A Most Improbable Story

This book is a "Big History" of the evidence regarding how we came to be. It briefly explores philosophical thought and how our past might affect our future. The text summarizes different perspectives, including the strengths and weaknesses of each. The genesis of our planet is explored, especially the circumstances that must exist for complex life to arise. This brief journey highlights the history of life, the emergence of simple lifeforms, and the evolution of complex creatures, including humans. The book concludes with a discussion of why other humanoids went extinct while our species achieved dominance. The author speculates on potentialities awaiting humankind and our planet.

Steven J. Theroux was Professor and Chair of the Natural Sciences Department at Assumption University in Massachusetts. He currently teaches Genetics and Biology of Cancer at Harvard University and Assumption University.

A Most Improbable Story

The Evolution of the Universe, Life, and Humankind

Steven J. Theroux

CRC Press
Taylor & Francis Group
Boca Raton London New York

CRC Press is an imprint of the
Taylor & Francis Group, an **informa** business

First edition published 2023
by CRC Press
6000 Broken Sound Parkway NW, Suite 300, Boca Raton, FL 33487–2742

and by CRC Press
4 Park Square, Milton Park, Abingdon, Oxon, OX14 4RN

CRC Press is an imprint of Taylor & Francis Group, LLC

ISBN: 978-1-032-21854-0 (hbk)
ISBN: 978-1-032-21851-9 (pbk)
ISBN: 978-1-003-27029-4 (ebk)

DOI: 10.1201/9781003270294

Typeset in Times
by Apex CoVantage, LLC

Contents

PART II   The Emergence of Life From a Lifeless World

Chapter 11   The Emergence of the Genus *Homo*—Earth's First Humans
Come Into Being 221

Chapter 12   The Genesis of Behaviorally Modern *Homo sapiens*—A
Cognitively Advanced Human That Can Reflect Upon Its
Existence 239

Preface

We are an impossibility in an impossible universe.

Ray Bradbury

Why are we here? It is a simple question, but humankind has been searching for the answer since we first appeared on the planet. In the quest to explain why we are here, many creation stories have been generated, and today, humanity has a creation narrative like no other in history. This modern, scientifically based account of how we came to be is every bit as engaging and exciting as the best novels, and the story is so grand that we may be tempted to believe it is a work of fiction. However, as incredible as it is, this account is not fictional. Rather, it is a saga based on scientific evidence, and it represents our best assessment of how we came to be.

Like many good stories, ours will take us to distance times and faraway places. It will introduce us to strange happenings and unfamiliar ideas. It will entertain and educate. In fact, as we explore our genesis, we will travel to the very beginning of time and the edge of the known universe and beyond. We will contemplate the behavior of subatomic particles, the structures of galaxies, and the function of the human brain. And throughout it all, we will reflect on how the natural world influenced who and what we are.

Humanity's modern creation story has fascinated me for a long time, but trying to understand it requires forays into many disciplines including physics, cosmology, chemistry, geology, earth science, evolutionary biology, environmental science, and philosophy, to name just a few. There is no individual who is an expert in all of these areas, and venturing into these foreign lands of intellectual thought can be scary and intimidating. However, if we want to gain a sense of the big picture that outlines how we came to be, we must move outside our comfort zone, and we must navigate intellectual realms that we rarely travel. In any journey of this kind, one benefits immeasurably from the wisdom and knowledge of talented guides, and in order to tell this story, I relied heavily on scientists at the vanguard of their disciplines. These individuals will lead us through the rugged and seemingly impenetrable territories that they are intimately familiar with, and I am personally grateful to them for allowing me to move outside my own disciplinary boundaries and into areas replete with astonishing vistas. It is to these guides that I dedicate this book because it is they who have made this adventure possible.

Like any traveler who is depending heavily on the information possessed by others, I must correctly understand and employ the data I am given, or I will end up hopelessly lost. I am grateful to the many individuals who helped me in my quest to do this, and I am indebted to them for assisting me when I fell off the intellectual trail. However, despite all of the help I received, it certainly is possible that I occasionally took an incorrect turn or ventured down the wrong alley. So, while I cannot take credit for discovering the knowledge that allowed us to undertake this adventure, I can and do take responsibility for any errors I may commit in the telling of it.

Despite the challenges that come with a transdisciplinary enterprise, I greatly enjoyed this undertaking, and during the trip, I came to better appreciate our place in the universe and what it means to be human. I hope this text, which I view as the travel log that I generated during my excursion, helps you do the same. So, let's get started, and let's learn more about our modern creation story, the greatest story ever told.

Acknowledgments

There are many individuals who helped make my presentation of this story possible. The idea to write this book arose out of the numerous conversations I had with my colleague, Hubert Meunier. Throughout this endeavor, Hubert provided encouragement and direction, and he kindly reviewed the entire manuscript. My daughter, Millie, who is trained in neuroscience and psychology, was the first to read the book in its entirety, and her enthusiasm, feedback, and encouragement for the project were invaluable and motivating. Georgi Georgiev reviewed the text, initiated many fascinating philosophical discussions, corrected several of my errors, and taught me quite a substantial amount of physics along the way. My colleague in biology, David Crowley, also reviewed the entire text, and he helped with the organization and writing of the document. I am also greatly indebted to my daughter, Katrice, for her detailed and extensive editorial suggestions. Kate is a trained editor and a fantastic writer, and she went through the manuscript several times. Her assistance greatly improved the writing of the document, and she helped make it clearer and better organized.

Others who contributed to the text include Allan Barnitt, my now-retired biology colleague, and Paul Douillard, a talented philosopher and educator, and my former Dean. Stuart Borsch kindly made use of his very unusual and diverse expertise in nuclear engineering and history to review the beginning of the manuscript, and Tom Slakovsky provided feedback on the chapters related to physics. Asling Dugan employed her expertise in microbiology to review the chapter on the evolution of life, Ed Dix reviewed the chemistry at the beginning of the book, and Jimmy Hauri and Kevin Hickey helped me with topics related to environmental science and earth science. In addition, Karolina Fucikova generously used her expertise in botany and the evolution of terrestrial life to provide feedback on these subjects, while the archeologist and biological anthropologist, Greg Landen, provided very helpful advice on the chapters related to the evolution of humans. In addition, my colleague from the history department, Tom Wheatland, who worked as a publisher earlier in his career, provided helpful guidance on how to navigate the book publishing world. Last, but not least, I am forever thankful for the support, encouragement, and patience of my wife, Sharon. I am grateful to all these individuals for what they did to shape the manuscript, but most of all, I am thanking them for the very many ways they enrich my life.

Finally, I am enormously appreciative of Chuck Crumly for supporting, encouraging, and assisting in the production of this text. He and his team at CRC Press/ Taylor & Francis have been a pleasure to work with, and I am fortunate to have had the opportunity to do so.

Introduction
Perspective Matters, But Perspective Isn't Everything

I believe everyone should have a broad picture of how the universe operates and our place in it. It is a basic human desire. And it also puts our worries in perspective.

Stephen Hawking

I think it's much more interesting to live not knowing than to have answers which might be wrong.

Richard Feynman

What was God doing before He created the Universe? Before He created Heaven and Earth, God created hell to be used for people such as you who ask this kind of question.

St. Augustine

How did our universe begin? Was there nothing, and then a Big Bang occurred, and our universe appeared? Or has our universe always existed, and then at some point, it expanded as described in the Big Bang theory? Or is it possible that a multiverse always existed, and eventually it gave rise to our particular universe? We don't have rational answers to these questions, and it is hard to imagine what such answers would sound like.

These questions are difficult to contemplate in part because our conceptual abilities are limited. As humans, we can't imagine something emerging from nothing. Our minds can't conceive of a mechanism that would allow that to happen. And we don't do any better when we try to imagine infinity. After all, what does it mean to say something has always been, for how can something exist if it never had a beginning?

While we may not be able to answer these questions, we do know that an event referred to as the "Big Bang" occurred 13.8 billion years ago, and with that "Big Bang," our universe began the process of adopting the characteristics it has now. To the best of our knowledge, this makes the Big Bang the single most propitious and important incident that ever occurred, for the Big Bang gave rise to time, space, and matter within our universe, and it profoundly influenced all that we are and ever will be. But what was the Big Bang, and why did it occur 13.8 billion years ago? What initially arose out of this astounding event, and how did it influence the formation of humanity? These questions have kept armies of thinkers occupied for many years and for good reason. If we want to understand why the universe and humanity have the traits they do, we need to begin by trying to understand the answers to these foundational questions.

DOI: 10.1201/9781003270294-1

THE SILENT "BIG BANG"

The first thing to know about the Big Bang is that it was silent.[1] That's right, the Big Bang did not involve a bang at all.

Sound is produced when energy generates a compression wave in a medium such as air or water, and at the moment of the Big Bang, there was no medium for energy to travel through. In fact, at the moment of the Big Bang, there was an unfathomable amount of energy in the form of radiation, but that's it. Nothing else existed. Not even an atom.

Why did this event occur? This is one of the most fundamental of all questions, and the scientific community's answer is simple: "we don't know." However, more than a few thinkers have put forward possible answers. Among them is Stephen Hawking, who suggests that gravity and quantum fluctuations were responsible for the Big Bang and the subsequent genesis of our universe. He writes in his best-selling book, *The Grand Design* [1]

> If the total energy of the universe must always remain zero, and it costs energy to create a body, how can a whole universe be created from nothing? That is why there must be a law like gravity. Because gravity is attractive, gravitational energy is negative: One has to do work to separate a gravitationally bound system, such as the earth and moon. This negative energy can balance the positive energy to create matter.

Hawking later goes on to state,

> Bodies such as stars or black holes cannot just appear out of nothing. But a whole universe can. Because gravity shapes space and time, it allows space-time to be locally stable but globally unstable. On the scale of the entire universe, the positive energy of the matter *can* be balanced by the negative gravitational energy, and so there is no restriction on the creation of whole universes. Because there is a law like gravity, the universe can and will create itself from nothing.

Hawking's ideas about how the universe began raise at least as many questions as they answer. For starters, how did the law of gravity come into existence? Hawking gives gravity the power of creation, but he fails to explain its origin, and he also does not explain why gravity exerts the force it does. Further, the energy in the known universe is thought to be fixed, and according to the first law of thermodynamics, it cannot be created or destroyed, but how did this initial fixed energy come to be? Put another way, why is there "gravitationally induced negative energy," and why is there "positive mass energy"? And why does the universe exhibit mathematically based predictable laws like the ones Hawking discusses? In short, what gave rise to the initial conditions of the universe, and why does our universe have the structure and regularity necessary to produce something rather than nothing?

To work his way around at least some of these dilemmas, Hawking brings to the forefront a theorem developed by mathematicians and physicists called "M theory." This theorem unites different versions of string theory, all of which attempt to explain how Einstein's theory of gravity could be compatible with quantum mechanical theories. Remarkably, Hawking suggests that while studying M theory, physicists have uncovered mathematical support for the idea that there may be as many as 10^{500} universes and that each of these universes may have its own laws.

Up until as recently as 1923, which is just a few generations ago, we believed that our galaxy was alone in the universe. However, Edwin Hubble's astronomical investigations revealed that the stars he was studying were too far away to be part of the Milky Way. Hubble realized that what he was looking at were constituents of another galaxy altogether. At that moment, after two million years of existence, humanity's understanding of our place in the universe changed profoundly. After all our time on the planet, we finally realized that there were in fact numerous galaxies and that ours was one component of an unimaginably vast universe. Today, we believe that there may have been as many as three trillion galaxies in our universe, but although we have come to accept that there are an incredible number of galaxies, it was not until very recently that we have been asked to wrestle with the idea that there may be 10^{500} universes.

10^{500} is a mind-bogglingly large number. As Hawking notes, it is so large that if you could count 1,000 universes per second, and if you started counting at the moment of the Big Bang, and you did not take coffee breaks, 13.8 billion years later you would have counted only 10^{20} of them. Therefore, for all practical purposes, M theory suggests that there are effectively an uncountable number of universes, and each of these may have its own laws of physics.

If M theory is correct, it is not surprising that we live in a universe like ours, because as living creatures, we must by necessity exist in a universe where the physical laws make it possible for life as we know it to evolve. Presumably, in the vast majority of the unfathomable number of "other universes," the physical laws that operate would prevent the emergence of life.

The idea that there are an uncountable number of universes has become known as the "multiverse concept," but again, invoking this hypothesis does not reveal why there should be gravity, energy, or quantum fluctuations in any of the supposed uncountable number of universes. The multiverse proposal also demands that we accept the idea that the M theorists' mathematical theorems accurately predict the structure of reality, and it stipulates that we do so without empirical data. In his text, *Not Even Wrong*, mathematician Peter Woit rejects this demand. He writes that M theory makes no testable predictions, and therefore it cannot be proven correct. He goes on to say that M theory is so far from acceptable that it is "not even wrong." To call it "wrong," he suggests, is to give it more standing than it deserves. Woit is not alone in his reservations, and no less a physicist than Richard Feynman, a Nobel Laureate, and a founding father of quantum theory, questioned the validity of M theory [2]. Feynman writes,

> I don't like that they're not calculating anything. I don't like that they don't check their ideas. I don't like that for anything that disagrees with an experiment, they cook up an explanation—a fix-up to say, "Well, it still might be true." For example, the theory requires ten dimensions. Well, maybe there's a way of wrapping up six of the dimensions. Yes, that's possible mathematically, but why not seven? When they write their equation, the equation should decide how many of these things get wrapped up, not the desire to agree with experiment. In other words, there's no reason whatsoever in

superstring theory that it isn't eight of the ten dimensions that get wrapped up and that the result is only two dimensions, which would be completely in disagreement with experience. So, the fact that it might disagree with experience is very tenuous, it doesn't produce anything; it has to be excused most of the time. It doesn't look right.

In addition to the aforementioned concerns, M theory is troubling because even otherwise reliable mathematically based models can give rise to incorrect predictions. For example, quantum mechanics and special relativity have been remarkably successful at predicting phenomena in our universe, but they are significantly limited in their ability to arrive at the correct masses of the elementary particles. According to Harvard physicist Lisa Randall, their equations suggest that the fundamental particles are "one ten-thousand-trillion times smaller than we would expect" [3]. Empirical data reveal the errors that these equations produce, and consequently, we don't suffer from the delusion that the physical particles of the universe are many, many orders of magnitude smaller than they are. But in M theory, where the empirical verification of a mathematical theorem is not possible, incorrect assumptions will go undetected.

Despite all the above, it remains possible that M theorists are correct and a multiverse exists, and it is possible that we cannot detect the many alternate universes because they have generated their own spacetime [4], with which we cannot interact [5]. It is also possible that other universes exist on the outskirts of our spacetime, and we have not yet encountered them [6], and as some have suggested, it may be that there are parallel universes within the multiverse that can only be accessed by somehow traveling through black holes [7]. But, while many things are possible, to move from the realm of philosophical speculation into the territory of empirical science, we need experimental evidence.

STRANGE BEHAVIOR IN THE SUBATOMIC WORLD

Having just argued that we can't accept ideas based solely on mathematical reasoning, I will now suggest that we also can't simply dismiss them because they seem odd. We have learned from the study of quantum theory that many strange and counterintuitive phenomena do occur in nature [8]. For instance, we know that because of quantum entanglement, manipulations that affect one photon can instantaneously affect an entangled photon, even if the partners are many miles apart. How these photons become entangled on the quantum level is not clear, and we do not see this type of entanglement in the macroscopic world. But, despite its strangeness, and notwithstanding Einstein's assessment that the phenomena seemed like "spooky action at a distance" [9], quantum entanglement can be empirically demonstrated.

There are plenty of other peculiar and mysterious phenomena that occur in the physical world. For instance, we know that matter has both particle and wave character. However, particles are discrete structures, whereas waves are spread out, so it is hard to reconcile how matter can exist as both. Stranger still, we know which of the two states matter assumes is influenced by whether it is being monitored. This was first demonstrated in a "double slit experiment." In these studies, a single electron was fired at a barrier with two slits in it, and once the electron moved past the barrier, it was allowed to strike a screen on the other side. Interestingly, if the electron was not being monitored, it exhibited wave-like characteristics, and it moved through both slits simultaneously, which suggested that the *single* electron could be in more than one place at one time (see Figure I.1A). However, a very different story emerged when the electron was being monitored. In that situation, the electron passed through one slit or the other, and when it hit the screen, it did so as a discrete single particle (see Figure I.1B) [10–12].

To explain these seemingly unfathomable observations, a group of physicists generated what is referred to as the Copenhagen interpretation of quantum mechanics. According to this view, in the absence of an observer, the electron does not have a definitive location, and it does not exist at any one point in spacetime. Thus, before an observer arrives, the electron can best be envisioned as a foggy cloud, and within that

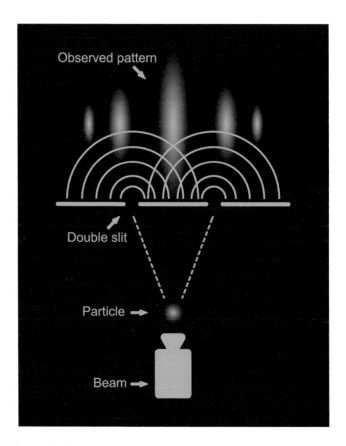

FIGURE I.1A In the absence of a detector, an electron behaves like a particle, and it moves through two slits at the same time. As a result, the photons generate constructive and destructive interference patterns, and they produce the pattern depicted.

FIGURE I.1 The double-slit experiment: A particle, such as an electron, has both particle and wave characteristics.

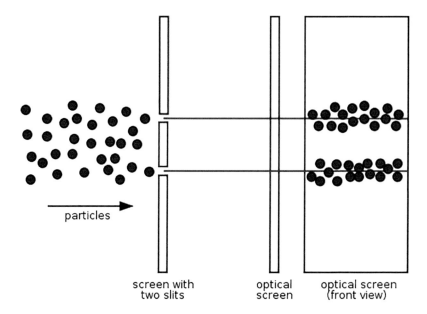

particles

screen with   optical   optical screen
two slits   screen   (front view)

FIGURE I.1B   However, if the electron is being monitored, then it behaves like a particle and it travels through one slit or the other, but not both. The result is that two distinct bands of equal intensity are formed. (From: Open source. Permission can be found at https://commons. wikimedia.org/wiki/File:Two-Slit_Experiment_Particles.svg.)

FIGURE I.1   (Continued)

cloud, there are different probabilities that the electron could undergo a change in its quantum state and appear as a particle-like entity. However, when an observer is present, the "interaction" between the observer and the electron alters the quantum state of the electron's probability cloud, and although we can't say exactly where the electron will appear, it ultimately does assume particle-like characteristics within spacetime [13].

After pondering ideas like those described earlier, Albert Einstein quipped that he could not accept the Copenhagen group's view of reality, because he liked to think the moon was there whether he was looking at it or not. Other exceptional physicists, such as Richard Feynman, also struggled to make sense of quantum theory, and he famously stated that "no one really understands quantum mechanics." But, if no one understands quantum mechanics, then no one truly understands many of the fundamental tenets that form the foundation of physics, and to quote Sean Carroll, "If nobody understands quantum mechanics, nobody understands the universe" [14].

In the end, the Copenhagen group's ideas have certainly caused consternation within rarified intellectual realms, and their ideas appear to be devoid of "common sense," but it is important to emphasize that their views are completely consistent with experimental observation. Other even more bizarre explanations for the experimental observations made by quantum physicists have also been generated, and among the strangest are those produced in 1957 by Hugh Everett. At that time, Everett was a graduate student at Princeton University, and as a relative newcomer to the field, he introduced the "many worlds" hypothesis, in which he suggested that it was possible to explain quantum mechanical phenomena, like those exhibited in the double-slit experiment, by assuming that every possible quantum mechanical manifestation of an object really does exist. Thus, according to Everett, electrons, atoms, people, and presumably even very large structures such as moons, planets, and stars can be found in the state we see, but they can also be found in other states as well. The obvious question that arises is, why can't we directly or indirectly see all these various manifestations of reality, and Everett's answer was that we can't perceive them because they each exist in a parallel universe.

Even though there is no compelling evidence for the existence of parallel universes, Everett's hypothesis is nevertheless thought provoking, because if he is correct, there are countless versions of you, and each is living his or her own unique life. However, if Everett is correct, even though you and your doppelgangers share much in common, you and they will never meet, nor will any of the many countless versions of you ever know with certainty that the others exist.

Ultimately, it is very unlikely that Everett's explanation would have put Einstein's mind at ease, and neither the Copenhagen interpretation of quantum physics nor the "many worlds hypothesis" is easy concept for most of us to accept. This fact led the mathematical genius and polymath Roger Penrose to suggest that both explanations are problematic at best and nonsensical at worse, and he argued that this is a good indicator that quantum mechanical theory and our best explanations of quantum mechanical phenomena are inchoate. According to Penrose, this is a result of our lack of understanding of the detailed relationship between gravity and quantum mechanics, and he postulated that once we fully understand the effect of gravity on quantum entities, we will understand why people and moons don't behave like photons and electrons [15].

Most scientists agree with Penrose's statement. Still, we know that bizarre and counterintuitive phenomena take place, and we know this because we have the mathematics *and* the empirical data to support our assertions. However, this is not the situation with M theory. M theory does generate strange predictions, but its proponents haven't produced supporting empirical data to back them up. Indeed, even the predictions derived from many of M theory's intrinsically pleasing mathematical theorems are often indirect. For some advocates of the multiverse hypothesis, this reality is not troubling. However, once we leave the realm of experimentally verifiable science, almost any hypothesis is potentially plausible, and if empirical data are not necessary, then there is no reason why we cannot support MIT physicist Max Tegmark's idea that there are universes based on every imaginable mathematical structure [16]. Additionally, we could, as Paul Davies points out in his book *Cosmic Jackpot*, postulate that there are universes that are non-mathematical and that there are universes based on every type of ethical system imaginable. Davies goes on to argue that if empirical evidence is not needed, you could even claim that the majority of universes within the hypothesized multiverse are computer-simulated facsimiles generated by an advanced civilization and that you are in just such a virtual universe [17]. But while it is possible everything around you is nothing more than a computer simulation, I wouldn't bet the farm on this solipsistic hypothesis.

Despite the lack of data and the seemingly extreme nature of the various manifestations of the multiverse hypothesis, some of M theory's promoters still argue that clues exist that support their beliefs. Once again, Max Tegmark comes to mind as an example of those in this group when he states,

> some of the fine-tuning (of the universe) appears extreme enough to be quite embarrassing—for example, we need to tune the dark energy to about 123 decimal places to make habitable galaxies. To me, an unexplained coincidence can be a tell-tale sign of a gap in our scientific understanding. Dismissing it by saying "We just got lucky—now stop looking for an explanation!" is not only unsatisfactory, but is also tantamount to ignoring a potentially crucial clue.

Tegmark is certainly correct: It is truly remarkable that the universe contains precisely the right amount of dark energy to allow for the existence of galaxies, planets, and people, and Davies estimates that the odds of our universe getting this amount of dark energy by chance are about one in 10^{120}. To give this number meaning, Davies points out that this would be like getting heads in a coin toss no fewer than four hundred times in a row [17]. However, the apparent fine-tuning of the universe does not prove the existence of countless universes. Some physicists are confident that there is a reason why dark energy has the value it does, and we just haven't discovered that reason yet. In addition, many theists use the very same data Tegmark presents to support the existence of an omnipotent creator [18, 19].

Other clues the multiverse proponents put forward are also not convincing enough to unequivocally justify the multiverse concept. For example, Tegmark and Lawrence Krauss argue that the analysis of the microwave pattern in the early universe strongly indicates that the universe is flat [16, 20], and according to inflation theory, if a multiverse exists, one would expect that to be true. However, physicist George Ellis remarks that the universe may only appear to be flat within the area we can study, and if we were able to see further, we may yet learn that the universe is not flat after all [21].

Interestingly, Tegmark acknowledges much of the criticism leveled against his multiverse hypothesis. Despite this, he writes, "I'll still happily bet my life savings on the existence of a multiverse!" [22]. Fortunately for Tegmark, it is very unlikely that anyone will be able to prove him definitively wrong anytime soon, and therefore he can rest assured that if he makes the bet, and is incorrect, no one will be collecting.

Ultimately, the claims of a multiverse are extraordinary, and as the saying goes, "extraordinary claims require extraordinary evidence." Someday, the mathematics and hypotheses that predict the existence of a multiverse may be corroborated by data obtained from the study of this universe. Krauss, for example, believes that the study of gravitational waves may reveal that our universe formed because of an inflationary transition within a region of space, and our universe may only represent one bubble that jutted off as a result of this vast, eternal process of cosmic inflation. He and others argue that local regions of inflationary transitions may continue to occur forever, and each time they do, a new universe may arise. Furthermore, Krauss believes that if the study of gravitational waves leads to this conclusion, we will know that other universes exist, even if we cannot detect them. Perhaps, but at this point, the compelling data Krauss hopes for are absent.

So, we can marvel at the intrinsic beauty and logic of some of the mathematical propositions suggesting the existence of countless universes, and we can use these ideas to creatively explain the seemingly unexplainable attributes of our universe. However, when we label aesthetically pleasing mathematical models as convincing evidence, we enter the world of mathematically based philosophy. Ultimately, since we don't know if the mathematics of these models is tethered to reality [23, 24], believing in the validity of the concepts produced by these mathematical analyses requires a significant act of faith.

THE "GOD HYPOTHESIS"

When it comes to explaining how the universe came to be, faith of a different sort has played a role in the minds of many. Long before M theory, multiverses, and all the related ideas, most people believed that the universe was here because "The Gods" or "God" made it so.

In the polytheistic world, there are elaborate and creative stories about how the Gods created the universe, and there are fascinating tales about the structure of the cosmos itself [25]. However, for sheer entertainment value, it's hard to beat what I will call, "it's tortoises all the way down," description of how the universe was constructed. In his book, *A Brief History of Time* [26], Steven Hawking recounts this humorous explanation of the structure of the universe as it was supposedly expounded by a "little old lady" engaged in a verbal exchange with a famous scientist. According to Hawking, the conversation unfolded as described:

> A well-known scientist (some say it was Bertrand Russell) once gave a public lecture on astronomy. He described how the earth orbits around the sun and how the sun, in turn, orbits around the center of a vast collection of stars called our galaxy. At the end of the lecture, a little old lady at the back of the room got up and said: "What you have told us is rubbish. The world is really a flat plate supported on the back of a giant tortoise." The scientist gave a superior smile before replying, "What is the tortoise standing on?" "You're very clever, young man, very clever," said the old lady. "But it's tortoises all the way down!"

In the Judaeo Christian Westernized world, the best-known creation story is found in the Book of Genesis. According to this text, God created the universe *ex nihilo* (from nothing) in six days. We learn in Genesis 1 that the famous, "let there be light" statement was uttered by God on the first day of creation, and then the water, sky, land, and vegetation—essentially Earth and the first life—were created over the next two days. It wasn't until day 3 that the Sun, Moon, and stars appeared, and the story reveals that humankind appeared on the scene on day 6.

As a science text describing the beginning of the universe, Genesis comes up short. For starters, the story suggests there was an Earth without a Sun or stars, and in this starless system, there was apparently both light and plants. You don't have to be an astronomer or a botanist to know that in our galaxy, if stars like our Sun were not present, there would be no significant amount of light on Earth, and Earth itself, if it existed, would be an unbearably cold rock utterly devoid of plant life. However, many religious traditions have moved past the idea of interpreting the book of Genesis literally, and this text, and the others that compose the Bible, are often used by believers as metaphorical and spiritual guides for understanding the relationship between humankind and its creator.

Throughout history, many of the most successful, productive, and prominent philosophers, scientists, and mathematicians accepted the idea that God created the universe, even if He may have done so in a way that differs from the literal description outlined in Genesis [19]. For more than a few of these individuals, their work was a way to gain insight into the nature of their creator God, and the urge to learn more about this God inspired much of what they did. The great Sir Isaac Newton was an example of just such an individual. Newton possessed one of the most impressive scientific minds humanity has ever encountered, and he used his impressive intellect to make seminal and groundbreaking contributions in mathematics, physics, and optics. Newton's thoughts produced the calculus, a description of the three of the laws of motion, the law of universal gravitation, as well as many additional noteworthy insights into the workings of the physical world [27], and with the possible exception of Albert Einstein, he is considered by many to be without equal in the scientific realm. However, despite being able to study the physical world like few have done, Newton reserved much of his intellectual energy for the study of Christian theology. Throughout his life, Newton sought to clarify his vision of the nature of God, and for him, science was both an oblation and a fantastic theological tool. Newton reflects these views in the introduction to his *Opticks*, where he states that the purpose of natural philosophy (i.e., science) is to determine causes from effects until we finally arrive at the "First Cause," which Newton believed was God [28].

As a result of his studies, Newton concluded that at the beginning of time, God created the laws that govern the universe. However, although one might think that an omnipotent and omniscient God would set up a universe that behaved exactly as it should, Newton suggested otherwise. He proposed that occasionally God had to do some work to keep things functioning as intended. For example, Newton argued that God needed to step in on a regular basis to carry out the seemingly mundane task of slightly adjusting a planet's unstable orbit. So for Newton, God physically tinkered with the functioning of the universe, and in Newton's mind, miracles that involved altering the physical functioning of the universe were a routine act for God, and this was a demonstrable fact [27].

Other prominent thinkers who were convinced that God created the universe include the pious and devout physicist Galileo Galilei and the philosopher and mathematician Rene Descartes. Galileo is famous in part for his opposition to the belief that Earth is the center of the universe, and while he had his doubts about the infallibility of papal proclamations, there was no apparent ambiguity in his belief that God created the universe and that God did so using laws He ordained and implemented at the beginning of time [29]. For Galileo, God always existed, and the universe we enjoy was a product of His handiwork.

Like Galileo, Descartes was a Roman Catholic who believed that God created the universe and that God, in the act of creation, generated the laws that governed our world. However, unlike Newton and Galileo, Descartes argued that the laws that God freely imposed were, like God himself, immutable [30]. From Descartes' perspective, once God established the laws that governed the universe, He no longer had to intervene. This belief led some to declare Descartes a deist, and Descartes's position that the laws of nature were unchanging, and that miracles did not require the alteration of these laws, necessitated careful exposition if he was going to survive and thrive in the Roman Catholic world of 17th-century Europe.

THE BELIEF IN GOD: IS THIS JUST A QUAINT SET OF ANACHRONISTIC IDEAS?

The fact that many of the greatest thinkers of the past adopted the hypothesis that God created the universe and the laws that govern it is often dismissed as a quaint intellectual cultural artifact of a distant time and place. According to some, Newton, Galileo, Descartes, and numerous other great minds of long ago could be expected to believe in a deity, because their society was theistic, and criticism of religious belief was vigorously discouraged and sometimes punished by torture and death. In addition, those who make this statement often argue that thinkers like Newton, Galileo, and Descartes were prone to theistic belief because they were not aware of the structure of the galaxy or the universe, and they did not know about quantum mechanics, general relativity, M theory, Darwinian evolution, and a great number of other scientific advances. Presumably, according to this crowd, if the great thinkers of the past lived today, and if they knew what we know now, they would be solidly ensconced among the atheists and the materialistic reductionists.

These claims are factually accurate on several counts. It is true, for example, that religious belief has waned over the last several hundred years, and because of this and the development of modern societies, religious proclamations are now criticized more openly and more frequently than in the past. One need not look further than 21st-century France, where approximately 40% of the population consider themselves to be atheists to find evidence of this [31]. At the beginning of the 18th century, it would have been very difficult to find anyone in that part of the world who openly aligned him or herself with an atheistic belief system. Today, you can meet people of this persuasion on every street corner.

The proponents of this argument are also correct in stating that other dimensions of society, culture, and history heavily influenced the ideas of these earlier modern thinkers. Nevertheless, despite their cultural milieu, the great intellectuals of the past still had reason to doubt the monotheistic God hypothesis. The idea that there may not be a God very likely crept into their minds, at least on occasion, for despite numerous theodicies, there has never been an issue more vexing and intractable to the monotheist, and none more likely to sow doubt about the existence of God, than the problem of evil [32]. And this problem is not new. In fact, "natural evils" (such as tsunamis, plagues, and volcanic eruptions) have been around since the dawn of humankind, and human moral evil was even more ubiquitous in centuries past than it is today [33, 34].

Implied in the assertion that religion is a quaint intellectual idea is the corresponding belief that modern science has largely wiped away religious belief among the most accomplished intellectuals. However, many modern intellectuals are deeply spiritual, including Max Planck and Albert Einstein, both of whom rank among the greatest physicists of the 20th century [35]. As it turns out, Planck and Einstein were often in conflict with many of the ideas expressed by the great religious traditions of their day, but they were deeply spiritual, nonetheless. Planck, for example, stated that he did not accept the God of Christianity, nor did he believe in a personal God [36]. Einstein also did not believe that God was a personal deity receptive to prayers and prone to initiating miraculous interventions, and for him, God was perhaps best described by the 18th-century Dutch rationalist, Baruch Spinoza, who argued for the existence of a panentheistic deity. However, despite their personal differences with many of the dogmas of the major religious traditions, Planck and Einstein did believe that there was a God-like process or entity that was deserving of their reflection, contemplation, and admiration. According to Einstein, his beliefs were based on "a knowledge of the existence of something we cannot penetrate, of the manifestations of the profoundest reason and the most radiant beauty, which are only accessible to our reason in their most elementary forms." In addition, for many of the world's greatest scientists, science was worth studying, because as Einstein suggested, it allows one to, "know His thoughts," and it gives an individual the opportunity, "to know how God created this world" [37]. Ultimately, Einstein's views resonated deeply with many of those of his 17th-century predecessor, Isaac Newton, and he was in agreement with one of his contemporaries who stated that "in this materialistic world of ours the serious scientific workers are the only profoundly religious people" [38].

UNCERTAIN REALITIES

SCIENCE AS A VEHICLE TO UNDERSTAND THE UNIVERSE

From the modern-day perches of science, theology, and philosophy, we can survey an intellectual landscape with plenty of room for contemplation. In reflection, some reject the idea that God exists, and they choose to place their faith in science. Within this group, many live with the hope that science will ultimately clearly illuminate the great mysteries of our universe.

Those choosing science as their primary epistemological approach have many reasons for doing so. Science has opened vistas into the nature of human existence that were not imaginable just 100 years ago, and unlike religious traditions, of which there are an almost endless number of variants, science has a universal quality to it. Scientists in Boston, Massachusetts, read the same journals and textbooks as their counterparts in Beijing, China, they accept most of the same basic facts and concepts, and they utilize the same methodologies and techniques to explore the world. There is something very comforting in this level of agreement regarding the nature of the universe and the best strategies by which to explore it. However, the fact that scientists often reach a consensus does not necessarily mean that their views are correct, and the troubling reality is that many scientific experiments are not replicable [39, 40]. Furthermore, occasionally new paradigm-shifting data do cause the scientific community to reject or substantially reshape its former understanding of the natural world [41], and to some, these sorts of changes are disturbing. Still, despite its limitations, science has the benefit of being self-correcting, and it has generated theories with significant predictive value. Consequently, the most rigorously tested scientific theories we have represent humanity's best explanations of natural phenomena.

In addition to providing new information about the nature of the universe, modern science and the political reforms that arose during the enlightenment made it possible for humans to experience a degree of freedom that was previously unimaginable. Today, in many parts of the world, individuals can articulate thoughts based on scientific principles, some of which would have been deemed heretical just a few hundred years ago. Because of these societal changes, people around the world feel free to question and reject some of the most troubling aspects of the canon of religious traditions, including the suggestion that an all-powerful being occasionally may give the righteous order to commit mass murder and genocide. In most modern societies, skeptics question without the worry that they will be punished and subjected to the cruelest and most dehumanizing forms of torture and execution imaginable. These advances are a noteworthy gift to humanity, and to the extent that science contributed to their formation, people are attracted to it.

Besides contributing to societal freedom, scientific and political advances also liberated many from the existential fear of the afterlife. For the agnostic and atheist, it is no longer necessary to worry about the doctrine of predestination, which states that they or their loved ones may be destined, from the moment time began, to suffer a horrific, never-ending existence after death. In short, for many religious skeptics, science provides relatively reliable information about the natural world and our existence within it, and it also offers an autonomy of thought and freedom from a God that is sometimes depicted as brutal, merciless, and capricious.

Lawrence Krauss—who is a self-proclaimed anti-theist—nicely presents these aspects of the scientific atheist's position in his text, *A Universe From Nothing* [20], when he writes,

> I would find little purpose living in a world ruled by some divine Saddam Hussein-like character, as my late friend Christopher Hitchens put it, who not only makes all the rules, but punishes those who disobey them with eternal damnation. I find living in a universe without purpose to be amazing, because it makes the accident of our existence and our consciousness even more precious—something to be valued during our brief time in the sun.

For the committed atheist or anti-theist, science is the only reasonable source of answers about our universe. For them, there is simply no other alternative.

SOME OF THE LIMITS OF CHOOSING SCIENCE TO EXPLORE REALITY

Still, despite its many virtues, science is not as powerful and liberating as some would like to believe. Immanuel Kant argued that we can only address the veracity of events that occur in the natural world, and he suggests that science, which analyzes events within the natural world, can neither prove nor disprove the existence of God, the soul, or any other supernatural entity that is, by definition, outside of the natural world. As a result, from a Kantian perspective, the conviction of the scientific atheist is misplaced.

Kant also pointed out that scientists encounter other limitations as well, since according to him, we cannot know with certainty the true nature of anything. Kant believed that our understanding of the world is partly the product of the activity of our minds. This activity allows us to perceive objects in space and time, it enables us to understand the world as a system in which causation occurs, and it helps us organize our ideas about what we perceive as reality. Similar views have been explored by modern-day investigators, some of whom essentially agree with Kant and argue that our minds actively construct an image of what is real in a process that involves "controlled hallucinations" [42]. But, if our minds do perceive and interpret the world through such a process, our view of reality may be askew and warped, and consequently, we may not be seeing the universe as it truly is.

That our minds can misinterpret reality is simple enough to demonstrate. If you have ever been on the remarkable simulator ride at Disney World called "Soarin," you know that your mind can trick you into believing that you are gliding at high elevations over various parts of the state of California, and although part of your mind knows that you are doing no such thing, other parts are nevertheless absorbing the incoming physical stimuli and arranging them in time and space so that you get the impression that you are really gliding through the California air like a bird.

A perhaps less entertaining example of how our brains create an inaccurate picture of reality occurs when we look at any apparently solid, motionless structure. For example, for most of our existence, we have believed that solid substances, such as rocks, consist of uniformly filled space, and we have considered the material that occupies that space to be static. However, we now know that solid matter is composed of atoms that are in ceaseless motion [43], and these atoms have their mass concentrated in an extremely small area called the atomic nuclei. Outside the nucleus there are electrons that typically appear within orbits approximately 100,000 times the diameter of an atomic nucleus [44]. Knowing this, a modern physicist views a rock as being made of highly dynamic structures that consist of mostly empty space, but this is certainly not an interpretation that a prehistoric human could have seriously entertained, and clearly, the physicist's understanding of a rock is not what we "see" when we glance down at one. So, can we trust our senses to tell us "the true nature of reality"? Of course, we know the answer is that we cannot. Furthermore, what is equally obvious and perhaps even more disturbing is the fact that whenever we think we understand something, we can't be sure that we really do. We live in a state of perpetual ignorance, and we never know whether there is a deeper state of reality that remains to be uncovered. In short, we live our lives knowing that we can never truly fully know something.

Thus, the work of modern science supports Kant's idea that our minds are limited in many ways. However, as Kant also suggested, the entity we call the mind does afford us some advantages, and it allows us to pull off a neat trick. Specifically, it allows us to generate new knowledge about the world independent of experience. But, here again, this ability comes with a catch, and there is evidence that suggests that when the mind is generating knowledge of the world, it may actively obscure, distort, or fail to pursue the true nature of some aspects of reality. Some of this evidence can be seen in the computer modeling done by Donald Hoffman, who demonstrated that there are instances in which simple models of a complex system outcompete those with more veridical and nuanced versions of reality. Hoffman argues that the simpler models are "more fit" and better able to compete for survival in part because seeking truth is an expensive endeavor that requires substantial time and energy. And given that time and energy are often in short supply, Hoffman argues that truth seeking can sometimes divert precious assets from endeavors that would more likely ensure survival and reproduction [45]. Furthermore, in more complex biological organisms, there are times when "the truth" can negatively impact psychological well-being, and this too can be detrimental. For example, in some situations, statistical realities paint a bleak picture of one's chances for survival. An individual in such a predicament whose mind overestimates the odds of success by ignoring parts of reality may be able to act more optimistically and confidently than the situation merits [46]. Ultimately, such a mind may survive even when the odds are against it, while other minds that possess a more accurate picture of the situation perish due to a lack of enthusiasm, confidence, or effort [47].

In short, some philosophical and evolutionary theories suggest that the pursuit of truth is not always a luxury we can achieve or afford, and they argue that our minds exist primarily to enhance our chances of survival and reproduction, not to discover the truth. If this is correct, we may often find it difficult to know the truth, even when we are engaged in seemingly logical pursuits such as those that characterize science.

USING THEOLOGY TO UNDERSTAND THE NATURAL WORLD—ATTRIBUTES AND SHORTCOMINGS

Theology is another epistemological approach that many employ to understand at least some aspects of the natural world. However, theological knowledge aspires to do much more than answer questions about nature, and it affects more than our understanding of the natural world. Theological beliefs structure many individuals' ethical systems, and they can impart inspiration to those who believe in their validity. Interestingly, theological beliefs can often spur successful empirical investigations into realms that are not directly addressed through revelation, and as mentioned earlier, Newton and Einstein's theological beliefs aided and inspired them in their quest to understand the realities of the natural world.

Religious beliefs also influence the structure of our societies, and on occasion, they may affect the survival of an individual and the faction to which he or she belongs by enhancing the unity and functioning of the group. In addition, religious beliefs may provide a form of psychological support that enhances the likelihood of survival by inspiring confidence [48, 49]. So, religious belief may have value to an individual and a group, but what religious belief does not provide is certain knowledge of what is true. Ultimately, religious doctrines are a matter of faith, and despite the assertions of those who argue that they are divinely inspired, the believer, using sense data, cannot prove even the most fundamental precepts of their faith, such as the existence of God. Of course, one's faith system could indeed be correct, on at least some matters, and it is often the case that the religious believer cannot be proven wrong [50–53]. Yet, despite the numerous bold and confident-sounding conclusions proclaimed by spiritual adherents, religion, like science, fails to provide the definitive proof about the nature of reality that many seek.

PHILOSOPHICAL LOGIC, YET ANOTHER WAY OF KNOWING

Given the nature of these disparate worldviews, it is easy to see why some atheistic scientists loathe the religious and why some religious individuals strongly dislike scientists. However, it is interesting that what both science and religion have in common is the inability to uncover the definitive "Truth," and neither can generate clear answers to simple questions such as "how did the universe begin and why does it have the characteristics it does." When asked, many scientists hold firm to their belief that the answers lie in understanding mathematics, quantum fluctuations, particle physics, and forces such as gravity. But when queried on why mathematics, quantum fluctuations, subatomic particles, and gravity exist, their response is like that of their theistic brethren, namely, that these conditions simply must exist. Like their theistic counterparts, they propose that there are some questions that cannot be answered, and therefore we must simply accept particular facts about the universe without understanding why they are true [54].

To circumvent the limitations of science and theology, there are some who have tried to establish a third camp in which they can ponder compelling ontological questions. Members of this group hedge their bets and look to philosophy—and more specifically to logic—for an answer. While admitting that a clear, detailed description of how the universe originated may never be uncovered, they nevertheless suggest that their approach may still be able to offer significant insights. For many of this disposition, the key to understanding what is true lies in the application of abduction, Bayesian reasoning, and Occam's razor. Those in this group have faith that ultimately, the simplest explanation is almost certainly the correct one. Of course, discerning the simplest explanation is itself not simple, and despite the raging debates about how the universe came to be, many in this third camp nonetheless believe the answer is apparent. For them, it is easier to believe that mathematics, quantum fluctuations, positive and negative energy, and gravity arose from nothing than it is to believe that an all-powerful God did. Linus Pauling, one of the greatest scientists of the 20th century, and the only person to be the sole recipient of two Nobel prizes, articulated this position when he appeared on the Phil Donahue show in the 1980s. *The Donahue Show* was a popular television program that regularly delved into the intellectual and spiritual realms, and when hosting Pauling, Donahue decided to ask his guest about his religious beliefs. Pauling responded that he was an atheist, and when he did, the audience let out an audible gasp of shock and disapproval. Initially, Pauling looked a bit taken aback by this response, but he proceeded to explain that, in his opinion, it made more sense to believe a simple system arose from nothing than it did to believe that a God, which is the equivalent of an unfathomably complicated system, did. However, despite his efforts, Pauling could not convince the audience to accept his ideas. In some ways, this is understandable because even if we could all agree on what the "simpler system" is, the attempt to apply logic to answer a question that human logic is not equipped to answer is itself an inherently flawed approach. Furthermore, if simple systems and simple answers are always better than more complicated ones, then we would predict that nothing would exist. Nothing, after all, is the simplest and least complicated system we can imagine, and logically, nothingness should give rise to more nothingness.

MOVING FORWARD

So, we limited humans ramble onward in our search for Truth and understanding. Atheistic scientists such as Hawking argue for a universe without a God, while their more theistically minded comrades such as Planck and Einstein deny the existence of a personal God but ponder the presence of a force that constructed the universe and continues to guide it. Meanwhile, there are many deeply religious people who have a robust and unshakable belief in the existence of a personal God that hears their prayers and regularly intercedes in the workings of the universe on their behalf.

Scattered liberally among the atheists and deeply religious are many who say they just don't know if God exists. The members of this crowd may at first appear to have taken the easy road but being agnostic can be arduous. Agnostics are by definition, "uncertain," and human nature does not tolerate uncertainty with ease. In many parts of our life, when uncertainty arises, problems ensue. In the face of uncertainty, the stock markets drop, stress hormones skyrocket, and people's minds do metaphorical gymnastics to fill in the knowledge gaps and minimize the unknowns. The reason we experience this discomfort is that we intuitively perceive that uncertainty is coupled with risk, and risk can be a dangerous thing.

Given all of this, it is not surprising that we are often disquieted by those who answer questions we deem fundamentally important by saying, "I don't know." To the scientist who fears an eternal existence with a genocidal and capricious God, not knowing whether God exists is far from soothing, and to the theist who believes that God will protect him and those he holds dear for eternity, answering "I don't know" to the question of whether God exists is also disconcerting. We all want to know, and we want to know with certainty, but certainty escapes us, and during our relatively short lifespans, many of us vacillate in our views. At various times, we may see ourselves as atheists, agnostics, or true believers. Indeed, even the most committed among us sometimes grapple with our deepest beliefs.

Perhaps one of the best-known examples of an individual who struggled with these concepts is Mother Teresa of Calcutta, India. Mother Teresa, who was recently canonized by the Roman Catholic Church, was an exceptionally virtuous nun and Nobel Peace Prize recipient. Many who knew her considered her to be a living saint, and she served the poor of India with energy, enthusiasm, and commitment. But despite her achievements and deeply rooted spiritual life, Mother Teresa spent more than five decades battling profound doubts about the existence of her own personal relationship with God, and at times, she even doubted whether God existed at all [55, 56]. This unusually devout woman died a theist, but her life was not one of unwavering faith.

A less pious example of an individual who struggled with his beliefs was Charles Darwin, the originator of the theory of evolution, and one of history's most prominent biologists. Early in life, Darwin was a believer in special creation, and when he set off on the *Beagle*, at the age of 20, to begin his trans-global sea voyage, he considered himself to be an orthodox Christian. Eventually, however, Darwin jettisoned his belief in special creation, and he developed the theory with which he is now associated. Ultimately, Darwin came to the conviction that the concept of special creation, while wrong, was pleasing to many people's imaginations, and he feared that many would obdurately cling to this belief despite the evidence against it. In his autobiography, he summarized his position when he wrote, "we can allow satellites, planets, suns, universe, nay whole systems of universes, to be governed by laws, but the smallest insect, we wish to be created at once by special act." Unfortunately, we now know that Darwin's fears were well founded, and despite overwhelming evidence, many still doubt his theory of evolution.

As Darwin aged, he began to question more than just the validity of special creation, and eventually, he developed doubts about Christianity in general. Darwin's views were conditioned by several factors including some Christian texts, which stated that non-believers, and those who did not accept the tenets of Christianity, would suffer eternal damnation. Darwin found this concept particularly troubling because his father, whom he described as the kindest man he ever knew, was a "freethinker," and so was his brother and many of his best friends. By the time he was an old man, Darwin had condemned the aforementioned Christian precept as "a damnable doctrine," and given this teaching, he questioned how anyone could wish Christianity to be true. Darwin also struggled with the question of natural evil, and he was particularly disturbed by the existence of unmitigated pain and suffering in the animal kingdom. About digger wasps (i.e., Ichneumonidae), which paralyze their prey and bring them back to the nest for their young to eat alive, he wrote to his colleague Asa Gray in May of 1860,

> With respect to the theological view of the question: This is always painful to me. I am bewildered. I had no intention to write atheistically, but I own that I cannot see as plainly as others do, and as I should wish to do, evidence of design and beneficence on all sides of us. There seems to me too much misery in the world. I cannot persuade myself that a beneficent and omnipotent God would have designedly created the Ichneumonidae with the express intention of their feeding within the living bodies of caterpillars [57].

At the end of his life, Darwin was convinced that evolution had shaped life on Earth, but by then, he was not at all certain that God had any role in the process. And regarding the "mystery of the beginning of all things," Darwin stated that he believed the puzzle to be insoluble by humankind. Ultimately, he concluded that he "must be content to remain an Agnostic" [58].

LIFE IN THE REALM OF UNCERTAINTY

Whether or not God and other forms of supernatural phenomena exist, most of us are confident that the universe exists and that it somehow came to be. Perhaps in our struggle to make sense of the universe and our existence, we are like a dog watching a rocket ship blast into space. The dog plainly sees the rocket lifting into the sky, and the sight is no doubt an awesome one, even to a dog. However, the dog cannot understand how or why the rocket is able to disappear into the heavens, and he will never comprehend the calculus and engineering that went into making the rocket what it is. When we stare out into the universe, or when we gaze inward and ponder our own existence, we know something about how that dog must feel. However, when we look out into the seemingly endless expanse of space and ponder its creation, we, unlike the dog, can strive to overcome our intellectual limitations. And while "The Truth" may be beyond our grasp, I agree with Gotthod Lessing's statement that "the aspiration to the truth is more precious than its assured possession" [59]. Ultimately, we are driven to find the answers to our questions; it is something we are apparently genetically programmed to do, and humanity will seek answers even when we know we may never find them.

Happily, despite the potential frustration of this endeavor, there is still much to learn from the scientific study of how we came to be. For the theist, the enterprise is useful, because if you believe that God created the universe, then there is much you can discover about God through the study of His creation. The atheist and agnostic also benefit from this exploration because understanding how we came to "our moment in the sun" is one of the most fascinating questions that can be asked. And regardless of whether one is a theist, atheist, or agnostic, the study of our universe and how we came to be will certainly provide valuable insight into the workings of our world and the nature of humanity.

In the end, we may have different perspectives about how best to find the answers to some of life's biggest questions, but while perspective matters, it isn't everything. Regardless of our personal beliefs, we can all marvel at the beauty and nobility of the endeavor to comprehend, and we can all learn to live with and celebrate the uncertainties and mysteries we encounter along the way.

It is in this spirit that we will explore some of the remarkable events that needed to occur for us to be here. Whether these happenings were orchestrated by God, the inevitable consequences of a multiverse, the result of a yet-to-be-discovered unifying theory of everything, or the outcome of simple dumb luck, they are nevertheless fascinating and worth pondering.

As we continue to consider some of the events that made our presence possible, it is important to note that we will not contemplate every apparently fortuitous occurrence. To do so is not possible, for we don't know, and we never will know, every event upon which our existence is contingent. There are also many incidents of uncertain importance, and even attempting to characterize all the crucial incidents would be an immensely overwhelming task. Therefore, this "Big History" book will focus on a small but salient group of phenomena that had to occur for our universe and humanity to exist. By doing so, we can more fully appreciate the wonder of life and the time we spend as sentient beings on this remarkable planet.

Having circumscribed the topics we will be exploring, we will continue by examining what we know about the formation of the known universe. We will begin this journey by investigating some of the consequences of the Big Bang, which was the epic event that generated our universe and the conditions necessary for life itself.

NOTE

1. Fred Hoyle coined the term "Big Bang," but he never intended it to be a literal description of what was happening at the moment time began. Instead, many believe that Hoyle, who supported the competing Steady State Hypothesis, generated the term largely as a derogatory moniker for an idea he did not like. Hoyle himself stated that the term was helpful in distinguishing the Steady State Hypothesis from the ideas being expressed by others with whom he disagreed.

REFERENCES

1. Hawking, S. and L. Mlodinow, *The grand design*. 2010, New York, NY: Bantam Books. 198 p.
2. Woit, P., *Not even wrong: the failure of string theory and the search for unity in physical law*. 2006, New York, NY: Basic Books. p. 320.
3. Sweeney, S., *Explaining the higgs*. Harvard Gazette, 2014.
4. Veneziano, G., *The myth of the beginning of time*. Scientific American, 2004. **290**(5): p. 30–39.
5. Deutsch, D. and M. Lockwood, *The quantum physics of time travel*. Scientific American, 1994. **270**: p. 68–74.
6. Linde, A., *The self-reproducing inflationary universe*. Scientific American, 1994. **271**: p. 48–55.
7. Popławski, N.J., *Cosmology with torsion: an alternative to cosmic inflation*. Physics Letters B, 2010. **694**(3): p. 181–185.
8. Musser, G., *Virtually reality*. Scientific American, 2019. **321**(3): p. 30–35.
9. Rotman, D., *Einstein's "spooky action at a distance" paradox older than thought*. MIT Technology Review, 2012. https://www.technologyreview.com/2012/03/08/20152/einsteins-spooky-action-at-a-distance-paradox-older-than-thought/.
10. A. Tonomura, et al., *Demonstration of single-electron buildup of an interference pattern*. American Journal of Physics, 1989. **57**(2): p. 117–120.
11. Markus Arndt, et al., *Wave—particle duality of C60 molecules*. Nature, 1999. **401**.
12. *The Universe*. Microscopic universe. History Channel. Aired on June 3, 2012.
13. Feynman, R.P., *The character of physical law*. The Messenger Lectures. 1965, Cambridge, MA: MIT Press. 173 p.
14. Carroll, S., *Even physicists don't understand quantum mechanics*. New York Times, 2019.
15. Folger, T., *If an electron can be in two places at once, why can't you?* Discover Magazine, 2005. https://www.discovermagazine.com/the-sciences/if-an-electron-can-be-in-two-places-at-once-why-cant-you.
16. Tegmark, M., *Parallel universes. Not just a staple of science fiction, other universes are a direct implication of cosmological observations*. Scientific American, 2003. **288**(5): p. 40–51.
17. Davies, P., *Cosmic jackpot: why our universe is just right for life*. 2007, Boston, MA: Houghton Mifflin. 242 p.
18. Pinker, S., et al. *Does science make belief in God obsolete*, 2007. www.templeton.org/belief.
19. Meyer, S.C., *Return of the God hypothesis: three scientific discoveries that reveal the mind behind the universe*. First edition. 2021, New York, NY: Harper One, an imprint of HarperCollins Publishers. 568 p.
20. Krauss, L.M., *A universe from nothing: why there is something rather than nothing*. 1st Free Press hardcover ed. 2012, New York, NY: Free Press. xix, 202 p.

21. Ellis, G.F., *Does the multiverse really exist?* Scientific American, 2011. **305**(2): p. 38–43.
22. Vilenkin, A. and M. Tegmark, *The case for parallel universes.* Scientific American, 2011. www.scientificamerican.com/article/multiverse-the-case-for-parallel-universe/.
23. Houston-Edwards, K., *Numbers game.* Scientific American, 2019. **321**(3): p. 35–40.
24. Hossenfelder, S., *Lost in math: how beauty leads physics astray.* First edition. 2018, New York, NY: Basic Books. xi, 291 p.
25. Gleiser, M., *The dancing universe: from creation myths to the Big Bang.* 1997, New York, NY: Dutton. xiii, 338 p.
26. Hawking, S., *A brief history of time: from the Big Bang to black holes.* 1988, New York, NY: Bantam Dell Publishing Group. 198 p.
27. Gleick, J., *Isaac Newton.* 2003, New York, NY: Pantheon Books. p. 272.
28. John D. Barrow and F.J. Tipler, *The anthropic cosmological principle.* 1986, Oxford: Oxford University Press. 706 p.
29. Sobel, D., *Galileo's daughter: a historical memoir of science, faith, and love.* 1999, New York, NY: Walker & Co. ix, 420 p.
30. Osler, M.J., *Eternal truths and the laws of nature: the theological foundations of Descartes' philosophy of nature.* Journal of the History of Ideas, 1985. **46**(3): p. 349–362
31. European Commission, *Special Eurobarometer.* 341 / Wave 73.1—TNS Opinion & Social, 2010 (Biotechnology, p. 381).
32. Mathewes, C.T., *Why evil exists,* in *The great courses; variation: great courses (compact disc).* 2011, Chantilly, VA: Teaching Co.
33. Kristof, N., *The best news you don't know.* New York Times, 2016.
34. Pinker, S., *The better angels of our nature: why violence has declined.* 2011, New York, NY: Viking. xxviii, 802 p.
35. Barrow, J.D., *The constants of nature: from Alpha to Omega—the numbers that encode the deepest secrets of the universe.* 2002, New York, NY: Pantheon Books. p. 368.
36. Heilbron, J.L., *The dilemmas of an upright man: Max Planck as spokesman for German science.* 1986, Berkeley: University of California Press. xiii, 238 p.
37. Jammer, M., *Einstein and religion: physics and theology.* 1999, Princeton, NJ: Princeton University Press. p. 288.
38. Einstein, A. and H. Allen, *The world as I see it.* 1949, New York, NY: Philosophical Library. p. 108.
39. Ioannidis, J., *Why most published research findings are false.* PLoS Medicine, 2005. **2**(8): p. 0696–0701.
40. Yong, E., *Bad copy.* Nature, 2012. **485**: p. 298–300.
41. Kuhn, T.S., *The structure of scientific revolutions.* Second Edition, ed. International Encyclopedia of Unified Science. Foundations of the Unity of Science, v. 2, no. 2; Variation: International Encyclopedia of Unified Science; v. 2, no. 2. 1970, Chicago: University of Chicago Press. xii, 210 p.
42. Seth, A., *Our inner universes.* Scientific American, 2019. **321**(3): p. 40–47.
43. Rovelli, C., *Reality is not what it seems: the journey to quantum gravity.* First American edition. ed. 2017, New York, NY: Riverhead Books. 280 p.
44. Cline, K., *Solid objects are mostly empty space.* Independent Record, June 6, 2012.
45. Hoffman, D., *What scientific idea is ready for retirement: truer preceptions are fitter perceptions.* Edge, 2015.
46. McNamara, J.M., P.C. Trimmer, and A.I. Houston, *It is optimal to be optimistic about survival.* Biol Lett, 2012. **8**(4): p. 516–519.
47. Mark, J.T., B.B. Marion, and D.D. Hoffman, *Natural selection and veridical perceptions.* Journal of Theoretical Biology, 2010. **266**(4): p. 504–515.
48. Kinnvall, C., *Globalization and religious nationalism: self, identity, and the search for ontological security.* Political Psychology, 2004. **25**(5): p. 741–766.

49. Ysseldyk, R., K. Matheson, and H. Anisman, *Religiosity as identity: toward an understanding of religion from a social identity perspective.* Personality and Social Psychology Review, 2010. **14**(1): p. 60–71.

50. Armstrong, K., *A history of God: the 4000-year quest of Judaism, Christianity, and Islam.* First American ed. 1993, New York, NY: A.A. Knopf: Distributed by Random House. xxiii, 460 p.

51. Ehrman, B.D., *How Jesus became God: the exaltation of a Jewish preacher from Galilee.* First edition. ed. 2014, New York, NY: Harper One. 404 p.

52. Armstrong, K., *The battle for God.* 2000, New York, NY: Knopf/HarperCollins.

53. Friesen, J.P., T.H. Campbell, and A.C. Kay, *The psychological advantage of unfalsifiability: the appeal of untestable religious and political ideologies.* Journal of Personality and Social Psychology, 2015. **108**(3): p. 515–529.

54. Carroll, S.M., *The big picture: on the origins of life, meaning, and the universe itself.* 2016, New York, NY: Dutton est 1852, an imprint of Penguin Random House LLC. p. 496.

55. Kolodiejchuk, B., *Mother Teresa: come be my light.* 2007, New York, NY: Doubleday. 404 p.

56. Barron, R., *Saint of light, saint of darkness.* 2016, Irondale, AL: National Catholic Register.

57. Darwin, C., *To Asa Gray.* Darwin correspondence project, 1860. DCP Letter 2814.

58. Darwin, C., *The autobiography of Charles Darwin, 1809–1882: with original omissions restored.* Norton library; Variation: Norton library. 1969, New York, NY: Norton. 253 p.

59. Livio, M., *Brilliant blunders: from Darwin to Einstein—colossal mistakes by great scientists that changed our understanding of life and the universe.* First Simon & Schuster hardcover edition. ed. 2013, New York, NY: Simon & Schuster. 341 p.

Part I

The Making of a Habitable World

1 In the Beginning
The Genesis of the Universe

The beginning is the most important part of the work.

Plato

He who chooses the beginning of the road chooses the place it leads to.

Kami Garcia

If you think this Universe is bad, you should see some of the others.

Philip K. Dick

What is time? If nobody asks me, I know; but if I were desirous to explain it to one that should ask me, plainly I know not.

St. Augustine

Regardless of how many universes there may be, and irrespective of how our universe arose, we know that the universe we live in can generate life. Why is this true? As we will see, for life to exist, our universe had to possess features that would lead to the production of galaxies populated by stars and planets. The qualities of our universe that made such galaxies possible were acquired during the Big Bang, and these attributes are more critical and precise than anyone living just a century ago would have imagined.

ONE HECK OF A BEGINNING

The first of the fortuitous events that we can clearly identify is the moment that our universe came into being. According to physicists, when this happened, all the baryonic and radiative energy (i.e., ordinary energy) that the universe would ever have appeared, and it existed within a point smaller than that occupied by an atom. This fact is worth pondering. In some ways, it seems like just another fun factoid—like saying that Fenway Park in Boston can accommodate 37,400 fans; however, a moment of reflection reveals that this assertion is perhaps one of the most fantastic ever to be made. To begin with, it implies that the energy of as many as several trillion galaxies was present in this infinitesimally small spot. Einstein provided insight into just how much energy this is when he determined that the energy (E) in the universe is equal to mass (m) times the speed of light squared (c^2). This now-famous equation indicates that the production of even a very small amount of matter requires a truly staggering amount of energy. In fact, it has been estimated that to produce 1 gram of matter (which is about the weight of a single M&M candy), you would need an amount of energy similar to that released by the atomic bomb

DOI: 10.1201/9781003270294-3

that destroyed Nagasaki, Japan. Ponder this for a while and then try to imagine the amount of energy necessary to create the ordinary matter of the universe. When doing so, consider the relative sizes of some of the objects in the known universe. You may look outside your window and see a large boulder and think it is colossal, or you may gaze out onto the massiveness of Earth-bound structures such as the Grand Canyon in awe-struck wonderment; however, on an astronomical scale, these structures are virtually nothing. NASA reports the weight of our Sun to be a whopping 4.3×10^{30} pounds or about 330,000 times that of Earth, but in terms of stellar heavy weights, our Sun is no big deal. It is just one somewhat larger than average star. Scattered throughout the universe are stars so enormous that envisioning their mass boggles the mind. One of the largest we know of is R136a1, which is hundreds of times more massive than our Sun [1]. While considering all of this, note that there are more stars in the universe than there are grains of sand on Earth. Now consider how much energy would be needed to generate just the ordinary stellar matter that is currently in our universe.

In addition to the energy needed to form the stars that now blaze, we must also consider the free energy in our cosmos. Much of the energy that existed at the moment of the Big Bang was present in the form of gamma radiation, which is a type of energy that we now encounter during nuclear explosions. At the moment the universe began, gamma radiation permeated all of the space that came into existence; however, as that space expanded, so did the gamma rays. Today, 13.8 billion years after the Big Bang, the cosmos is a very large place indeed, and the high-energy gamma rays produced by the Big Bang have been stretched into low-energy microwaves. These microwaves permeate all of the incredibly large cosmos, and they form what is referred to as the cosmic microwave background (CMB) radiation.

The existence of CMB radiation was first predicted in 1948 by Ralph Alpher [2], and this energy was initially detected in 1965 by Arno Penzias and Robert Wilson [3–5]. Today, this radiation has an almost perfectly uniform distribution throughout the cosmos, and its existence constitutes some of the most compelling evidence for the idea that our universe is expanding and unfolding as described in the Big Bang and inflation theories. Keep in mind that all this free energy that permeates the universe was also concentrated in the subatomic universe that formed at the moment of the Big Bang.

In addition to the cosmic background radiation, there is of course radiative energy that has been emanating from the gigantic nuclear explosions we call stars. For more than 13.5 billion years, some of the baryonic matter in these nuclear furnaces has been converted into radiative energy and then released into the vastness of space. In some instances, near the end of a large star's life, the mass-to-energy conversion process is so spectacular that the discharge from the dying star can be seen through large swaths of the observable universe. In fact, during some supernovae, the amount of energy released is similar to the estimated total amount of energy the Sun will liberate in its entire nine-billion-year lifetime. At the moment a star goes supernova, it is among the most energetically intense objects that ever existed. But, as impressive as all this matter and energy is, there was yet still more energy present at Big Bang. Much more.

We now know that the Big Bang also eventually led to the production of very large amounts of ill-defined and strangely behaving dark matter. Some have speculated that this matter now exists in dimensions of space that we cannot perceive, while others have suggested that it consists of primordial black holes that formed when dense regions of fundamental particles collapsed during the first second of cosmic time [6]. Whatever its nature, we know that for the galaxies to maintain their integrity, this mysterious dark matter must exist, and in total, dark matter constitutes approximately 85% of the total matter in the observable universe.

Having pondered this, you will no doubt arrive at the conclusion that there was indeed a very large amount of concentrated energy at the point that initiated the Big Bang. But together, radiative energy and the energy stored in the form of baryonic matter constitute a mere 5% of the mass energy of the universe, and dark matter accounts for another 27%. Remarkably, the remaining 68% of the mass energy in the universe consists of enigmatic dark energy.

Today, physicists believe that dark energy permeates the entire cosmos, and many argue that dark energy is a property of space itself [7]. However, despite dark energy's ubiquity, we are not sure how it came into existence, and although we know that baryonic and radiative mass energy has negative gravitational potential (i.e., it is attractive) while dark energy appears to have positive gravitation potential (i.e., it is repulsive), we do not understand all the other ways in which these forms of energy differ.

Many cosmologists are confident that dark energy was present at the moment of the Big Bang, and this energy or something like it (e.g., "a vacuum energy") powered a faster-than-light inflationary period. This inflationary period is believed to have lasted only 10^{-32} seconds [8], but some estimate that during this exceptionally ephemeral moment, two points within an atomic radius were separated by about four light years, which is the approximate distance between our Sun and the next star [6]. At first glance, this inconceivable rate of expansion seems to violate the rule that nothing can move faster than the speed of light; however, during the inflationary period, space itself was being created, and the rate of space creation immediately after the Big Bang was much faster than the speed of light.

After the initial inflationary burst, the rate of expansion of our universe slowed, but it did not stop. However, given the way the universe did expand, it is very unlikely that all the universe's current supply of dark energy was present at the moment of the Big Bang. If it had been, the universe's expansion rate would have been much higher. So, to make sense of this, some have suggested that dark energy is being created *de novo* as space is enlarging. If this is the case, and if the universe is a closed system, then perhaps dark energy does not obey the first law of thermodynamics (which states that energy cannot be created or destroyed), and perhaps dark energy arises constantly as a result of non-random fluctuations in quantum fields [9, 10].

The existence of a process that violates the first law of thermodynamics would certainly shake up the apple cart of ideas within the realm of physics. However, most physicists believe that dark energy is likely conserved, and they are confident that once we better understand dark energy, and the nature of the universe, we will see that dark energy conforms to the first law of thermodynamics as well as the other laws of physics.

As of this writing, however, we know very little about dark energy, and we certainly do not fully understand the nature of the universe. Nevertheless, if energy is defined as the ability to do work and make things happen, then dark energy is apparently aptly named, because this force appears to make the entire universe expand, and its nature is cloaked in a seemingly impenetrable mystery.

After contemplating all the aforementioned, it's now time to try to imagine the nearly incomprehensible amount of energy that was present in the moment that marked the birth of our universe. If you have a hard time doing so, you are not alone. In some ways, envisioning this is harder than imagining that the universe is structured upon a layer of never-ending "tortoises all the way down" (see the Introduction for more on this). However, unlike the protagonist in the tortoises all the way down the story, the physicists who generated the Big Bang theory have gathered a large amount of empirical data to support the idea that there was indeed a Big Bang that contained an unimaginable amount of energy. It is possible that some of the details of this theory will change with the ongoing acquisition of information, but if the data that support it continue to withstand scrutiny, and if this information was interpreted correctly, then the Big Bang truly was an utterly spectacular event that possessed a magnificence, power, and grandeur that is impossible for us to fully comprehend and appreciate.

SCALAR FIELDS AND THE REMARKABLE UNFOLDING OF THE UNIVERSE

While most physicists believe that both the Big Bang and inflation occurred, the best minds in this discipline still cannot definitively explain *why* they did so. We also don't know why the Big Bang event had the exact energy that it did, and not more or less. What we do know is that the universe didn't really begin with a big "explosion" because, as I mentioned in the Introduction, explosions depend on the presence of pressure differentials, and at the moment of the Big Bang, there was no space, time, or matter, and therefore, no pressure differentials. So, when space came into existence, why did it expand, and why did this process form our observable universe?

In an attempt to answer these questions, some have postulated that scalar "inflation fields" made it possible for the universe to enlarge. Andrei Linde suggests these fields are analogous to electrical potential fields in that they appear only if the potential between two poles is asymmetrical [11]. According to Linde, a tremendous potential energy asymmetry existed within a scalar field, and as it dissipated the incredibly rapid process of inflation unfolded and space was created.

Physicists also suggest that, as the potential energy of the inflation scalar field moved toward its minimum, some of the field energy was converted into elementary particles, and as the energy of the field began to fill the expanding volume of space, the universe quickly cooled. As a result, after a mere 0.01 seconds, the temperature of the universe was 10^{11} degrees Celsius—down from a temperature of at least 10^{15} degrees at the initial moment of the Big Bang [12]. Further, after 0.1 seconds, the temperature of the nascent universe had dropped to 3×10^{10} degrees Celsius, and 1 second after the Big Bang the temperature of the universe was "just" 10 billion (1×10^{10}) degrees Celsius [13].

THE ESTABLISHMENT OF THE COSMOLOGICAL CONSTANT, Λ (LAMBDA)

As was mentioned earlier, after the initial burst of expansion, the rate of inflation of the universe slowed, but it did not stop. Had it done so, the universe would have been unstable, and gravity would have eventually gathered matter into a point like that from which it began. Einstein realized this, and when he was pondering this issue in 1917, he concluded that after expanding for some period, the universe eventually stabilized. To explain how a static, stable universe could exist, Einstein argued for the existence of a "cosmological constant" that could "correct" or exactly compensate for the gravitational force that would tend to cause the universe to contract. He also suggested that this cosmological constant, which he called "lambda," wasn't just a fudge factor. Rather, he viewed it as describing a property of spacetime. Yet Einstein did not speculate as to how this property arose, nor did he describe the nature of the physical processes that generated this putative force [14]. He did, however, suggest that the magnitude of this expansion force increased as a function of distance, so two points that were relatively close together, for example, two points in our solar system, would not be significantly affected by lambda, but two locations at the opposite ends of the universe would be profoundly impacted.

Despite the cosmological constant's initial utility, Einstein quickly retracted the suggestion that his equations should contain this value once he realized that the universe was in fact not a static entity. The Belgium cosmologist and Jesuit priest, Georges Lemaitre, was among the first to propose that this was so, and the American physicist Edwin Hubble eventually provided empirical data that clearly indicate that most of the galaxies in our universe are indeed receding from us because the universe itself is expanding [15].

By 1998, investigations of type Ia supernovae by Brian Schmidt, Adam Riess, and Saul Perlmutter revealed an even more dramatic reality. They found that not only was the cosmos expanding, but surprisingly, it was doing so at an increasing rate [16–18]. The force that causes this accelerated expansion has been labeled "lambda," and so decades after Einstein's death, the cosmological constant has returned. But this time it is not being used as a mathematical tool to describe a force that maintains a static universe. Instead, lambda is viewed as a representation of the energy of empty space, and presumably, it symbolizes the dark energy that is driving the accelerating expansion of space itself.

The magnitude of the repulsive dark energy in newly forming space has profound implications regarding the fate of our universe. We now know that, if the density of energy in this newly materializing space had been larger, the rate of expansion would have overwhelmed the force of gravity, and the planets, stars, and galaxies would have failed to form. In fact, if lambda had been stronger, even atoms themselves would have failed to coalesce. On the other hand, if lambda had been weaker, the universe would have eventually contracted, and it would have done so before intelligent life could evolve [12]. We obviously live in a universe with stars, planets, and life, and we now know that all this was possible because, in our universe, lambda has a value of 7×10^{-30} g/cm^3 [19]. Given that the force of lambda could, in theory, have been very, very different, why does lambda have the specific value that we observe? Speculation abounds [19, 20], but in short, we simply don't know.

After considering this issue, the American Nobel Prize-winning physicist Steven Weinberg asserted that, if there is only one universe, it seems incredibly unlikely that it would just happen to have the value of lambda that it does. Therefore, to explain the seemingly inexplicable, Weinberg postulated that the value of lambda we observe strongly suggests that a multiverse exists [12, 21]. Presumably, within a multiverse, virtually all the other universes have different constants for the value of lambda. But again, other than anthropic arguments, there is no compelling evidence for the existence of a multiverse, and furthermore, even if one grants that a multiverse does exist, the question remains: Why does space in the various universes of the multiverse have dark energy to begin with? As was mentioned earlier, some say that the space in each universe has this property because of quantum mechanical phenomena, but this just furthers the question, and we must then ask why quantum mechanical phenomena imprint physical traits on a potentially infinite number of universes.

Regardless of what the answers to these questions are, the realization that our universe has a very small nonzero value for lambda helps us understand how our universe evolved, and it also allows us to predict what our universe may look like in the distant future. Indeed, the work of Adam Riess reveals that, for most of the first eight billion years, the mass of our universe was sufficient to slow the rate of the universe's expansion. However, as the cosmos continued to enlarge, the amount of space increased while the energy density of that space remained constant. As a result, the matter in expanding space was diluted, and after about eight billion years, an equilibrium state was achieved in which the gravitational force of matter was equal to the repulsive force of lambda. However, this static universe, which was similar to the one Einstein envisioned when he proposed the cosmological constant, was short-lived. As space continued to expand and matter continued to become more dilute, the repulsive energy of space increasingly overwhelmed the gravitational attraction of matter, and the rate of expansion of the universe began to accelerate exponentially. This increasing rate of expansion, which causes the distant galaxies to recede from us at ever-increasing speeds, has been going on now for more than five billion years.

Will the rate at which galaxies separate from each other eventually slow? We don't know for sure, but the answer appears to be no. If this is the case, in 100 billion years, the matter located in distant parts of the universe will move away from us so quickly that it will no longer be possible to detect light emanating from it. When that day comes, future cosmologists in our galaxy will no longer be able to obtain empirical evidence to prove that they are a part of a universe that contains trillions of galaxies. Instead, these scientists will find data consistent with the idea that our galaxy and a few relatively nearby sister galaxies are alone, and unless they are somehow familiar with the work of cosmologists who lived many billions of years earlier, they will remain forever ignorant of the true composition of our universe. In the absence of our current information, some of these future cosmologists may speculate that countless galaxies do indeed exist, but they will be in a position similar to the multiverse proponents of today: They will have an interesting idea, but they will have difficulty in providing compelling evidence to support their proposition.

FUNDAMENTAL FORCES ESTABLISH THE FABRIC
OF THE UNIVERSE

The birth of the universe was an incredibly busy period. In addition to inflating, cooling, generating the first subatomic particles of matter, and ultimately obtaining a value of lambda conducive to galaxy formation, during the first millionth of a second, the universe also generated the four fundamental forces that govern every aspect of our world. These all-important determinants of physical reality are gravity, the electromagnetic force, the strong nuclear force, and the weak nuclear force.

Gravity is by far the weakest of the four forces, but it is pervasive, and it is important since it binds matter together. The electromagnetic force consists of the electric force produced by the interaction of charged particles and the magnetic force generated by charged particles in motion. This force is very large in comparison to gravity, and it governs most of the chemical properties of atoms and molecules. Interestingly, when the attractive force of gravity exactly counterbalances the repulsive electromagnetic force that acts on atoms, structures such as planets can form.

The remaining two nuclear forces are not those we typically think much about; however, they are essential for the structure and behavior of the universe. The strong nuclear force binds subatomic particles called quarks, and this allows for the formation of protons, neutrons, and atomic nuclei. Meanwhile, the weak nuclear force acts on subatomic particles named fermions, and it is manifest in the exchange of force carrier particles that operate over very small distances. This weak nuclear force governs the decay of some forms of unstable atoms, and it can initiate the nuclear fusion reactions that occur in stars. Processes influenced by this force, such as the fusion of two protons to form deuterium (which consists of one proton and one neutron), make it possible for stars to produce the massive amounts of energy they need to exist.

Each of the four fundamental forces has unique characteristics, but most physicists believe that, at the moment of the Big Bang, there was only one force, and very soon after the four individual forces were derived from this one. However, despite intense efforts, attempts to explain how all four were once united have been unsuccessful. Demonstrating the common link between the four fundamental forces remains the goal of those who work on the so-called "Theory Of Everything" (TOE).

In addition to trying to demonstrate how the four forces interrelate, physicists are also attempting to explain why the fundamental forces possess specific, and sometimes perplexing, traits. For example, we know that the electric and magnetic forces are two variations of the electromagnetic force, but even though the electrical force can be converted into a magnetic force and vice versa, these two manifestations of the electromagnetic force have very different properties. The electric force, for instance, can easily generate positive and negative charges, but the magnetic force is not separable in the same way. Instead, physicists have found that whenever a magnetic dipole is split in two, each of the resulting smaller magnets is also a dipole, and interestingly, to date, no naturally occurring monopoles have been identified. And while we don't know why the electric and magnetic forces differ in this way, it is fortunate that this dichotomy exists, because if we lived in a universe where the electrical force could not be easily separated into opposite charges, protons and electrons would not readily form, and atoms, as we know them, would not be part of our universe.

THE IMPORTANCE OF MATTER AND ANTIMATTER ASYMMETRY

Physicists describe many other fascinating processes that took place during the very early moments of the universe. Among these are interactions between the subatomic particles produced during the Big Bang. These particles initially existed in the form of matter and antimatter; however, upon interaction, matter and antimatter annihilate each other in a burst of pure energy [22]. Had the balance of matter and antimatter after the Big Bang been exactly equal, all of the subatomic particles in the cosmos would have converted back into energy, and the universe today would consist of nothing but pure energy distributed throughout vast amounts of empty space. Fortunately for us, the ratio of matter to antimatter was 1,000,000,001 to 1,000,000,000 [23], so after matter and antimatter finished annihilating each other, the tiny fraction of matter that remained was sufficient to form the estimated 10^{78} atoms of our observable universe [24]. Why was there an asymmetry in the balance of matter and antimatter? To date, the answer to this question remains elusive. Whatever the cause, what is clear is that soon after the Big Bang, there was an unimaginably large amount of mass that came into existence and then disappeared in what could only have been one very spectacular blast.

GETTING THE DENSITY OF MATTER RIGHT: THE VALUE OF Ω (OMEGA)

The fact that all but 0.5 billionth of the matter initially produced after the Big Bang was eliminated due to the presence of antimatter is worth taking a moment to contemplate. What would the universe be like if, after the interaction with antimatter, significantly more matter had remained? Well, we know that in a cosmos with a gravitational force like that of our universe, if the density of the remaining matter had been significantly higher, matter would have accreted, and a "Big Crunch" would have ensued. This process would have prohibited the formation of stars and all that arises from them. It is also possible that the universe would act like a yo-yo, expanding in a violent "Big Bang" and then accreting in an equally ferocious "Big Crunch." Alternatively, had there been significantly more matter in our universe after the Big Bang, our universe may have simply collapsed into a series of black holes.

It is also worth asking what the universe be like if, after the matter–antimatter interactions, even less matter remained? In this situation, the density of matter could have been too low for accretion to occur, and as a result, there would once again be no planets, stars, or galaxies. In this case, the matter of the universe would have drifted apart and spread out over the vast expanse of space. Such a universe would have been a very dull place, completely devoid of higher-order structures [23]. Luckily, the density of matter in our universe is neither too high nor too low for the purposes of galaxy building.

THE IMPORTANCE OF SYMMETRICALLY DISTRIBUTING MATTER: THE VALUE OF Q

While the density of matter was certainly critical to the formation of galaxies, so too was the symmetry of the distribution of that matter. Had a perfectly symmetrical universe arisen, all of space would have had an equal density of material, and there would be no regions where the concentration of matter was high enough to exert a gravitational pull on other nearby matter. Such a universe would contain nothing more than small particles floating uniformly through an otherwise empty void. In a perfectly symmetrical universe, hydrogen atoms could not have congregated into massive gas clouds, and there could have been no accretion of these gas clouds, and no planets or stars. A universe with a perfectly homogenous distribution of matter would quickly become cold and dark, and it would be structurally uninteresting.

Fortunately, a slight asymmetry in the matter distribution yields a completely different outcome. In this situation, the force of gravity becomes all-important, and as small clusters of hydrogen and helium gas form, the gravitational influence of pockets of matter grows, the asymmetries in the distribution of matter are amplified, and large-scale structures, such as galaxies, eventually appear. Only in a slightly asymmetrical universe does life become a possibility [23].

What is the magnitude of this asymmetry, and what caused the asymmetrical distribution of matter? The first of these two questions can be answered with confidence: It appears that the degree of asymmetry was exceedingly small. In fact, analysis of the background radiation from a period near the birth of the universe indicates that the distribution of matter was nearly perfectly homogenous (see Figure 1.1) and that only one part out of 100,000 was out of place. But, as it turns out, these miniscule anomalies in the distribution of matter were enough to initiate galaxy formation.

As for the cause of these diminutive irregularities, that remains an enigma. It may be that quantum fluctuations generated the asymmetries, and then these tiny perturbations were extended and distributed smoothly across the universe during the inflationary period when space expanded faster than the speed of light. The current size and uniformity of the universe suggest that an inflationary process did in fact

FIGURE 1.1 Cosmic microwave background radiation of the nascent universe. Small differences in the uniformity of the radiation distributed throughout the universe are displayed as variations in color. (This image was generated by NASA using the Wilkinson Microwave Anisotropy Probe (WMAP). European Space Agency, CC BY-SA 4.0 <https://creativecommons.org/licenses/by-sa/4.0>, via Wikimedia Commons.)

evenly distribute the initial asymmetries of the atom-sized universe, and as a result, only slight irregularities in the distribution of matter are now present in the immense cosmos of today.

EXISTING IN A THREE-DIMENSIONAL UNIVERSE

After the initial events of the Big Bang, our universe eventually emerged with three dimensions of space. It is possible, however, that this situation could have been different. For instance, it is conceivable that our universe could have possessed fewer or greater numbers of spatial dimensions. Had our universe contained just two spatial dimensions, the complexity of structures that could form would pale in comparison to that possible in our world. For example, in a two-dimensional universe, spherical planets would not exist, and complex three-dimensional structures like those found in us would not be possible. And, while it is more difficult to envision, we know that a world with four spatial dimensions would have also been a disaster. In our three-dimensional universe, the force of gravity decreases at a rate equal to the square of the distances between the interacting objects, but in a universe with four spatial dimensions, gravity would decrease as a function of the cube of the distances. As a result, this four-dimensional world would not be able to maintain stable, gravitationally induced orbits, because if an object, such as a planet, moved toward the structure it orbited, the force of gravity would increase much more quickly than it does in our world, and the orbiting object could quickly be dragged into the larger structure and destroyed. Furthermore, in a world with four spatial dimensions, if an orbiting object moved slightly away from a more massive structure, the sudden decrease in the force of gravity exerted by the larger object would make it easy for the satellite to break free, and as a result, it could be jettisoned into the vastness of space. Due to these phenomena, the existence of stable planetary systems like our solar system would be rare and short-lived in a four-dimensional universe.

In a world with four spatial dimensions, problems at the subatomic level would also become manifest because even the orbits of electrons would be precarious. At the atomic level, the positively charged protons repel each other, as do the negatively charged electrons, and in our three-dimensional universe, the strength of the repulsion of like-charged particles varies as a function of the square of the distance between the charges, i.e., electrical repulsion also obeys the inverse square law. But, in a world with four spatial dimensions, the magnitude of the like-charged repulsion would once again vary as a function of the cube of the distance between the interacting particles, and in such a universe, electrons circling a nucleus of protons and neutrons would frequently spiral inward toward the more massive nucleus, or alternatively, they could escape from their orbits.

The problems of living in a world with four spatial dimensions would not be limited to the existence of unstable solar systems and atoms. In a universe with an extra spatial dimension, stars would also be unstable. The integrity of a star depends on a delicate dance between the outward pressure generated by the force of the star's nuclear reactions and the inward pressure generated by the force of gravity acting on the star's mass. If the force of gravity was to vary by the cube of the distance between the interacting matter, the gravitational force would more easily overwhelm the outward pressure generated by the nuclear explosions, and the star would quickly collapse upon itself. As a result, in a four-dimensional world, stars would have significantly shorter life spans, and biological evolution, if it occurred, would be much more constrained [23].

TIMING IS RELATIVE, AND TIMING MAKES A DIFFERENCE

The dimension of time is difficult to comprehend, and it is likely that many of us feel as St. Augustine did when he stated that, when he was not questioned about time, he had a sense that he understood it, but when asked to define it, he was at a loss. However, it may indeed be difficult to define time, because according to Einstein and many other physicists who study quantum gravity, time is not a fundamental property of the universe [25–27]. Instead, time may emerge from the interactions among quantum particles [28], and our sense of time may arise when we compare the order and number of times one regularly occurring event takes place relative to another.

As a young man, Galileo was among the first to think of time in this way, and he reportedly developed this insight while he was pondering how to accurately measure time in a world that lacked precision timepieces. According to legend, while he was sitting in a Catholic Mass in the Pisa Cathedral in Italy, Galileo noticed that a chandelier was oscillating at what appeared to be a regular interval. This observation led him to realize that there was an order in which regularly occurring events took place, and the number of times one regularly occurring event transpired was directly proportional to the number of times another regularly occurring event ensued. In particular, Galileo noticed that his heart, which he assumed was beating regularly, contracted a specific number of times before the chandelier moved through its arc.

Galileo's insight had great practical value, and it led to the development of much better clocks, but Galileo knew that the schema he generated to precisely measure "time" was not directly measuring something like length, width, or height. Instead, Galileo understood that he was defining time as the relationship that arose between regularly occurring processes.

Ultimately, if time is relational, then although it may not be a fundamental property of the universe, it is still important, because a change in the order or number of times one process occurs relative to another could alter the universe. For example, if the processes that occur during a supernova were to take place more quickly relative to the oscillation of a pendulum, supernovae might have produced fewer heavy elements. On the other hand, if the expansion of the universe after the Big Bang had occurred more slowly relative to the rate of the oscillation of a pendulum, the temperature of the universe would have decreased more gradually, and more of the hydrogen produced during the Big Bang would have been transmuted into heavy elements [23]. In the end, if either of these sets of relationships had been different, our universe would be fundamentally different, and these changes would preclude the development of our universe as we know it.

Before moving on, it is worth mentioning that the scenarios described earlier represent two of the countless ways that the intrinsic relationships between regularly occurring physical events in the universe could be transformed. Ultimately, there is an exceedingly large number of changes of this type that could potentially alter our universe.

THE ARROW OF TIME

Yet another critical aspect of our universe's character is that "time" within our cosmos appears to move unidirectionally, and this is true even though all but one of the laws of physics can unfold equally well in the forward and backward directions. The one law that does not work bidirectionally is the second law of thermodynamics, which states that within closed systems, entropy (or disorderliness) always increases. Interestingly, this one exception may be what imposes directionality upon time [28].

Physicist Sean Carroll nicely explains why entropy tends to increase in a closed system by describing why entropy increases in places such as your coffee mug. He notes that if you mix milk and coffee together in your mug, the molecules in the two liquids tend to remain intertwined and in a disorderly state because there are a very large number of ways in which the molecules that compose milk and coffee can be mixed, but there are only a very limited number of ways in which these same molecules can be unmixed [10].

Of course, it is possible that all the milk molecules could, just by chance, separate from all the coffee molecules, but this is an extraordinarily low probability event. In fact, Carroll suggests you would likely have to wait longer than the current age of the universe to see it happen. Given this, all the events that we perceive as moving forward in time may appear to do so because they are moving toward a greater state of disorder, and to quote Carroll, "the arrow of time is simply the tendency of systems to evolve toward one of the numerous, natural, high-entropy states." However, it follows that if our universe unfailingly moves toward a higher state of entropy, it must have started in a very low state of entropy. Consequently, when our universe began, its energy and mass must have been arranged in a highly symmetrical pattern.

To say that the universe's matter and energy were highly symmetrically distributed soon after the Big Bang may, at first, seem to be an argument against the idea that the universe began in a low entropy state. After all, when milk and coffee molecules are symmetrically distributed, the entropy (i.e., the disorder of the coffee and milk molecules within the system) is said to be high. However, in a universe where gravity exists, there is only one way to get a very highly ordered distribution of matter and energy, and that is to space that matter and energy in a manner that prevents it from clumping. To do that, the matter and energy must be symmetrically distributed throughout space.

As it turns out, the matter and energy in our universe were nearly perfectly symmetrically distributed at the time of the Big Bang, but as was mentioned earlier, a very small degree of asymmetry did exist. This wee bit of unevenness eventually caused matter to clump, structures to form, and the cosmos to evolve, and currently, the state of asymmetry, disorder, and entropy in our universe continues to increase. Consequently, the cosmos continues to evolve, and the arrow of time remains decidedly one directional.

Today, as we live out our lives in a time-directional universe, we are left with a truly puzzling question: "why did our universe start in such a highly-ordered state"? Our universe could have started in a highly disordered state, and in such a place, time and complex structures would not exist [29].

Interestingly, the odds of our universe starting in a highly ordered state appear to be fantastically small, and if there is only one universe, the probability that we would be here is therefore infinitesimal, at best [30]. However, as I mentioned earlier, some have suggested that there are an astoundingly large number of universes, and they believe that our universe's initial configuration happened to be, just by chance, a highly ordered one [10, 31, 32]. If so, this would be one more example of an exceedingly unlikely trait that our universe just happens to possess.

THE CRITICAL VALUES OF MANY OTHER FUNDAMENTAL CONSTANTS ALSO INFLUENCE THE NATURE OF OUR UNIVERSE

It is not just the existence, character, and magnitude of the forces mentioned earlier that are critical for the formation of a universe like ours. The values of many cosmic constants are also of great significance [24, 33, 34]. For example, the ratio formed by the mass of the electron divided by the mass of a proton is called beta (β), and it governs the structure of the chemical world. The value of β is 1/1836, and for the molecules in our universe to function as they do, β must have a small value. If the mass of an electron were significantly increased, the atomic nucleus would not be able to constrain its electrons as effectively as it does now, and under these conditions, John Barrow speculates that highly structured molecules such as DNA would either fail to form, or in the event that they did, they would cease to engage in processes essential for life [24].

The fine structure constant, α, is also critical in constructing a universe like ours. This constant is a measure of the electromagnetic force that governs the interactions between photons and charged particles. This dimensionless constant has a value of 1/137, and it is influenced by other constants including the speed of light in a vacuum (c), the elementary charge constant (e), Plank's constant ($\hbar$), and the electric constant (ε_0). In our universe, for stars to burn hot enough to initiate a nuclear reaction, β cannot exceed 0.005 α^2. Consequently, if any of the components of the fine structure constant were changed, a universe without stars may emerge.

The aforementioned examples certainly do not constitute an exhaustive description of alternations in the fundamental constants that could change the nature of reality. But hopefully, they do illustrate the fact that the values of fundamental constants matter, and these constants set the parameters for life in our universe.

Why do the fundamental constants of the universe have the values they do? This is certainly an intriguing question, and to generate an answer, Einstein tried to develop a unified field theory that would characterize the relationships between the fundamental forces, elucidate the essential physical constants, and reveal the values those constants assume. During part of a series of letters he exchanged with his friend and former student Ilse Rosenthal Schneider, Einstein wrote:

> With the question of the universal constants, you have broached one of the most interesting questions that may be asked at all. There are two kinds of constants: apparent

and real ones. The apparent ones are simply the outcome of the introduction of arbitrary units but are eliminable. The real [true] ones are genuine numbers which God had to choose arbitrary, as it were, when He designed to create this world [24]

Ultimately, Einstein hoped to develop a unified field theory that would allow us to calculate the values of nature's "real [true]" constants as precisely as we wished.

Einstein's belief that such a theory would one day come to be was a product of his scientific and spiritual outlook. He indicated that he had a strong desire to determine if the universe itself depended on the precise values of the true constants, and if it did, he wondered whether God had any choice in assigning values to these constants. He mused that if God was going to generate a universe like ours, perhaps the values He could assign to the true constants were constrained.

Einstein spent a considerable portion of his life in pursuit of a unified field theory, and while many believe he could have spent his last years more productively, Einstein, having already made a name for himself, felt that it was his duty to undertake this grand task. In his view, he was in a position to do so, and he believed that younger, less established physicists, could not engage in this high-stakes endeavor without risking their careers [35]. However, despite a sustained effort, Einstein never succeeded in developing a grand unifying theory. Many others tried to carry on where he left off, and Steven Weinberg, Abdus Salam, and Sheldon Glashow shared the 1979 Noble Prize for demonstrating how the electromagnetic force and the weak nuclear force could be united, but despite the application of techniques and approaches that ranged from pseudoscientific numerology to pure thought, mathematical reasoning, and sophisticated particle physics, to date, no one has succeeded in developing a theory that unites all four fundamental forces.

Today, not everyone believes that we will develop a theory that describes why the fundamental constants have the values that they do. In fact, some question whether there really are unchanging fundamental constants at all. Among the skeptics are John Barrow and John Webb, both of whom suggest that the so-called "constants" may vary in different sections of the universe or at different times in the universe's history [33, 36, 37]. In addition, as was mentioned earlier, many believe that a multiverse exists, and each universe within the multiverse has different values for its constants.

Whether the fundamental constants have varied, or do vary, within our particular universe remains a matter of speculation and debate, but the National Institute of Standards and Technology lists more than 300 fundamental physical constants that can be used to describe our current universe (see the NIST web page at http://physics.nist.gov/cuu/Constants/Table/allascii.txt). The study of dark matter and dark energy may reveal additional constants, and as of today, the actual number of "real [true]" dimensionless constants remains a mystery.

In some ways, the fundamental forces and the fundamental constants of our universe are modern science's equivalent of Plato's forms. They exist "out there," and they are the basis for all that there is, but like Plato's forms, these forces and constants have the character of a metaphysical enigma.

THE BIRTH OF THE UNIVERSE'S PRIMORDIAL ATOMS

Up until now, we have focused largely on the events that occurred in the first one-millionth of a second after the Big Bang. And as we have seen, after this very short period, the universe was off to a good start.

However, one-millionth of a second after the Big Bang, matter was comprised primarily of free-floating protons, neutrons, and electrons. A lone proton forms the nuclei of the most common isotope of hydrogen, but immediately after the Big Bang, the universe was still too hot for the strong nuclear force to join protons and neutrons, and therefore nothing larger than a hydrogen nucleus (with its one proton) could exist. This situation changed within a few minutes when the plummeting temperature of the universe made it possible for deuterium nuclei (which contain one proton and one neutron) and helium nuclei (which have two protons and two neutrons) to begin to form [13].

Five minutes after its birth, the universe's temperature was down to a billion degrees Celsius. By then, 75% of the atomic nuclei formed hydrogen, and most of those remaining formed the nuclei of helium atoms. The first atoms and most of the other elements in the periodic table did not form until later in the universe's history.

Due to their early production, hydrogen and helium nuclei have often been referred to as primordial raw materials. These nuclei formed the basis for the formation of hydrogen and helium atoms, and these atoms, in turn, formed the primary constituents of stars and the raw material for the genesis of the far less common heavier elements. On a more personal level, it is worth noting that hydrogen nuclei are found in the molecules that compose us, and they are present in every cell of our bodies. It is therefore fair to say that we are composed in part by 13.8-billion-year-old relics of the Big Bang, and in a very real way, we are a modern-day walking embodiment of that event.

THE FIRST FIVE MINUTES OF THE UNIVERSE—AN INFLUENTIAL TIME PERIOD INDEED

The first five minutes of the universe was the most important five minutes in the last 13.8 billion years. During this fleeting period, gravity, the electromagnetic force, the strong nuclear force, and the weak nuclear force came into being; the precise characteristics and values of each of these forces were determined; matter was generated; the physical constants of the universe were set in place; the ratio of matter to antimatter was established; and a specific and limited number of spatial and temporal dimensions were weaved into the fabric of our cosmos. Furthermore, during this five-minute period, the expansion of space began, and the density and distribution of matter were determined. It was truly an utterly remarkable time.

For the theist, the Big Bang and the five minutes that followed can be viewed as evidence that God exists. The universe, with all its grandeur and mystery, is a creation worthy of a God, and the intellect required to engineer our cosmos would be God-like. The story of the creation of the matter and energy of the universe, and a description of how the creation was set on a course that would one day produce life forms capable of pondering the creator is one that could form its own sacred text.

From the theist's perspective, learning about the genesis of the universe is an awe-inspiring opportunity to gain insight into how God works, and to quote Einstein,

> the scientists' religious feeling takes the form of a rapturous amazement at the harmony of natural law, which reveals an intelligence of such superiority that, compared with it, all the systematic thinking and acting of human beings is an utterly insignificant reflection.

For the atheist who also does not accept the precepts of M theory and the notion of a multiverse, the probability of a universe like ours coming into existence is infinitesimally small, and our presence is either proof that we are luckier than anyone could have imagined, or it strongly suggests there is some underlying reason why the universe has the properties it does. Given the extreme improbability of our existence, members of this group would no doubt feel better if Einstein's dream of a grand unifying theory were realized, and they would clearly be happy if such a theory led to a relatively simple explanation for why our universe has the properties it does.

In some respects, the aftermath of the Big Bang is easier for atheistic M theory advocates to embrace. For them, it's not surprising that our universe has the characteristics necessary for creatures like us to emerge. After all, in a multiverse with 10^{500} universes, the odds that there is one universe like ours may indeed be a statistical inevitability, and therefore, sooner or later we had to come to be. But even for this group, a great mystery remains: What animates the laws of the multiverse and what makes the multiverse possible? For the M-theorists, this question may forever be shrouded in mystery and the unknown.

Finally, for the spiritual and multiverse agnostics who believe that there is no grand unifying theory, the reason for the Big Bang and the five-minute period that came after it remains a most profound enigma, and it is indeed a most compelling and interesting riddle.

Having finished our short tour of a few of the important cosmic features that arose out of the Big Bang, we are now able to shift our focus to the more complex systems that arose from this event. Ultimately, we will end with a discussion of our favorite complex system, namely, ourselves. But it is still a long journey from the Big Bang to *Homo sapiens*. So, we will start by looking at some comparatively simple, but important, inanimate by-products of the Big Bang. One of the most significant of these is stars. Within 200 million years of the Big Bang, our universe started to generate these cosmic plasma gas balls, and while these enormous spheres of hydrogen may seem rather dull, it turns out they are anything but. Stars are the universe's cosmic wombs. They are where the elements necessary for life arose, they are the sites from which life-giving energy radiates, and they are the structures that generate black holes, which, as we will see in Chapter 3, are the organizing centers of life-supporting galaxies. It is these all-important astronomical marvels that we will examine next.

REFERENCES

1. Crowther, P.A., et al., *The R136 star cluster hosts several stars whose individual masses greatly exceed the accepted 150 M ⊙ stellar mass limit.* Monthly Notices of the Royal Astronomical Society, 2010. **408**(2): p. 731–751.

2. Alpher, V., *Ralph A. Alpher, Robert C. Herman, and the cosmic microwave background radiation*. Physics in Perspective, 2012. **14**: p. 300–334.

3. Larson, R.B. and V. Bromm, *The first stars in the universe*. Scientific American, 2001. **285**(6): p. 64–71.

4. Penzias, A.A. and R.W. Wilson, *A measurement of the flux density of CAS A At 4080 Mc/s*. Astrophysical Journal Letters, 1965. **142**: p. 1149–1154.

5. Dicke, R.H., P.J.E. Peebles, P.G. Roll, and D.T. Wilkinson, *Cosmic black-body radiation*. Astrophysical Journal, 1965. **142**: p. 414–419.

6. Garcia-Bellido, J. and S. Clesse, *Black holes from the beginning of time*. Scientific American, 2017. **317**(1): p. 38–43.

7. Krauss, L., *Cosmological antigravity*. Scientific American, 2002. **280**: p. 52–59.

8. Tate, K., *Cosmic inflation: how it gave the universe the ultimate kickstart* (infographic). Space.com, 2014. https://www.space.com/25075-cosmic-inflation-universe-expansion-big-bang-infographic.html.

9. Carroll, S.M. and J. Chen, *Spontaneous inflation and the origin of the arrow of time*, 2004 (http://arxiv.org/abs/hep-th/0410270).

10. Carroll, S.M., *The cosmic origin of time's arrow*. Scientific American, 2008. **298**: p. 48–57.

11. Linde, A., *The self-reproducing inflationary universe*. Scientific American, 1994. **271**: p. 48–55.

12. Weinberg, S., *Facing up: science and its cultural adversaries*. 2001, Cambridge, MA: Harvard University Press. p. 306.

13. Weinberg, S., *The first three minutes: a modern view of the origin of the universe*. 1977, New York, NY: Basic Books. x, 188 p.

14. Livio, M., *Brilliant blunders: from Darwin to Einstein—colossal mistakes by great scientists that changed our understanding of life and the universe*. First Simon & Schuster hardcover edition. ed. 2013, New York, NY: Simon & Schuster. 341 p.

15. Hubble, E., *A relation between distance and radial velocity among extra-galactic nebulae*. Proceedings of the National Academy of Sciences, 1929. **15**: p. 168–173.

16. Riess, A., et al., *Observational evidence from supernovae for an accelerating universe and a cosmological constant*. Astronomical Journal, 1998. **116**: p. 1009–1038.

17. Riess, A.G. and M.S. Turner, *The Expanding Universe. From slowdown to speed up*. Scientific American, 2008, https://www-scientificamerican-com.ezp-prod1.hul.harvard.edu/article/expanding-universe-slows-then-speeds/.

18. S. Perlmutter, et al., *Measurements of Omega and Lambda from 42 high-redshift supernovae*. Astrophysical Journal, 1999. **517**: p. 565–586.

19. Steinhardt, P.J. and N. Turok, *Why the cosmological constant is small and positive*. Science, 2006. **312**: p. 1180–1183.

20. Barrow, J.D. and D.J. Shaw, *The value of the cosmological constant*. General Relativity and Gravitation, 2011. **43**(10): p. 2555–2560.

21. Weinberg, S., *Anthropic bound on the cosmological constant*. Physical Review Letters, 1987. **59**(22): p. 2607–2610.

22. Burhidge, G. and F. Hoyle, *Anti-matter*. Scientific American, 1958. **198**: p. 34–39.

23. Rees, M.J., *Just six numbers: the deep forces that shape the universe*. 2000, New York, NY: Basic Books. x, 173 p.

24. Barrow, J.D., *The constants of nature: from Alpha to Omega—the numbers that encode the deepest secrets of the universe*. 2002, New York, NY: Pantheon Books. p. 368.

25. Rovelli, C., *Reality is not what it seems: the journey to quantum gravity*. First American edition. ed. 2017, New York, NY: Riverhead Books. 280 p.

26. Sorli, A., *The biggest misunderstanding of 20th century science is that time is the 4th dimension of space*. General Science Journal, www.gsjournal.net/Science-Journals/Research%20Papers/View/1148.

27. Folger, T., *Newsflash: time may not exist*. Discover Magazine, 2007. https://www.discovermagazine.com/the-sciences/newsflash-time-may-not-exist.

28. Rovelli, C.S.E., et al., *The order of time*. 2018, New York, NY: Riverhead Books. 240 p.

29. Lightman, A., *Benedict Cumberbatch meets Albert Einstein in Carlo Rovelli's new audiobook*. New York Times, 2018.

30. Meyer, S.C., *Return of the God hypothesis: three scientific discoveries that reveal the mind behind the universe*. First edition. ed. 2021, New York, NY: Harper One, an imprint of Haper Collins Publishers. 568 p.

31. Earman, J., *The "past hypothesis": not even false*. Studies in History and Philosophy of Science Part B: Studies in History and Philosophy of Modern Physics, 2006. **37**(3): p. 399–430.

32. Brown, H.R. and J. Uffink, *The origins of time-asymmetry in thermodynamics: the minus first law*. Studies in History and Philosophy of Modern Physics, 2001. **32**(4): p. 525–538.

33. Barrow, J.D. and J.K. Webb, *Inconstant constants: do the inner workings of nature change with time?* Scientific American, 2005. **292**(6): p. 56–63.

34. Barrow, J.D. and F.J. Tipler, *The anthropic cosmological principle*. 1986, Oxford: Oxford University Press. 706 p.

35. Isaacson, W., *Einstein: his life and universe*. 2007, New York, NY: Simon & Schuster. p. 675.

36. J.K. Webb, et al., *Indications of a spatial variation of the fine structure constant*. Physical Review Letters, 2011. **107**: p. 191,101.

37. Webb, J., *Is life on Earth due to a quirk in the laws of physics?* The Conversation, November 3, 2011.

2 The Life and Death of a Star

Throughout human history, stars have inspired many dreams and more than a few wishes, but ultimately stars are important because without them the universe would be devoid of life. However, before the universe could assemble stars, it first had to generate atoms. So, we will begin our discussion about the life and death of a star by examining the formation of the smallest atoms. Following this, we will contemplate how these small atoms led to the genesis of stars. Then, in the next several chapters, we will examine how the stars of our universe produced the heavy elements needed for the formation of planets and living creatures, and we will investigate how the death of some of the universe's first stars led to the generation of new stars, Earth-like planets, black holes, and life-supporting galaxies.

FORGING ATOMS AND SETTING THE STAGE FOR THE BIRTH OF A STAR

Within five minutes of the Big Bang, charged atomic nuclei existed, but the first neutral atoms didn't appear until about 380,000 years after the universe began. It took that long before the temperature of the cosmos had dropped to the point at which electrons could be held in close association with atomic nuclei [1, 2]. The period when neutral atoms began to materialize in the universe is called "the recombination era"; however, the term "recombination" is misleading because, prior to this point, atomic nuclei and electrons had never "combined" to form a neutral atom. So, it has been suggested that perhaps it would have been more fitting to call this "the combination era."

DOI: 10.1201/9781003270294-4

In any case, before "recombination," electrons and atomic nuclei formed a universe that was soup-like, and the photons that moved through space were scattered aimlessly by these charged particles. However, after recombination, short-wavelength, high-energy photons moved through space largely unfettered, and the universe went from opaque to clear. In addition, during the recombination event, new high-energy photons were generated in a process called decoupling. These nascent photons joined those that had existed previously, and they also dispersed unencumbered in all directions.

As time went on, the universe continued to expand, and as it did, the short wavelengths of electromagnetic radiation that filled the cosmos expanded as well. Today, almost 14 billion years later, this energy now exists as long-wavelength electromagnetic radiation. To detect this energy, look at the static that appears on your television: about 1% of that noise is generated by the cosmic background radiation that filled the newborn universe.

THE DARK AGE OF THE UNIVERSE

As the universe continued its ceaseless enlargement, the afterglow of the Big Bang and the recombination era faded, and the universe entered a dark age that lasted for hundreds of millions of years [3]. During this period, there were no discrete light sources within the cosmos, and the universe was a black and foreboding place consisting mainly of nebulae composed primarily of molecular hydrogen. These clouds were established because the matter of the early universe had a slightly asymmetrical distribution (see Chapter 1), and that allowed some material to begin to exert a gravitational pull on other nearby molecules. In between these gas nebulae, there was nothing but vast voids of nearly empty space.

The dark age of the universe began to come to an end when massive clumps of matter started to form within nebulae. As the size of the clumps grew, their gravitational force increased, and when clouds collided, they began to acquire angular momentum. Over time, the hydrogen molecules within the nebulae bumped into each other with increasing frequency, and as they did, the kinetic energy within the hydrogen nebulae was converted into thermal energy. This caused the rapidly moving disc of gas to flatten out, and eventually, the center of the cloud became so massive and pressurized that its temperature approached 14 million degrees Fahrenheit. The molecular clouds' attainment of this temperature was a watershed event within our universe because, at this temperature, four hydrogen nuclei can be transmutated to a helium nucleus, and in that process, matter is converted into energy. The conversion of matter to energy is what being a star is all about, and until its death, a star's existence is marked by its ability to convert massive amounts of matter into energy.

To initiate the all-important transmutation of hydrogen to helium, protons and electrons are converted into neutrons in a process governed by the weak nuclear force. Once the neutrons exist, one proton and one neutron can interact to form a deuteron, but to get that to happen, the high-temperature particles must slam into each other with enough force that they end up approximately one femtometer (10^{-15} meters) apart. At that distance, the strong nuclear force, which only acts over extremely short gaps, can bind subatomic particles and stabilize them.

Once a proton and neutron are linked, the strong nuclear force of the bound pair produces allows yet another proton to join the nucleus. The formation of the triad then makes it possible for the final neutron to join the group, and that completes the helium nucleus.

Remarkably, during this transmutation, about 0.7% of the mass involved is transformed into pure energy. Within the gas cloud, this energy flows outward, and as it does, it offsets the crushing gravitational force generated by the tremendous mass of congregating hydrogen. Eventually, an equilibrium is established between the explosive outward force of the nuclear reactions and the imploding inward force of gravity, and with that, a star is born.

THE IMPORTANCE OF THE HYDROGEN NUCLEAR BINDING EFFICIENCY (ε)

What is particularly noteworthy about the initiation of the fusion process is its exquisite dependence on the magnitude of its nuclear binding efficiency (ε) [4]. This binding efficiency is said to have a value of 0.007, since 0.7% of the bound matter is converted into energy during the fusion of the hydrogen nuclei. We now know that if ε were significantly smaller in value, it would not have been possible to stably join a proton to a neutron, and the pathway to helium synthesis would have shut down. If our universe had a low ε, there would be no blazing stars powered by giant nuclear furnaces. Instead, there would only be collections of superheated protons and neutrons, and these structures would radiate energy for a small fraction of the lifespan of a nuclear-powered star. In addition, if the nuclear binding efficiency of hydrogen in our universe was low, there could be no elements heavier than those produced during the Big Bang, since the synthesis of most heavy elements depends on stars' ability to carry out the efficient nuclear fusion of hydrogen. In short, if the universe had a low ε, it would have forever been a dull place utterly devoid of life [4].

Interestingly, the situation in our universe would not improve if the hydrogen nuclear binding efficiency were significantly greater than 0.007. This is true because in a high ε universe, the protons produced at the moment of the Big Bang would quickly fuse [4, 5], and consequently, free hydrogen nuclei would have been rapidly depleted.

STARS ARE THE HEAVY ELEMENT GENERATORS OF THE UNIVERSE

Fortunately, in a world with an ε of 0.007, hydrogen-based stars, and these magnificent structures generate heavy elements while simultaneously releasing heat, visible light, and other forms of energy. The process of stellar element production is essential for increasing the complexity of the universe, and surprisingly, these nucleosynthetic reactions are significantly influenced by the existence of atomic resonance states, which are in turn influenced by the value of ε and a large number of other forces and constants [6].

The first example of just how critical atomic resonance states are to the formation of elements such as carbon and nitrogen came from the work of Fred Hoyle and William Foley. In the 1950s, Hoyle knew that to produce carbon (which has six protons), stars needed to fuse three helium nuclei with two protons apiece. However, Edwin Salpeter, a contemporary of Hoyle's, determined that it was extremely unlikely that three helium nuclei would collide in one instant, and he suggested that it was much more probable that carbon was synthesized in a two-step process. According to Salpeter, the first stage would involve two helium nuclei colliding to form an unstable beryllium nucleus that contained four protons and four neutrons, and then in the second stage, a third helium nucleus would join the unstable beryllium nuclei to form carbon [7].

Salpeter's hypothesis, however, had a potential problem because beryllium nuclei have a fleetingly short half-life of 10^{-16} seconds, and as a result, they don't persist long enough to collide with large numbers of helium nuclei. Hoyle realized that, for carbon to be produced in abundance, the efficiency of the second step of Salpeter's pathway would have to be high. But what would increase the efficiency of this reaction? At first, the answer wasn't obvious; however, Hoyle recognized that some high-efficiency mechanism for the fusion of helium and unstable beryllium nuclei must exist, because if it didn't, the universe would have much less carbon than what is observed, and carbon-based creatures like us wouldn't be here.

Ultimately, Hoyle's answer to the carbon synthesis paradox was to propose that stellar-based carbon must exist in an excited, high-energy, resonance state. Specifically, he suggested that in stars, carbon must have a naturally occurring energy state that is equal to the sum of the energy in the thermally excited helium and beryllium nuclei. Hoyle made this suggestion because he knew that if a naturally occurring product exists at an energy level that is equal to the sum of the energy levels of the reactants, the reaction that generates that product would proceed very efficiently. Having generated this prediction, Hoyle precisely calculated the energy level at which excited stellar carbon would need to exist to account for the abundance of carbon in our universe, and he determined that value to be 7.65 million electron volts [7].

Predicting the existence of an energy state that would speed up the synthesis of carbon is one thing, but gathering empirical data that verify the existence of that predicted resonance state is another. So, to move his carbon resonance hypothesis out of the realm of speculation, Hoyle convinced a group of nuclear physicists led by Willy Fowler to run the experiments necessary to test his prediction. Remarkably, the data generated by Fowler and his associates indicated that Hoyle was right, and thermally excited carbon did indeed have a previously undocumented high-energy resonance state at 7.68 ± 0.03 million electron volts [8]! As it turns out, most of the carbon nuclei in this excited resonance state revert back to beryllium and an alpha particle, but some release energy in the form of gamma radiation, and then they relax to form carbon in its non-excited ground state configuration [9].

The confirmation that carbon has a resonance state that facilitates its production was great news for anyone trying to understand why carbon is so abundant in the universe. However, what immediately became apparent to those studying nucleosynthesis was that if oxygen, which is typically produced by fusing carbon to helium, also had a resonance state that sped up its production, most of the carbon in the universe would quickly be transmuted to oxygen. As it turned out, Fowler and his colleagues determined that oxygen does not have such a resonance state, and therefore, we live in a universe where carbon is relatively plentiful.

Fortunately, the absence of a resonance state that promotes the synthesis of oxygen from carbon and helium did not limit the evolution of life, and this is due in part to the fact that the ratio of carbon to oxygen in our observable universe is approximately 1:2 [10]. So, even without an energy state to facilitate its production, oxygen is plentiful, and that's good news for us water-loving aerobes.

The apparent fine-tuning necessary for the nucleosynthesis of large amounts of carbon was surprising to many, and after some reflection, Hoyle came to view the existence of his previously unpredicted carbon resonance state as a form of revelation. In 1981, he stated that

> A common sense interpretation of the facts suggests that a super intellect has monkeyed with physics, as well as with chemistry and biology, and that there are no blind forces worth speaking about in nature. The numbers one calculates from the facts seem to me so overwhelming as to put this conclusion almost beyond question [11].

As impressed as Hoyle was by the existence of this carbon resonance state, Stephen Hawking and Leonard Mlodinow suggest that modern computer models of nucleosynthesis would have struck Hoyle as even more amazing [12]. They point out recent data indicating that "a change of as little as 0.5 percent in the strength of the strong nuclear force, or four percent in the electrical force, would destroy either nearly all of the carbon or all oxygen in every star" [10, 12]. Perhaps not surprisingly, even with the computer simulations of nucleosynthesis, not everyone is as stunned as Hoyle was by the apparent fine-tuning that is necessary for element production [13], but regardless of one's opinion, it is hard not to take notice of the incredibly high degree of precision required for stellar alchemy to occur.

ROUNDING OUT THE PERIODIC TABLE

After elucidating the surprising pathway to the synthesis of carbon, Hoyle, Fowler, and their colleagues outlined how most of the other elements in our universe were generated [14]. With the publication of their groundbreaking work, humanity was finally able to understand how carbon and many of the other naturally occurring heavier elements came to be.

As for the elements lighter than carbon, Hoyle and Fowler suggested they were produced largely outside of stellar cores. Hydrogen nuclei account for approximately 73% of the visible mass in the universe, and as was mentioned earlier, these nuclei were generated during the Big Bang. Helium nuclei, the second most abundant substance in the universe, accounts for the vast majority of the remaining visible matter, and like hydrogen nuclei, most of the helium nuclei in our universe were produced during the Big Bang. In fact, it is estimated that 83–96% of the helium that currently exists was generated during that event. The remaining helium was largely produced by the nuclear fusion of hydrogen within stellar cores.

Most of the remaining small nuclei (i.e., lithium, beryllium, and boron) were generated when cosmic rays, which are essentially high energy protons, fragmented larger atomic nuclei that were present in the gaseous remains of supernovae [15].

In addition, some of these small nuclei were likely produced during the ephemeral period between the Big Bang and the inflationary expansion of the universe, when the temperature of the cosmos was still hot enough to generate them.

STARS AS PLANET MAKERS AND GALACTIC CONSTITUENTS

Stars are efficient generators of heavy elements, and they also play a critical role in forging planets and many of the other structures in a star system. Most of the material that forms the planets comes from the nebular debris that did not get swept up into the star itself. In addition, the initial nuclear explosion, which occurs during the birth of a star, contributes a significant amount of material to the non-stellar accretion disks that give rise to the planets, moons, comets, and asteroids of a star system. Many of these structures remain gravitationally locked to the star around which they formed, and consequently, most of the stars in our galaxy likely have numerous planets and moons as well as a plethora of comets and asteroids in their neighborhoods [16].

During the 200 million years that followed the Big Bang, countless stellar systems appeared in our universe. Eventually, groups of stars and sub-stellar bodies formed galaxies, and these galaxies coalesced into clusters, superclusters, and filaments. Over a period of billions of years, the light produced by these galactic conglomerations permeated the universe, and as a result, the dark age of the early cosmos slowly receded.

Eventually, the number of galaxies in the universe grew to at least two trillion [17, 18], which means that at one point in the history of our universe, there were approximately 286 galaxies for every person alive on Earth as of this writing. Over time, however, many of these galactic entities merged, and the distinct galaxies that now remain are distributed within the observable universe of about 96 billion light years [19]. Ninety-six billion light years is a lot of space, so many of the galaxies that exist today are located far from their neighbors, and in the case of our own Milky Way galaxy, one would have to travel about 2.5 million light years just to get to Andromeda, which is one of our closest galactic neighbors [20–23].

THE DEATH OF A STELLAR GIANT

The first stars within the newly forming galaxies were essentially monstrous collections of hydrogen that carried out the mundane task of releasing energy while transmuting hydrogen into helium. However, toward the end of their lives, the nature of their activity changed dramatically. Over many of the first 100 million years of their existence, the raging internal battle between nuclear fusion and gravity was locked in a stalemate. But this eventually had to end, because while the amount of hydrogen a star can fuse is finite, gravity exists as long as there is mass to generate it. Consequently, in this battle, gravity was destined to prevail. So, when the first stellar denizens of the galaxies burned through most of their hydrogen, the relentless force of gravity began to gain ground. In the presence of dwindling hydrogen, gravity compressed the core of the star, but the fight was not over yet. Once gravity pushed the atomic nuclei closer together, the temperature of the stellar core increased until it reached about 180 million degrees Fahrenheit, and at that point, there was so much thermal energy in the star that the helium nuclei within the structure collided violently enough to overcome much of the protons' electrical repulsion. As a result, the helium nuclei came very close together, so close that they fell under the influence of a strong nuclear force. Once that happened, the star began to transmute helium nuclei into heavier elements such as beryllium and carbon, and it continued to release enough energy to counter the crushing gravitational force that existed within the star. This new round of nucleosynthesis bought the star more time, but the moment eventually came when the star started to run out of helium. Even then, the largest stellar giants were still okay because their cores were able to heat up to even higher temperatures, and consequently, they began to fuse their remaining elements into still heavier elements.

As a result of these processes, many of the early large stars had successive, onion-like layers of nuclear reactions occurring within their cores. In each stratum, the remaining hydrogen, helium, carbon, or other heavier nuclei were undergoing the process of nuclear fusion and transmutation. As the stars carried out these fusion reactions energy continued to be released, but importantly, the amount of energy being released per fusion eventually began to decrease as the size of the atomic nuclei increased. This is true because the strong nuclear force that holds the protons together in the nucleus decays exponentially as a function of distance; however, the electrostatic repulsion of the protons decays as a function of the inverse square of the distance. As a result, the electrostatic repulsion of the protons continued to grow significantly when large nuclei formed, but the release of energy when the protons interacted via the strong nuclear force did not increase proportionately. Ultimately, this meant that the largest stars started to have difficulty making nuclei larger than iron-56, because when a star started making nuclei that big, the nucleosynthesis process required more energy than was released as a result of nuclei fusion. So, when copious amounts of iron and other large nuclei started to form in the core of gargantuan stars, the end was near, and the star's fate was sealed. Without the ability to liberate enough nuclear energy to fend off the large and overwhelming force of gravity, the contest between the explosive outward force of nuclear fusion and the relentless crushing force of gravity was soon over, and gravity emerged the undisputed victor.

In stars that had a mass greater than four to five times that of our Sun, the ability of the gravitational force to overtake the nuclear force resulted in gravity compacting the stellar core so tightly that the temperature within it exceeded 18 billion degrees Fahrenheit. Under these conditions, the electrons and protons were pushed together with enough force to produce high-energy neutrons and neutrinos. Some of these high-energy particles can pass unrestricted through planets and other solid objects, but even they could not traverse the ultra-compact, ultra-high-density stellar core that gravity generated at the end of a large star's lifespan. As a result, many of the neutrons and neutrinos of the dying star collided with its core, and then they ricocheted back up through what remained of the stellar giant. As neutrons and neutrinos made their way through the expiring star, they superheated its gases, and this triggered the ejection of large segments of stellar mass. In some of the biggest stars, the discharged debris was easily more massive than ten of our Suns, and the temperature of this discarded wreckage could reach more than a billion degrees. By the time this stellar rubble was freed from the star, it typically had a velocity equal to approximately 10% that of the

FIGURE 2.1   An image of a supernovae explosion in the center of our galaxy.

speed of light, and the remains of the once seemingly indestructible star were ultimately scattered far and wide across hundreds of light years of space (see Figure 2.1).

If the dead star had a mass approximately four to eight times that of our Sun,
what remained after the supernova explosion was a dead "neutron star." The neutron
core is only about 12 miles in diameter, but a teaspoon of this entity weighs about a
billion tons, and the dead core has a total gravitational pull about two billion times
that of Earth. Despite their astonishing mass and density, the force of the supernova
explosion is so great that it usually leaves the neutron core rotating at speeds up to
43,000 times per minute.

The formation of a spinning neutron core is certainly a dramatic way for a large
star to end its life, but the largest stars in the universe end their existence in an even
more spectacular manner. When truly giant stars run out of nuclear material to fuse,
their mass is so prodigious that even the ultra-dense neutron core concedes defeat to
gravity. When this happens, the core of the dead star collapses until it has a density
so great that nothing can escape from it. These structures, which we commonly call
black holes, profoundly warp time and space, and in their light-absorbing darkness,
they mark the final resting place of a once luminous stellar giant.

Within each galaxy, there are likely very large numbers of these stellar grave sites.
Some of these black holes are relatively small, and they have masses that are only
several times that of our Sun. Others, such as the supermassive black holes that exist
in the center of most galaxies, can be billions of times more massive than our Sun [24].

THE SUPERNOVA: A BEACON IN THE NIGHT
AND AN IMPRESSIVE ALCHEMY FACTORY

A supernova is a notable event by any standard, and it has a grandeur that is in
keeping with that of the universe itself. During their last days, some dying stars
can release more than ten times the amount of energy our Sun will liberate in its

ten-billion-year lifespan. In fact, a large supernova can release more energy than that emitted by some spiral galaxies over an entire year. When a supernova occurs, it can outshine billions of stars, and the death of such a large star is so spectacular that even when it occurs in a faraway galaxy, it can still be detected in our little Earth-bound outposts many billions of light years away [25].

In addition to being extraordinary because of the amount of energy they release, supernovae explosions are special because they can seed the galaxy with large quantities of heavy elements such as gold (atomic number 79), lead (atomic number 82), and the largest of all naturally occurring nuclei, uranium (atomic number 92).

Over the billions of years that the Milky Way has been in existence, there have likely been over 100 million supernovae [25], but today the rate of supernova explosions in our galaxy is lower than it was billions of years ago, and currently, there are only about one or two of these events in our galaxy per century. Still, our universe is an exceedingly large place with a remarkably large number of galaxies, and it is estimated that with every second that passes, there is a supernova explosion somewhere in our universe [26].

THE KILONOVA: THE ULTIMATE ALCHEMY FACTORY

As a heavy element producer, the supernova is only outdone by the collision of two neutron stars [27]. Such a collision is called a kilonova because it can release about 1,000 times the energy of a standard supernova.

Humanity witnessed its first kilonova on August 17, 2017, when photons and gravitational waves emanating from a collision between two neutron stars 130 million light years from Earth arrived at our planet. Analysis of this incident revealed that when the neutron cores crashed into each other, they were moving at close to the speed of light [28], and the explosion they generated when they collided produced an 8,000-degree fireball with a diameter of about 5.6 billion miles [29]. This unfathomable pileup released about 100 million times as much energy as our Sun, and astronomers estimate that this one incident produced a mass of gold that was 40–100 times greater than the mass of Earth. This same kilonova also produced many other heavy elements including an amount of uranium that had a mass 10–30 times greater than that of our planet [29]. Collectively, the kilonovaes that have occurred within our galaxy may have generated an amount of gold equivalent in mass to that of 100 million Earth-like planets, and it is likely that a large fraction of the heaviest naturally occurring elements in the universe were produced by this type of collision.

Due to all the supernovae and kilonovae activity within the relatively young universe, large swaths of space became polluted with heavy elements, and ultimately more than 90% of this stellar debris contributed to the formation of new stars and planets. Many of the second- and third-generation stars that formed from this debris were considerably smaller than their predecessors, and due to their relatively diminutive stature, they burned through their nuclear fuel at a much more leisurely rate. Today, it is estimated that there are between two and seven new stars that join the Milky Way each year, and on average, these new galactic constituents have a mass that is about half that of our Sun [2, 30–32].

THE BIRTH OF OUR SOLAR SYSTEM

Our own solar system is among those that owe their existence to older stars that lived and died in the past. In fact, our Sun and the rest of our solar system formed from the elemental ash that was released during ancient supernovae and kilonovae. In addition, it is thought that a supernova may have generated a shock wave that jump-started the accretion process that ultimately formed our solar system about 4.57 billion years ago [33].

Our Sun is the largest object in our solar system, and like other stars, it is primarily composed of hydrogen and helium. However, because our Sun is a second- or third-generation star, it has a substantial amount of heavy elements, and in total, about 1.69% of our Sun's mass consists of elements such as carbon, neon, and iron [34]. All totaled, the heavy elements of our Sun have a mass that is about 5,577 times greater than that of Earth.

Due to its proximity, the Sun is by far the brightest star we can see, but even among the other stars in the galaxy, our Sun stands out. In terms of its luminosity, it ranks in the top quintile [35, 36], and to generate this level of energy output, the Sun must convert about 9.4 billion pounds of matter into energy every second. That is equivalent to converting the mass of more than 116,000 fully loaded, 18-wheel Mack trucks into pure energy every single second of every single day. And it is worth noting that our Sun has been doing this nonstop for over five billion years.

Some of the immense amount of energy that is emitted from the Sun travels to Earth as photons of light, which once free from the Sun's outer edge, manage the 93-million-mile journey to Earth in a little more than eight minutes. However, getting from the core of the Sun, where most of the photons are produced, to the Sun's surface is usually not a quick process, and as the photons move through the various layers of gas, they are constantly absorbed and readmitted by solar plasma. Typically, this results in a photon bouncing around the interior of the Sun for tens of thousands of years, and by some estimates, most of the light that is falling on us today was generated in the Sun's core sometime between 10,000 and 170,000 years ago.

The total amount of energy the Sun is releasing is not something that we can fully grasp but pondering just how much energy the Sun does liberate helps us understand why an object 93 million miles away can power almost all forms of life on our rocky little planet. In fact, despite its great distance, our Sun delivers so much energy to Earth that life on our planet only needs to capture about 1.2% of it to survive, and in total, more than a third of the incident sunlight simply bounces off clouds, the atmosphere, or Earth, and then it drifts back into space [37].

It is very fortunate that much of the incoming energy from the Sun does indeed radiate back into space because if significantly greater amounts were retained, the temperature of the planet would skyrocket, and life on Earth would be compromised or destroyed. Currently, much of humanity is worried about just such a problem, and as we continue to release heat-trapping molecules into the environment, we are capturing more of the Sun's energy within our atmosphere. Ultimately, this may threaten our very existence.

STARS AS THE ENERGY SOURCE FOR BIOLOGICAL EVOLUTION

I mentioned earlier that many of the first stars in the universe were relatively short-lived giants, which, after a few million years or so, were gone from the scene [2]. These first galactic citizens left important legacies, including heavy elements and black holes, but for life to develop, the universe needed more than just large, heavy-element-generating, ephemeral stars. For life to come into existence, the universe also needed galaxies with long-lived stars that could consistently liberate energy for billions of years [38]. Our Sun is an example of just such a star, and at 4.57 billion years of age, it is solidly ensconced among the middle-age stellar denizens of our galaxy. During our Sun's long life, it has provided the steady supply of the energy necessary to support life, and it has powered almost four billion years of evolution on our planet.

Unfortunately, all good things must come to an end, and our all-important Sun will also one day meet its demise. When this happens, Earth, and the life on it, will cease to exist. But for now, we will focus on the living, and we will next examine the role galaxies had in the generation of life.

REFERENCES

1. Alpher, V., *Ralph A. Alpher, Robert C. Herman, and the cosmic microwave background radiation.* Physics in Perspective, 2012. **14**: p. 300–334.
2. Larson, R.B. and V. Bromm, *The first stars in the universe.* Scientific American, 2001. **285**(6): p. 64–71.
3. Loeb, A., *How did the first stars and galaxies form?* Princeton Frontiers in Physics; Variation: Princeton Frontiers in Physics. 2010, Princeton, NJ: Woodstock. xiii, 193 p.
4. Rees, M.J., *Just six numbers: the deep forces that shape the universe.* 2000, New York, NY: Basic Books. x, 173 p.
5. Weinberg, S., *Facing up: science and its cultural adversaries.* 2001, Cambridge, MA: Harvard University Press. p. 306.
6. Meyer, S.C., *Return of the God hypothesis: three scientific discoveries that reveal the mind behind the universe.* First edition. ed. 2021, New York, NY: Harper One, an imprint of HarperCollins Publishers. 568 p.
7. Livio, M., *Brilliant blunders: from Darwin to Einstein—colossal mistakes by great scientists that changed our understanding of life and the universe.* First Simon & Schuster hardcover edition. ed. 2013, New York, NY: Simon & Schuster. 341 p.
8. Dunbar, D.N.F., et al., *The 7.68-Mev state in C^{12}.* Physical Review, 1953. **92**(3): p. 649–650.
9. Wolchover, N., *The Hoyle state: a primordial nucleus behind the elements of life.* Scientific American, 2012.
10. Oberhummer, H., *Stellar production rates of carbon and its abundance in the universe.* Science, 2000. **289**(5476): p. 88–90.
11. Hoyle, F., *The universe: past and present reflections.* Engineering and Science, 1981. **45**: p. 8–12.
12. Hawking, S. and L. Mlodinow, *The grand design.* 2010, New York, NY: Bantam Books. 198 p.
13. Davies, P., *Cosmic jackpot: why our universe is just right for life.* 2007, Boston, MA: Houghton Mifflin.

14. Burbidge, E.M., et al., *Synthesis of the elements in stars*. Reviews of Modern Physics, 1957. **29**(4): p. 547–650.
15. Viola, V. and G. Mathews, *The cosmic synthesis of lithium, beryllium and boron*. Scientific American, 1987. **256**(5): p. 38–45.
16. Cassan, A., et al., *One or more bound planets per Milky Way star from microlensing observations*. Nature, 2012. **481**(7380): p. 167–169.
17. Christopher J. Conselice, et al., *The evolution of galaxy number density at z < 8 and its implications*. Astrophysics arXiv:1607.03909 [astro-ph.GA], 2016.
18. Castelvecchi, D., *Universe has ten times more galaxies than researchers thought*. Nature, 2016. https://doi.org/10.1038/nature.2016.20809.
19. Bars, I.T.J. and F. Nekoogar. *Extra dimensions in space and time*. Multiversal journeys; Variation: Multiversal journeys. [Internet Resource; Computer File] 2009; 1 online resource (xiv, 217 p.): illustrations.]. Available from: http://site.ebrary.com/id/1035588 Materials specified: ebraryhttp://site.ebrary.com/id/10355881.
20. Uson, Juan, Stephen Boughn, and J. Kuhn, *The central in Galaxy Abell 2029: an old supergiant*. Science, 1990. **259**: p. 539–540.
21. Wilford, J., *Sighting of largest galaxy hints clues on the clustering of matter*. New York Times, 1990.
22. van Dokkum, P.G. and C. Conroy, *A substantial population of low-mass stars in luminous elliptical galaxies*. Nature, 2010. **468**(7326): p. 940–942.
23. Ignasi Ribas, et al., *First determination of the distance and fundamental properties of an eclipsing binary in the Andromeda galaxy*. Astrophysical Journal, 2005. **635**: p. L37–L40.
24. Scharf, C., *The benevolence of black holes*. Scientific American, 2012. **307**(2): p. 34–39.
25. Marschall, L.A., *The supernova story*. 1988, New York, NY: Plenum Press. xvii, 296 p.
26. Kasen, D., *Stellar fireworks*. Scientific American, 2016. **314**(6): p. 36–43.
27. Redd, N.T., *Neutron stars: definition & facts*. Space.com, 2017. https://www.space.com/22180-neutron-stars.html.
28. Betz, E., *Dawn of an era—astronomers hear and see cosmic collision*. Astronomy, October 16, 2017, https://astronomy.com/news/2017/10/astronomers-hear-and-see-cosmic-collision
29. Overbye, D., *LIGO detects fierce collision of neutron stars for the first time*. New York Times, 2017.
30. Wanjek, C., *Milky Way churns out seven new stars per year, scientists say*. NASA, 2006. https://www.nasa.gov/centers/goddard/news/topstory/2006/milkyway_seven.html
31. Robitaille, T.P. and B.A. Whitney, *The present-day star formation rate of the Milky Way determined from spitzer-detected young stellar objects*. Astrophysical Journal, 2010. **710**(1): p. L11–L15.
32. Diehl, R., et al., *Radioactive 26Al from massive stars in the galaxy*. Nature, 2006. **439**(7072): p. 45–47.
33. Williams, J.P., *The astrophysical environment of the solar birthplace*. Contemporary Physics, 2010. **51**(5): p. 381–396.
34. Zeilik, M.A. and S.A. Gregory, *Introductory astronomy & astrophysics*. Fourth edition ed. Boston, MA: Brooks Cole, 1997. p. 322.
35. Lada, C., *Stelluar multiplicity and the initial mass function: most stars are single T*. Astronomical Journal, 2006. **640**: p. L63–L66.
36. David A. Aguilar and C. Pulliam, *Most Milky Way stars are single*. Harvard Smithsonian Center for Astrophysics (News Bulletin), 2006. www.ncbi.nlm.nih.gov/pubmed/11759586.
37. Peter Schlosser, S.P., Mingfang Ting, Jason Smerdon, *Solar radiation, Earth's atmosphere, and the greenhouse effect*. Columbia University—EESC 2100 Spring, 2008 course notes published online.
38. Natarajan, P., *The first monster black holes*. Scientific American, 2018. **318**(2): p. 24–29.

3 The Milky Way
A Goldilocks Galaxy

The missing link in cosmology is the nature of dark matter and dark energy.

Stephen Hawking

Of all the conceptions of the human mind, from unicorns to gargoyles to the hydrogen bomb, the most fantastic, perhaps, is the black hole.

Kip Thorne

It is really striking that the Milky Way, containing us, lands smack-dab in the sweet spot of supermassive black hole activity.

Caleb Scharf

Stars generate large amounts of heat, light, and elements, but it takes a galaxy to create life. And not just any galaxy will do. Life-supporting galaxies must be able to generate immense numbers of giant stars that can produce vast amounts of heavy elements, and they must have long-lived stars that release energy gradually over billions of years. Galaxies that might give rise to creatures like you and I must also stop the remnants of stellar giants from leaving their boundaries after a supernova explosion, and they must limit how much stellar debris gets blown out of their boundaries by winds that emanate from their central supermassive black holes. In addition, life-supporting galaxies must have space that is relatively free from the gravitational tides and lethal energy that is found in their centers.

WHY DARK MATTER MATTERS?

In total, it has been estimated that 90% of galaxies cannot support life as we know it, and in some instances, this is because the galaxy lacks the optimally sized dark matter-induced gravitational sink necessary to efficiently catalyze the star formation process [1–4].

Dark matter appears to be particularly effective as a gravitational sink because it concentrates baryonic matter (i.e., normal visible matter), but it does not interact with radiation, which is defined here as any particle or photon that travels at or near the speed of light. In the early universe, baryonic matter was segregated due to intense radiation, but dark matter was not. As a result, dark matter formed the gravitational cauldrons that shaped baryonic matter into the stars of the earliest galaxies.

Often, when would-be galaxies did not have enough dark matter, they failed to thrive. In some cases, these proto galaxies faded away before they could form, and in other instances, all that materialized was a dwarf galaxy. Within our universe, dwarf galaxies are, on average, about 1% as massive as the Milky Way, and these

DOI: 10.1201/9781003270294-5

FIGURE 3.1   An artist's rendition of the invisible dark matter halo (shown in blue) that is believed to surround the Milky Way. (From: https://commons.wikimedia.org/wiki/File:Dark_matter_halo.png.)

galactic lightweights cannot effectively retain the heavy elements liberated by their dying stars. Their diminutive stature also insinuates that they lack the size required to generate safe havens for living creatures [5].

Occasionally, a newly forming galaxy was overwhelmed with dark matter, and in these cases, the extremely large gravitational forces generated by the dark matter prevented the gases in proto stars from cooling in an efficient and uniform manner. Typically, this triggered non-isotropic cooling and the formation of numerous dwarf galaxies, or in some instances, it led to the generation of a starless system [6–9].

Among the approximately two trillion galaxies, a small percentage are in the "Goldilocks range" in terms of their size and mass. These galaxies usually have the equivalent of at least 300 billion solar masses of dark matter that encase their visible baryonic matter in a halo-like configuration [10]. Such systems can form stars at an accelerated rate, and they can retain the chemical elements and other remnants of the centralized large stars that go supernova.

Not surprisingly, when it comes to dark matter, the Milky Way is in the "Goldilocks range." It has about ten times more dark matter than visible, baryonic matter, and its dark matter is distributed from the inner region out to more than 650,000 light years from the galactic center [11, 12]. Most of this dark matter is concentrated in a spherical halo (see Figure 3.1), and the Milky Way obviously has enough dark matter to form star systems with large amounts of heavy metals.

BLACK HOLES ARE AT THE HEART OF OUR EXISTENCE

Soon after they started to take shape, large galaxies began to generate black holes and supermassive black holes. By helping to anchor the center of the galaxy and agitating the material within it, these foreboding structures ultimately influenced the fate of a galaxy's life-supporting stars and planets. To get a sense of the role that these structures had in the genesis of life, we need to spend some time considering their origin, structure, and function.

The first black holes likely arose from the death of Brobdingnagian, population III stars early in the life of the primordial galaxy. These giants are estimated to have been at least 100 times more massive than our Sun, and some have suggested that their masses could have been a million times that of our Sun [13]. These galactic megastars were composed almost exclusively of hydrogen and helium, and as a result, they burned hot and fast, and they also generated enormous amounts of ultraviolet (UV) light. This radiation may have helped to reionize hydrogen atoms (i.e., it may have helped to break apart the hydrogen atoms that formed 380,000 years after the Big Bang), and after a few million years, many of these behemoths exploded as supernovae.

When the largest of the early universe's population III stars did meet their early demise, the intense radiation that existed at this time in cosmic history likely helped create a unique set of conditions that allowed for the "direct collapse" of the star [14–16]. This event concentrated matter within a giant star, and in turn, may have produced some of the universe's first large black holes.

To get an idea of just how dense black holes are, consider that for Earth to form a tiny black hole, all its 13 trillion pounds of mass would have to be condensed into a diameter of about 0.7 inches. For our Sun to do the same, its current diameter of about 870,000 miles would have to shrink to 3.7 miles. When a point in space becomes this dense, its gravitational force is so great that not even light can escape it.

Over time, the individual black holes of a galaxy consolidated into more massive structures. This process of black hole consolidation was first observed on September 14, 2015, when a pair of gravitational wave detectors in Washington State and Louisiana recorded the collision of two black holes. These fated galactic structures had been circling each other for eons, but when they finally merged at nearly half the speed of light, a black hole with a mass 36 times that of our Sun collided with the smaller structure that was 29 times as massive as the Sun. The merger reverberated across the universe, and during this unification process, three solar masses were converted into energy almost instantaneously. At that moment, the new, larger black hole released more energy than all the stars in a hundred galaxies, and this resulted in gravitational waves that traveled through space for 1.3 billion years before finally arriving at Earth on a late summer morning in 2015 [17]. In the four years following this historic event, humanity found evidence of dozens of black hole mergers, including one that took place 16.9 billion light years from Earth, and by 2023, our technology may allow us to detect the merger of a pair of black holes at the rate of one per hour [18].

The merging of black holes occurred for billions of years in proto galaxies all over the universe, and with this, the black holes in the center of galaxies became larger and more massive. The mammoth black holes that resulted continued to grow by devouring the disk of matter around them, which typically occupied an area many times larger than our solar system. Once matter within a disk transited the black hole's event horizon, its fate was sealed, and it disappeared from the cosmos. In a given year, a black hole similar to one in the center of our galaxy can easily devour a mass equivalent to several times that of our Sun, and a black hole such as SMSS J215728.21–3602151.1, which is 12 billion light years away and 20 billion times as massive as our Sun, can consume a mass equivalent to that of our Sun every two days [19].

Surprisingly, not all the material that comes under the influence of a black hole will be immediately consumed by it, and around each of these galactic monsters, some matter manages to skirt the event horizon and escape. To do so, the matter must enter the highly warped spacetime around the black hole on a trajectory that sends it into orbit around the giant gravity pit. Once there, this matter is at least temporarily safe from consumption, but its integrity is inevitably condemned, and the circling matter is ultimately shattered by the black hole's tremendous gravitational tides. As this process unfolds, the debris field produced becomes so dense that not even a black hole can consume it all at once, and as the material orbits at relativistic speeds, the galactic pileups that result from the colliding debris heats the area to millions of degrees. In addition to being aglow with thermal energy, the perimeter of a black hole also releases highly destructive X-rays and gamma rays deep into the surrounding galaxy. The net effect of this spewing energy is that the border of a black hole isn't black at all, and as various forms of radiation move away from the event horizon, the area shines with an intensity that is beyond belief. In the case of the largest of these entities, the energy that is released can outshine that generated by 700 trillion of our Suns [19].

Up until recently, our understanding of a black hole event horizon was based largely on data generated by theoretical physicists. However, in 2019, astronomers managed to capture a long-sought-after image of one of these structures in a galaxy called Meisser 87, which is 55 million light years away (see Figure 3.2). This

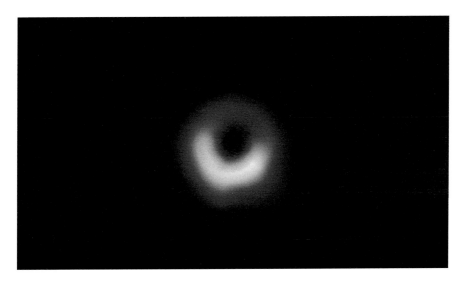

FIGURE 3.2   The event horizon that formed around a black hole located in the center of the galaxy M87 [20].

amazing picture contains an outline of a black hole with a mass 6.5 billion times that of our Sun, and we know that this structure, which is one of the largest black holes discovered, generates jets of energy that extend 5,000 light years into space. With this image, humanity can finally visualize an entity that is both a foundational generator of galactic structure and, at the same time, one of the most destructive forces in the known universe.

The aforementioned story illustrates how black holes liberate an outrageous amount of radiation by moving matter through curved spacetime, but there are still other means through which black holes can release even more energy. One of these involves spinning.

Throughout the universe, objects spin, and one important fundamental characteristic of our universe is that it conserves angular momentum. Angular momentum for a point mass is equal to the mass of the object, times the perpendicular distance to its rotational axis, times its velocity. So, when the mass of a super large star is condensed, the object's angular momentum is conserved, but the perpendicular distance to the mass' rotational axis decreases. As a result, the velocity at which the mass spins must increase. This is similar to what happens when an ice skater is spinning and she brings her arms closer to her body: Her angular momentum is conserved, but her arms' perpendicular distance to the rotational axis decreases, and as a result, she spins faster.

However, a black hole is no ice skater, and when objects that are spinning are condensed to the point at which they form a black hole, the velocity of their spin can reach phenomenal speeds. Our Sun, for instance, rotates around its axis once approximately every 26 days, which means that it is spinning around its axis at the impressive speed of about 4,400 miles per hour [21]. However, if the mass of the Sun were to be condensed to the point at which it formed a black hole, the principle of conservation of angular momentum suggests that it would rotate approximately once every 0.0001 seconds. This implies that the rate of the condensed Sun's rotation would increase about 9 billion-fold. However, the actual rate at which an object can rotate is limited by nature, and once a black hole spins at a speed close to that of light, its centrifugal force is so great that additional material cannot approach it. As a result, the interaction of the black hole with the universe around it ultimately slows its gyration down, and this keeps the black hole's maximal rotational speed below that of the speed of light.

With regard to a black hole's self-preservation, the huge centrifugal forces they produce are beneficial, because if there were no limits on how fast a black hole could spin, it would ultimately rotate so quickly it would rip itself apart. But, while black holes are not spinning faster than the speed of light, they are spinning at relativistic speeds, and at that rate, the charged particles in their accretion disks generate highly twisted magnetic fields. They also create charge differentials between the poles and the area near the equator, and these can be in the neighborhood of 1,000 trillion volts. As a result, the energy output generated by the spin of a black hole can approach that of 100 billion Suns [22], and this energy can help launch particles and gas from the accretion disk in a jet-like stream across hundreds of millions of light years of space.

In total, the energy liberated by a black hole results in a truly unimaginable force, and a black hole's ability to convert matter into energy is unparalleled within the universe. In fact, within a black hole's accretion disk, approximately 28% of the mass is converted into pure energy. This makes black holes almost 50 times as efficient as stars when it comes to converting matter into energy, and some supermassive black holes easily pump out more energy than all the stars in a hundred galaxies [22].

At the beginning of this chapter, there is a quote from Kip Thorne stating, "of all the conceptions of the human mind, from unicorns to gargoyles to the hydrogen bomb, the most fantastic, perhaps, is the black hole." This is a particularly apt statement, because it is truly impossible for us to fully grasp the mass, power, and destructive ability of a supermassive black hole. Despite our best efforts, how can we fully comprehend an object that may be tens of billions of times more massive than the Sun and have a density so enormous that it can puncture space itself. It is also hard to fully grasp how these structures can spin at speeds approaching that of light, outshine all of the billions of stars in hundreds of galaxies, and emit jets of intense electromagnetic radiation that stretch across tens of millions of light years [23]. So, a star-eating, supermassive, black hole is literally like nothing else we know of, and it would be the ultimate science fiction monster, if it weren't for the fact that it really does exist.

TRAFFIC CONTROL WITHIN THE ACCRETION DISK AND THE GROWTH OF A SUPERMASSIVE BLACK HOLE

A supermassive black hole's accretion disk is an unparalleled source of energy, but the energy in the outer edge of the disk can push matter away and slow the growth of the black hole [24, 25]. Given this, it is not clear how some black holes quickly grew into supermassive black holes.

One possible explanation is that early in the history of our universe, the density of stars and stellar gases was likely very high, and under these conditions, a small number of black holes may have merged within the center of a cluster of thousands of stars. In this environment, the dense interstellar gases would have absorbed much of the radiation produced by superheated material within accretion disks, and the gravitational pull of the thousands of nearby stars may have kept the growing black hole moving about in a highly dynamic state. As a result, the formation of the matter repulsing accretion disks may have been stymied by continuous reorientation of the black hole itself [26].

Supermassive black holes may have also formed as a result of galactic mergers (see Figure 3.3). During this process, the central black holes in each galaxy may have circled each other in a cosmic dance that extended for eons. However, when they finally did unite, the resulting fireworks were beyond description, and their joining was one of the most disruptive events to ever occur in the universe. Any life in the vicinity of two merging black holes didn't stand a chance, as entire planets would have been obviated.

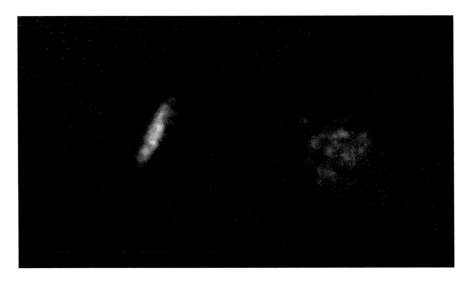

FIGURE 3.3   A depiction of the impending merger of two galaxies.

CURRENT AND UPCOMING GALACTIC MERGERS

The dynamic processes that shaped galaxies and their supermassive black holes continues, and as you read this, our Milky Way galaxy is merging with two dwarf galaxies called Canis Major and Sagittarius [27]. As impressive as these near-term colossal mergers may be, in the distant future, our galaxy will partake in an even more striking amalgamation when it collides with our largest nearby galactic neighbor, Andromeda. Currently, these two mega-galaxies are 2.5 million light years apart, and they are careening toward each other at approximately 20 miles per second, so it will take another 4.5 billion years or so before they finally do converge [28].

The new galaxy that will form, which astronomers have christened "Milkomeda" [29], will be truly enormous. The Milky Way, with its 100–400 billion stars, is already so vast that it takes light, traveling at six trillion miles a year, approximately 130,000 years to go from one end to the other [12]. And while the scale of our home galaxy is impressive, Andromeda is even more imposing. Weighing in at approximately twice the mass of the Milky Way [30], this giant has about one trillion stars spread out over an area of 260,000 light years. When "Milkomeda" comes to be, it will be one of the largest galaxies in the known universe.

Fortunately for our solar system, when these two galaxies do merge, our Sun will likely be located far away from the gargantuan supermassive black hole that will reside in the center of the newly formed galactic entity [31]. Nevertheless, the fusion of Milky Way and Andromeda may still portend bad news for our home world, and some theorists believe that when the merger occurs, supernovas will light up Earth's sky and potentially damage or destroy what remains of our planet [32].

However, until the merger, we in the Milky Way will experience the effects of the relatively small supermassive black hole that is centered approximately 27,000 light years away from our Sun. This structure, which astronomers refer to as Sagittarius A*, has a mass equivalent to 4.1 million Suns [33], which is still plenty large enough to significantly warp the surrounding space. As a result, all of the constituents of the galaxy rotate around the galactic midpoint [34], and as you read this, our solar system is orbiting Sagittarius A* at approximately 483,000 mph. At that speed, we will complete one trip around the galaxy's perimeter in about 225 million years, but given that *H. sapiens* have only been in existence for about 300,000 years, our species has witnessed merely 0.1% of the galactic orbital that our solar system travels [35]. As we will see in the next chapter, what we encounter on the remaining 99.9% of the voyage has the potential to influence the fate of humanity for better and for worse.

In addition to potentially creating existential angst, the rotation of billions of star systems around the Milky Way agitates the galaxy's components, and this enhances the interactions needed to keep the Milky Way in an active, star-forming state. Moreover, Sagittarius A* itself can shake up the components of the Milky Way through the tremendous amounts of energy it releases. Consequently, it is safe to say that the size, shape, and chemical composition of our galaxy have all been heavily influenced by the supermassive black hole that exists at its center.

THE ACTIVITY OF THE MILKY WAY'S SUPERMASSIVE BLACK HOLE APPEARS TO BE IN GOLDILOCKS RANGE

Recently, we have learned that the vigor of a supermassive black hole can vary greatly. Approximately 10% of the larger galaxies in the universe appear to have extremely dynamic and active supermassive black holes, and through the process of pulling material into their accretion disks, these insatiable goliaths are responsible for approximately 80% of the energy output of their galaxies [24, 25]. The winds and blast waves that the most active supermassive black holes produce can be more than 170 million miles per hour, and these potentially disruptive energy emissions can extend outward for more than 1,000 light years. In some of the more active supermassive black holes, the energy released by the material piling up in the accretion disk is so great that it can literally blow some of the elements and dust within a galaxy into intergalactic space. The most active supermassive black holes also heat a large fraction of the elements found in their galaxy, which can prevent this material from cooling enough to condense into stars. Furthermore, any heavy element-rich star systems that do manage to form in a galaxy with a very active supermassive black hole have to contend with the stray radiation that the galactic monsters release. As a result, it is unlikely that complex life could form anywhere near a very active supermassive black hole.

While some galaxies are burdened with unusually vigorous supermassive black holes, others have supermassive black holes that are atypically passive. In such places, these docile behemoths do not release enough energy to incite efficient churning of the interstellar gases and debris. As a result, the gases and elements near the center may not be moved to the outer realms of the galaxy, and while there may be extremely large numbers of gigantic stars in the middle of these galaxies, there are likely to be few heavy element-rich stars and planets in the potentially life-supporting, quiescent, outlying regions of the galaxy.

Luckily for us, Sagittarius A* appears to be in the middle range of supermassive black hole activity [36], and although Sagittarius A* has a mass equal to four million times that of our Sun, it is merely 5% as large as the supermassive black hole that is in the center of Andromeda. This makes the Milky Way's central supermassive black hole unusually small for a galaxy that is as large and old as our own [36].

Why the supermassive black hole at the center of our galaxy has the vigor and size that it does remain a mystery. Perhaps the number, size, and position of the first stars limited its growth and activity, or perhaps the various galaxies that merged and became part of the current Milky Way had central massive black holes that were relatively small and quiet. But regardless of how it came to be, Sagittarius A* has stirred up the galactic gas and debris of the Milky Way without blasting most of this material into intergalactic space. This, no doubt, has contributed to our galaxy's ability to support creatures like us.

A GIANT AWAKENS

While the Sagittarius A* appears to be relatively quiet now, it could increase its activity level, and there is evidence that it has done so from time to time. Some of these data that support this contention consist of X-rays that were found bouncing off stellar clouds approximately 300 light years from our galaxy's supermassive black hole. Additional evidence that Sagittarius A* can increase its activity was found in 2010 when a team of astronomers discovered a pair of faint plumes of radiation that emanate from the center of the galaxy and stretch more than 25,000 light years into space

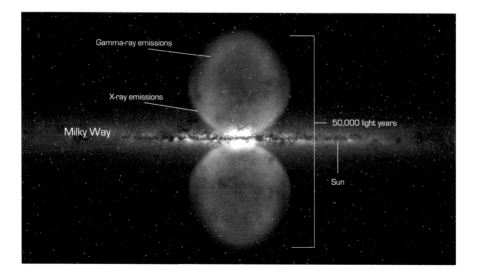

FIGURE 3.4 Fermi bubbles that radiate from the center of the Milky Way. In this image, the X-ray and gamma ray bubbles that emanate from the center of the galaxy are depicted. The location of the Sun, which can be found within the plane of the galaxy, is also depicted [36]. (From: NASA's Goddard Space Flight Center, Public domain, via Wikimedia Commons.)

(see Figure 3.4 from [36]). These structures, which are called Fermi bubbles, contain radiation that spans the electromagnetic spectrum, and they likely reflect a period during the last 100,000 years during which there was either a burst in star-making activity or a significant increase in the activity of our galaxy's supermassive black hole [22, 36–38]. If Sagittarius A* did increase its activity, we are not sure why it would have done so. However, some have speculated that the ingestion of a group of stars or the merging of two similar size black holes could have triggered an increase in the black hole's activity [37, 39].

Fortunately, at 27,000 light years from the Milky Way's central supermassive black hole, we are relatively safe from much of its destruction. Still, if there were to be a significant increase in the activity of this menacing giant, the radiation released could have a negative impact on life, even life ensconced on a planet tens of thousands of light years away. So, although Sagittarius A* is currently relatively quiet, astronomers such as Caleb Scharf argue this could change at any point [22, 36], and this fact may contribute to the vulnerability of life on planets like ours.

TO BE OR NOT TO BE

Ultimately, gravity can create stars and planets or destroy them, and the interplay between gravity's creative and destructive potential is complex. However, throughout large swaths of the universe, the creative character of gravity was more significant than its destructive potential. If this were not true, the galaxies themselves would not exist. However, of all the galaxies in our universe, how many of them harbor life-supporting stars and planets? As I mentioned earlier, it appears that about 90% are inhospitable to life because they do not contain the proper amount of dark matter, but of the remaining 10%, how many of these have a supermassive black hole that allows for the development of living creatures? Given the large number of variables that can influence the dynamic interactions between a black hole and its galactic constituents, it is hard to generate an answer to this question. But, the highly contingent nature of galactic development suggests that only a small subset of 200 billion or so potentially life-supporting galaxies are actually capable of hosting organisms like you and me.

STARS, BLACK HOLES, AND THE CONSTANTS OF THE UNIVERSE

Supermassive black holes are nature's most efficient transformers of matter into energy, and they are the universe's largest and most destructive agents. They are our universe's ultimate "bad boys," and if our universe could have a nightmare, it would no doubt involve supermassive black holes. However, like all other phenomena in the universe, supermassive black holes are influenced by the physical constants of the universe itself, and given the central role of black holes in shaping potentially life-supporting galaxies, it is worth taking a moment to consider how small alterations in the fundamental forces would reshape stars, black holes, galaxies, and our universe.

Black holes are essentially gigantic electromagnets and gravity pits, and therefore, they are particularly sensitive to the electromagnetic and gravitational constants of our universe. Physicists have defined the electromagnetic force that holds atoms together divided by the gravitational force generated by those same atoms as the constant "N," and in our universe, the value of N is incredibly large: in fact, it has a value equal to 1,000,000,000,000,000,000,000,000,000,000,000,000 (i.e., 1 with 36 zeros after it) [40]. The enormity of this number indicates that, when relatively small amounts of matter are interacting, the force that dominates is electromagnetic, but when the amount of matter is extraordinarily large, it becomes necessary to consider how gravity affects interacting particles as well. This is the case because, as the amount of matter increases, the overall net charge of the matter remains relatively constant, and typically, even in very large objects, the positive charge of the protons roughly balances the negative charge of the electrons. However, as objects become more massive, the gravitational interactions do not cancel out; rather, they summate. As a result, in extremely massive objects such as stars and black holes, the gravitational force becomes dominant, and gravity ultimately determines the fate of these entities. Consequently, altering the value of N would have significant effects on the behavior of objects both small and large. If, for example, the value of N was smaller, it would be easier to overcome the electrostatic repulsions that normally limit the fusion of atomic nuclei, and as a result, stars would form from less mass. In such a universe, stars would be more diminutive, and their lifespans would be greatly attenuated. In addition, the density of the stars would be greater, and they would not have to be as massive to implode into black holes. One consequence of this is that stars might not survive long enough to nurture the development of complex life, and if their numbers greatly increased, they would be more likely to interact and less likely to host planets with stable orbitals. Furthermore, there would likely be more black holes clustered in the center of a newly forming galaxy, and these structures would tend to merge and form larger and more voracious supermassive black holes. In short, in a universe with a smaller N, our galaxy would be a very different place and it would be far less amenable to the formation of complex life.

Here we find yet another example of how the fundamental features of the universe, which were acquired during its creation, affect the subatomic behavior of atoms and simultaneously set the stage for the genesis of amazingly large structures such as stars, supermassive black holes, galaxies, galactic superclusters, and galactic filaments. Here again, it is clear that at the universe's genesis, the nature of the large-scale structures that would come into existence billions of years later was determined.

THE NATURE OF A SUITABLE HOME

At this point, we have contemplated a few of the conditions that are apparently necessary to form a galaxy that can host complex life. However, even in seemingly hospitable galaxies, there is no guarantee that complex life will emerge. In fact, the construction of a potentially life-supporting galaxy is just the first of many steps necessary for the actual formation of life. In the remaining two chapters, we will begin to investigate some of the additional conditions that are likely required for potentially life-supporting stars and planets to form. As we will see, the journey from a potentially life-supporting galaxy to a potentially life-supporting planet is a protracted one that is full of surprises and demands.

REFERENCES

1. Strigari, L.E., et al., *A common mass scale for satellite galaxies of the Milky Way.* Nature, 2008. **454**(7208): p. 1096–1097.
2. Larson, R.B. and V. Bromm, *The first stars in the universe.* Scientific American, 2001. **285**(6): p. 64–71.
3. Loeb, A., *How did the first stars and galaxies form?* Princeton Frontiers in Physics; Variation: Princeton Frontiers in Physics. 2010, Princeton, NJ: Woodstock. xiii, 193 p.
4. Cho, A., *Complex life may be possible in only 10% of all galaxies.* Science, 2014. www.science.org/content/article/complex-life-may-be-possible-only-10-all-galaxies.
5. Piran, T. and R. Jimenez, *Possible role of gamma ray bursts on life extinction in the universe.* Physical Review Letters, 2014. **113**(23): p. 231102.
6. Perrotto, T.J. and W. Clavin, *Herschel measures dark matter required for star-forming galaxies.* NASA News, 2011.
7. Ciardullo, Robin, George Jacoby, and H. Dejonghe, *The radial velocities of planetary nebulae in NGC 3379.* Astrophysical Journal, 1993. **414**(2): p. 454–462.
8. Robert Minchin, et al., *A dark hydrogen cloud in the Virgo cluster.* Astrophysical Journal, 2005. **622**: p. L21–L24.
9. Christy A. Tremonti, et al., *The origin of the mass-metallicity relation: insights from 53,000 star-forming galaxies int the Sloan Digital Sky Survey.* Astrophysical Journal, 2014. **613**: p. 898–913.
10. A. Amblard, et al., *Sub-millimetre galaxies reside in dark matter halos with masses greater than 300 billion solar masses.* Nature, 2011. **470**: p. 510–512.
11. Iocco, F., M. Pato, and G. Bertone, *Evidence for dark matter in the inner Milky Way.* Nature Physics, 2015. **11**(3): p. 245–248.
12. Randall, L., *Dark matter and the dinosaurs: the astounding interconnectedness of the universe.* First edition. ed. 2015, New York, NY: Ecco, an imprint of HarperCollins Publishers. xv, 412 p.
13. Lemonick, M.D., *The first starlight.* Scientific American, 2014. **310**(4): p. 38–45.
14. Agarwal, B., et al., *New constraints on direct collapse black hole formation in the early universe.* Monthly Notices of the Royal Astronomical Society, 2016. **459**(4): p. 4209–4217.
15. Smith, A., V. Bromm, and A. Loeb, *Evidence for a direct collapse black hole in the Lymanαsource CR7.* Monthly Notices of the Royal Astronomical Society, 2016. **460**(3): p. 3143–3151.
16. CFA-Communications, *Just-so black holes.* Harvard Gazette, 2016.
17. LIGO-Scientific-Collaboration-and-Virgo-Collaboration, *Observation of gravitational waves from a binary black hole merger.* Physical Review Letters, 2016. **116**(061102): p. 061102–1 to 061102–16.
18. O'Callaghan, J., *The gravitational-wave "revolution" is underway.* Scientific American, 2019. www.scientificamerican.com/article/the-gravitational-wave-revolution-is-underway/.
19. Overbye, D., *A very hungry black hole is found, gorging on stars.* New York Times, 2018.
20. Doeleman, S., *Focus on the First Event Horizon telescope results.* Astrophysical Journal Letters, 2019. https://iopscience.iop.org/journal/2041-8205/page/Focus_on_EHT.
21. Editor, *Space math—how fast does the Sun spin?* NASA spacemath.gsfc.nasa.gov.
22. Scharf, C.A., *Gravity's engines: how bubble-blowing black holes rule galaxies, stars, and life in the cosmos.* First edition. ed. 2012, New York, NY: Scientific American/Farrar, Straus and Giroux. ix, 252 p.
23. Natarajan, P., *The first monster black holes.* Scientific American, 2018. **318**(2): p. 24–29.
24. Tombesi, F., et al., *Wind from the black-hole accretion disk driving a molecular outflow in an active galaxy.* Nature, 2015. **519**(7544): p. 436–438.

25. Nardini, E., Reeves, et al., *Galaxy evolution. Black hole feedback in the luminous quasar PDS 456.* Science, 2015. **347**(6224): p. 860–863.
26. Alexander, Tal and P. Natarajan, *Rapid growth of seed black holes in the early universe by supra-exponential accretion.* Science, 2014. **345**(6202): p. 1330–1333.
27. Naeye, R., *The newest closest galaxy.* Sky & Telescope, 2003. https://skyandtelescope.org/astronomy-news/the-newest-closestgalaxy/.
28. Roeland P. van der Marel, et al., *First Gaia dynamics of the Andromeda system: DR2 proper motions, orbits, and rotation of M31 and M33.* Astrophysical Journal, 2019. **872**(1).
29. Krauss, L., *What does the future of the universe hold?* Smithsonian Magazine, 2014. www.smithsonianmag.com/science-nature/what-does-future-universe-hold-180947977/.
30. Watkins, L.L., N.W. Evans, and J.H. An, *The masses of the Milky Way and Andromeda galaxies.* Monthly Notices of the Royal Astronomical Society, 2010. **406**(1): p. 264–278.
31. NASA, *Astronomers predict Titanic collision: Milky Way vs. Andromeda.* NASA Science, 2012. https://science.nasa.gov/science-news/science-at-nasa/2012/31may_andromeda.
32. Wong, J., *Astrophysicist maps out Milky Way's final collision.* News at University of Toronto, 2000.
33. Overbye, D., *Black hole picture revealed for the first time.* New York Times, 2019.
34. Ghez, A., et al., *Measuring distance and properties of the Milky Way's central supermassive black hole with stellar orbits.* Astrophysical Journal, 2008. **689**: p. 1044–1062.
35. Fraknoi, A., *How fast are you moving when you are sitting still?* The Universe in a Classroom, 2007. NASA, https://nightsky.jpl.nasa.gov/docs/HowFast.pdf
36. Scharf, C., *The benevolence of black holes.* Scientific American, 2012. **307**(2): p. 34–39.
37. Editor, *Hubble survey confirms link between mergers and supermassive black holes with relativistic jets.* Hubble Site, 2015.
38. Fox, A.J., et al., *Probing the fermi bubbles in ultraviolet absorption: a spectroscopic signature of the Milky Way's biconical nuclear outflow.* Astrophysical Journal, 2015. **799**(1): p. L7.
39. Felicia Chou, et al., *Hubble discovers that Milky Way core drives wind at 2 million miles per hour.* 2015, NASA, https://spacemath.gsfc.nasa.gov/sun/4Page1.pdf
40. Rees, M.J., *Just six numbers: the deep forces that shape the universe.* 2000, New York, NY: Basic Books. x, 173 p.

4 The Galactic Cradle

We live in a modest system, a galaxy called the Milky Way. If we named every star in the Milky Way and put them in the Hollywood telephone directory and stacked those telephone directories up, we'd have a pile of telephone directories 70 miles high.

John Rhys-Davies

Far out in the uncharted backwaters of the unfashionable end of the western spiral arm of the Galaxy lies a small unregarded yellow Sun. Orbiting this at a distance of roughly ninety-two million miles is an utterly insignificant little blue green planet whose ape-descended life forms are so amazingly primitive that they still think digital watches are a pretty neat idea.

Douglas Adam

If you want to see a black hole tonight, just look in the direction Sagittarius, the constellation. That's the center of the Milky Way Galaxy and there's a raging black hole at the very center of that constellation that holds the galaxy together.

Michio Kaku

Finding a galaxy suitable for life is not easy, and even within potentially life-supporting galaxies, there are likely only very limited domains that are capable of hosting complex life forms. These areas form galactic cradles, and within them, there is the possibility of finding a rare star and an even rarer planet that can support the type of complex life with which we are familiar.

DANGER WAITS IN THE CENTER OF THE GALAXY

As with all real estate, what matters most is location, location, location. As I mentioned in the previous chapter, one region of the Milky Way that is particularly foreboding is the center. If you were asked to pick out a spot for a new home, this would be one of the last places you would want to consider. To begin, the center of the galaxy is where the nightmarish supermassive black hole is located. Some of the stars that circle supermassive black holes do so at the breath-taking speed of 27 million miles per hour [1], and any star system that ends up too close to a black hole is literally devoured [2]. Furthermore, in the center of the galaxy, many of the stars exist as binary pairs. Occasionally, these duets get close enough to the supermassive black hole that they are ripped apart, and one of the two is literally ejected from the galaxy while the other remains trapped in orbit around the black hole. The banished star and its associated planets are doomed to travel through intergalactic space at near relativistic speeds, utterly alone and never to return [3, 4]. Some have suggested that life on a planet being towed through the universe by an ejected star may survive and that it may even be able to seed life in a distant galaxy [5], but such a planetary encounter with a black hole would be precarious to say the least.

DOI: 10.1201/9781003270294-6

Being in a region where most stars exist as pairs also has other potential problems. For example, a planet circling a binary pair would be heavily influenced by the gravitational turbulence of both stars, and as a result, it may not be able to maintain a stable orbit. As we will see in the next chapter, this would make the genesis and survival of life particularly difficult. To make matters worse, in the center of the galaxy the scorching radiation coming from supermassive black holes can rip apart a planet's atmosphere, and even planets thousands of light years away may be affected by the jets of radiation that occasionally stream out from the black hole's accretion disk, especially if they are in the direct path of these emissions.

Furthermore, in the center of the galaxy, the threat of atmosphere-destroying radiation does not just emanate from the supermassive black hole. In this relatively crowded region, there are giant blue stars that are at least 16 times as massive as our Sun, and they too emit enormous amounts of radiation. If one of these goliaths is in the neighborhood, the energy it releases could decimate the atmosphere of nearby planets and compromise the integrity of biologically important molecules such as DNA and RNA. To add to the peril, the behemoth stars in the center of the galaxy live a relatively short life, and when they eventually explode, they have an enormous impact on their crowded neighborhood. This is especially true of the stars located in the 200 or so globular clusters within the center of our galaxy. Among the densest of these is M15, which is estimated to have about 30,000 stars within a 22 light-year domain [6]. When stars in these regions go supernova, they rip apart their stellar neighbors and vaporize nearby planets. Even planets 30 light years away from the explosion could lose their atmospheres, and any life that gained a foothold on these worlds would be wiped out [7].

THE RELATIVELY TRANQUIL GALACTIC SUBURBS OF THE MILKY WAY

In comparison to the crowded and dangerous neighborhoods that exist in the center of the Milky Way, our Sun has a much more pleasant existence. At some 27,000 light years from the Sagittarius A*, it resides in the galactic suburbs, in an area called the Orin Arm [8] (see Figure 4.1). In our realm, the Sun cruises around the galactic perimeter at the relatively leisurely rate of 483,000 miles per hour, and in our neck of the Milky Way, stars are few and far between. Within 20 light years of our Sun, there are only about 150 other known stars [9], and the closest of these is the red dwarf Proxima Centauri, which is located more than four light years away [10].

To get a feel for just how much room there is between our Sun and our nearest stellar neighbor, it helps to consider how long it would take the fastest man-made object to reach Proxima Centauri. At the time of this writing, the title of the speediest object ever created by humanity goes to NASA's Juno spacecraft, which on September 8, 2014, was reported to be traveling 83,400 mph relative to Earth [11]. By our Earthly standards, that's certainly impressive, and at that speed, Juno is moving more than 100 times faster than a 0.22 long rife bullet when it exits the muzzle of a gun. Nevertheless, even at Juno's speed, it would take more than 34,000 years of nonstop movement to reach Proxima Centauri.

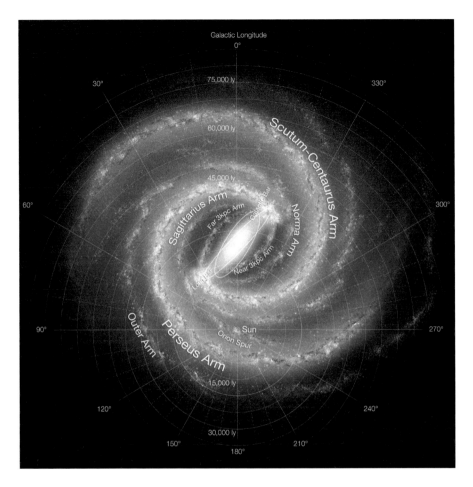

FIGURE 4.1 An artist's depiction of the Milky Way. The location of the Sun is noted. (From: NASA/JPL-Caltech/ESO/R. Hurt, Public domain, via Wikimedia Commons.)

Surprisingly, even though it is the closest star to us, Proxima Centauri is invisible to the unaided eye. Its diminutive status and low intensity ensure that it stays out of sight. There are, however, thousands of other larger stars, such as the blue-white supergiant Deneb, that exist within our galactic region, and many of these can be seen with a simple upward glace of the night sky. But, despite their visibility, these stellar luminaries are not even remotely close. In fact, Deneb is so far away that light that left that giant more than two and a half millennia ago will only be arriving at Earth later this evening [12]. It is worth noting that if there are intelligent beings on a planet that orbits Deneb, and if they can detect the radiation that left Earth, they will have to wait another 600 years or so before that radiation, traveling at the rate of 671 million miles per hour, allows them to learn of the birth of Jesus of Nazareth. It takes a long time to get the latest news from our stellar neighbors in our neck of the galaxy.

THE GALACTIC SUBURBS HAVE ELEMENTAL GRAVITAS

In addition to being rather spacious and uncrowded, the suburbs of our galaxy also have many stars and planetary systems that are enriched in heavy elements. Early in the Milky Way's history, the first gaseous giants burnt through their nuclear fuel and then exploded. When they did, the ash that composed their remains, which was enriched in heavy elements, drifted about in great nebulous clouds, largely in the center and suburban regions of the galaxy. Ultimately, the vestiges of these giants were swept up and incorporated into new stars, and as a result, the chemical composition of the more recent stellar additions to the galaxy differs fundamentally from the old stellar heavyweights that preceded them.

Further out in the Milky Way, in the rural regions of the galaxy, stars diminish in number, and many of the stars that do exist in the galactic hinterlands were formed from gaseous clouds deprived of significant amounts of heavy element debris. Consequently, they, and the planets that orbit them, are likely lacking in many of the raw materials needed for life. It is therefore probable that the outlying regions of our galaxy are desolate and inhospitable to life.

All of this suggests that, within the Milky Way, there are limited areas where complex life can exist, and Charles Lineweaver and his colleagues estimate that the bio-friendly zone consists of an annular region located approximately 23,000 to 29,000 light years from the center of the galaxy [13]. Others assign different boundaries [14], but it seems likely that most of the habitable zones of the galaxy lie somewhere between the dense and dangerous center and the remote and heavy element-barren outer regions [15].

SIZE MATTERS

By definition, our Sun is a star that exists within a habitable zone of the galaxy, but it also has a number of other attributes that make it unusual. For starters, our Sun is a type G dwarf, meaning that it is larger than most stars. Yet, despite its bulk, our Sun emits less than 10% of its energy in the UV range. Massive stars, and especially the 4% or so of stars that are significantly larger than our Sun [16, 17], typically radiate much more of their energy in the UV region of the electromagnetic spectrum. This form of radiation can superheat or destroy planetary atmospheres, and even if a planet manages to maintain its atmosphere, the UV light bombarding it poses a threat to the integrity of biologically important molecules such as DNA and RNA. This makes life near large stars improbable.

Given that large stars aren't ideal energy providers for complex life, one may suspect that small stars might be better suited for the job. As it turns out, stars that are 0.08–0.8 times the mass of our Sun are very common in the Milky Way, and indeed, they are very common in the universe as a whole. Within our galaxy, these small stars compose about 90% of the stellar objects [17], and the most frequently found subgroup within this constituency is that of the M dwarfs. Stars in this subclass are 8–45% as massive as our Sun, and they account for about 76% of the stellar objects in our galaxy. However, like giant stars, M dwarfs are less than optimal when it comes to supporting complex life. Their relatively low energy output greatly narrows the area around them that is habitable, and if a planet resides outside the highly constricted habitable zone, that bodies' water, if it had any, would either boil off or exist as ice. Furthermore, because M dwarfs are small, the habitable zone around them is very close to the star itself, and as a result, the orbit of planets within the habitable zone is often tidally locked. This means that the planet is gravitationally locked to its nearby star, and consequently, its rate of spin around its axis is the same as its rotational period around the star. Due to this phenomenon, one side of an orbiting planet is constantly exposed to the star and its light, while the other always faces away. Consequently, one face of the planet has a never-ending day, while the other exists in perpetual darkness. Under the constant heat of the star, water, and the atmosphere on the side of the planet facing the star, would likely boil away, while the dark side of the planet would be sealed in a deep freeze. The most probable place to find life on these planets would be in the very limited area between the two sides, where the light meets the darkness, but this would be a very limited zone, and diverse environments like those found on Earth would be hard to come by. Ultimately, the absence of such environments would impede the evolution of complex life.

Due to their proximity to an M dwarf, planets orbiting these stars would also potentially be strongly affected by solar flares. These stellar outbursts are generated by magnetic fluctuations within the star, and they can release energy in the form of X-rays and UV radiation. These forms of electromagnetic energy combine with the high-energy particles that solar flares also jettison, and when they rain down upon a planet, they rip apart its atmosphere and damage life [18]. Further, these solar flares, and the starspots that they are associated with, are likely to have significant effects on the climate of any orbiting bodies, so even if the atmosphere and life on a planet somehow survive the initial assault, the longevity of that life would be questionable, given that essential ecosystems on the orbiting globe would be subjected to regular disruption [19].

There are situations that may mitigate some of the negative consequences associated with living on a planet or moon that orbits a large or small star. Life forms living in the habitable zone of a large star may develop resistance to UV light, or they may evolve in a situation that does not depend on starlight for energy. For example, life on planets bombarded by UV light may evolve underground or underwater, and they may use geothermal or chemical energy to power their metabolism. However, it is hard to imagine highly intelligent life forms thriving in these circumstances. On the other end of the size spectrum, planets orbiting a small star such as an M dwarf may be able to maintain a moderate temperature, even when tidally locked, if they have a very dense, self-renewing atmosphere rich in carbon dioxide, or if they generate climates with atmospheric circulation patterns that allow the warm air produced on the warm side of the planet to enter the dark side and vice versa [20, 21]. Furthermore, given that the most intense solar flares and starspot activity occurs early in an M dwarf's very long lifespan, if life were to originate after these stellar events attenuate, it would have a better chance at long-term survival. Still, despite all the potential workarounds, life that evolved within the habitable zone of a very large or very small star would be difficult to initiate and maintain, and there are many advantages to living in a habitable zone created by a star with a radiation output and size that is like that of our Sun.

BEING MADE OF THE RIGHT STUFF IS IMPORTANT

Our Sun is near ideal in size, and it is in a galactic habitable zone, but it is also composed of a mix of unusual chemical elements. On average, it has 25% more heavy elements than do other similar size stars, and it has more metals than about two-thirds of the stars in its size cohort [15, 22]. This matters, in part because Sun's composition reflects that of the planets around it. As a result, if a star has few heavy elements, then the planets that accrete near that stellar body tend to also have few heavy elements. Any rocky planets that were generated in the habitable zone of a star deprived of heavy elements would tend to be small, and because of their diminutive stature and weak gravitational field, they would have trouble holding on to their atmospheres. In addition, without substantial amounts of metals such as iron and radioactive uranium, these planets would have weak magnetic fields and their internal heat sources would be limited. Additionally, even small decreases in the amount of planetary heavy elements such as oxygen and sulfur can matter a great deal. Water, for example, which contains the heavy element oxygen, makes rocks in Earth's mantle softer, and this helps them slide, a process that is necessary for plate tectonics and Earth-like ecosystems. Likewise, sulfur is necessary for our planet to exist in its current state, in part because it helps the iron core in the center of our planet remain partly liquefied and this is necessary for Earth to generate its life-protecting magnetic field. Consequently, a decrease in the amount of oxygen or sulfur on a planet could lead to the attenuation or elimination of its plate tectonics and magnetic field and that would profoundly affect the development of life [23].

Clearly, a solar system that exhibits a paucity of heavy elements is going to be different from ours. On the other hand, if a star has extraordinarily high amounts of heavy elements, that too is problematic. In this situation, many of the planets may be very large, Jupiter-size spheres [13]. These giants, with their immense gravitational pull and dense atmospheres, are not amenable to life as we know it. In addition, some of these gigantic planets may form far from the star and then, later in their life, migrate inward toward the center of their solar system. Such roving monsters would be bad news for any Earth-size globe that did manage to form in a habitable zone because, when the planetary giants move inward, they often destabilize the orbits of smaller Earth-size worlds, or else they collide with them and remove them from existence.

Even if a heavy element laden, Earth-size planet did manage to escape being knocked out of orbit or swept up by their larger sister planets, it would still be less than ideal for life because the massive quantities of the metals it possesses would likely be toxic, at least to complex life as we understand it. Furthermore, if the planet was particularly rich in some elements, such as carbon, it may have an atmosphere that traps massive amounts of heat [15, 22]. In this environment, greenhouse effects could generate temperatures that not even bacteria could tolerate. Finally, large amounts of heavy elements may end up producing globes overstocked in one or more heavy elements. All these worlds would be dramatically different than Earth and, in these places, intelligent life like that found on Earth would likely not be possible.

Fortunately for us, the pre-solar nebula that generated the chemical characteristics of our Sun also produced an orbitally stable, life-sustaining planet from the chemical mix. How did this pre-solar nebula arise and why was it different from many of the others that produced Sun-sized stars? To put the question another way, why did the elemental soup that formed our solar nebula exist in the Goldilocks state? Like many of the questions I have posed, the answer is not known with certainty, but some have speculated that our solar nebula arose in a region of the Milky Way that was close to the center, and that over time this gas cloud migrated out into the habitable zone of our galaxy [24, 25]. In such a scenario, the pre-solar nebula would have been likely to incorporate the heavy element enriched chemical remains of supernovae simply because there were more supernovae closer to the center of the galaxy. Another possibility is that the pre-solar nebula formed outside of the center of the Milky Way, but by chance, it just happened to be in an area that was augmented by the remains of two or more large supernovae. The odds of this are small, but there are many aspects of our existence that appear to be low probability events [26, 27]. Regardless of how the pre-solar nebula acquired its mix of heavy elements, what matters is that our solar system is chemically unusual and apparently quite special. Because of the specific stellar nebula that formed our solar system, we have the 26 elements needed for Earth-bound life, and the abundance of these elements is in a range that is suitable for the development of complex life forms like those we see around us.

CONSISTENCY COUNTS

The Sun has a number of other traits that contribute to its ability to support life. For starters, the Sun's energy output is much more consistent than that of many similar stars [28–30], and during a repeating 11-year cycle, our Sun's energy yield varies just a mere 0.1% [19]. This stability is essential for complex life on Earth, most of which could not tolerate the significant fluctuations in the high-energy radiation, cosmic rays, and infrared thermal energy that accompany more robust alterations in solar output. In fact, a relatively small change in thermal insolation could send many complex life forms into a tailspin from which they could not recover. We know, for example, that during the last ice age, when the planet's temperature decreased by 12.6°F, mass extinctions occurred around the globe, and today, climate scientists are warning that a 7.2°F increase in our globe's average temperature will be catastrophic for many Earth-bound species, including humans.

There are some organisms that are unusual in that they display exceptional resiliency when placed under thermal stress. In the laboratory, bacteria are routinely suspended in a glycerol solution and frozen at -4°F, and although they don't replicate or even undergo any significant rate of metabolism at that temperature, they can nevertheless remain viable for many years. There are also some animals, such as the tardigrades, that can enter a form of hibernation and survive even in the fierce coldness of space. On the other end of the spectrum, there are organisms, such as extreme hyper-thermophilic bacteria, that can endure temperatures exceeding that of boiling water. But surviving is different than flourishing, and in these punishing environments, all that even the most robust life forms can do is endure.

Given that even the best-adapted organisms have difficulty thriving in extreme situations, it's not surprising that for most plants and animals, prolonged bouts of moderate cold or mild heat are intolerable. For most creatures, low temperatures can cause the formation of ice crystals that disrupt cellular function, while high temperatures can generate enough kinetic energy to literally rip apart the cell's membranes and substructures. As a result, the range over which most plants and animals can prosper and complete their life cycles is limited to a narrow window of about 28–122°F [31, 32]. There are a few rare exceptions that can operate outside of this range, such as Emperor Penguins, which can complete their life cycle at temperatures below -40°F; however, these outliers can only function at one end of the temperature gradient, not both, so even they can't tolerate rapid fluctuations in temperature. Therefore, for complex life forms like those found on Earth to prosper, they must reside on a planet near a constant and consistent flow of heat. For life forms like us, a star's thermal consistency matters a great deal.

There are other important traits that a life-supporting star must exhibit, and one of these likely includes traveling in a path that is safe from cosmic dangers. Despite appearances, our Sun is doing a great deal of traveling. To start with, relative to the cosmic background radiation, our solar system, and indeed our entire galaxy, is moving uniformly at about 1.3 million miles per hour in the direction of the constellations Leo and Virgo. We don't know with certainty what is pulling our galaxy and a number of others in this direction, but it is likely that another large group of galaxies is exerting this effect, and the collective attractive force of these giant gravitational titans is aptly referred to as "the great attractor" [33–35]. The idea that we, as part of the Milky Way, have traveled more than 31 million miles since this time yesterday means that we are experiencing different areas of the universe daily, although looking up in the sky from our home deeply ensconced within the Milky Way, the universe doesn't look much different to us from one day to the next.

As was mentioned previously, our solar system is also rotating around the galactic core at a speed close to 500,000 miles per hour. This sojourn is not necessarily free from danger, and as any traveler knows, the art of successfully getting from one place to another often depends on finding the right path. As it turns out, our solar system is on what appears to be a slightly elliptical, near-ideal route around the galaxy. This is not typical, and 93% of the star systems within 130 light years of ours have more eccentric orbitals [22]. Consequently, some neighboring star systems occasionally plunge toward the center of the galaxy where they are more likely to encounter other stars, debris, and supernovae. As they take this dive toward the heart of the galaxy, they are also likely to be perturbed by the gravitational forces that operate in this crowded region.

Our Sun's ability to avoid the busy center of the galaxy makes life more probable within our solar system, but life on Earth also benefits from the fact that our elliptical path, and that of the nearby galactic spiral arms, intersect on average only about once every 140 million years. This is a good thing because like the center of the galaxy, spiral arms also contain enormous clouds of hydrogen gas, sizeable amounts of galactic dust, and a relatively high density of stars [36]. With regard to our own planet, Alexander Pavlov and his colleagues suggest that, when Earth is transiting these spiral arms, the amount of debris in the stratosphere accumulates, and they postulate that the buildup of molecular rubble may greatly enhance the amount of sunlight reflected back into space. If they are right, the temperature on Earth may have dropped precipitously during some of our voyages through the stellar arms, and an albedo effect triggered by these movements may have even initiated one or more of the "snowball Earth" eras, during which the entire globe was encased in ice [37]. So, it may be quite fortunate that we don't cross paths with spiral arms often.

There are still other dangers that await us as we move about the galaxy, some of which appear to be related to our position within the plane of the solar system's orbital ellipse. We know that, as our solar system circles the Milky Way, it resembles a carousel horse on a merry-go-round because, like a carousel horse, our solar system moves up and down through the plane of its disk. In fact, we move about 228 light years north of the mid-plane to about 228 light years south of the mid-plane once every 70 million years or so [38], and this implies that we cross the center of our galactic disk about once every 35 million years. Some evidence indicates that the bombardment of Earth by comets and asteroids increases about every 35 million years and Harvard physicists Lisa Randall and Matthew Reece suggest that the cause may be related to our transit through the center of the orbital ellipse. They hypothesize that a concentrated disk of dark matter lies in the mid-plane, and as we transit this dark matter disk, its gravity rattles our solar system's Oort cloud [38]. The Oort cloud, which begins in a remote region past Neptune and extends approximately one light year into space, is like the solar system's junkyard, and it contains an estimated 10^{12}–10^{13} remnants such as planetoids, comets, and asteroids that were produced early in the life of our star system [39]. Due to its great distance from the center, the Sun has relatively little gravitational control over the objects in these distant hinterlands. As a result, tidal forces generated by cosmic entities such as dark matter disks may be sufficient to knock some of the Oort cloud's comets and asteroids out of their orbitals, and once disturbed, the Oort cloud could send its debris hurling about in various directions. Some of these objects may end up being jettisoned into interstellar space, never to be seen again, but others may be directed toward the center of the solar system where they may encounter planetary bodies such as Earth [40, 41]. If this is true, then our solar system's oscillating movement through the galactic mid-plane may have triggered several mass extinctions in our planet's past, including the one that led to the demise of the dinosaurs 66 million years ago [38, 42].

While it is possible that the mid-plane of our galactic orbital has a dangerous band of dark matter, some believe that the most perilous part of our solar system's oscillation through its orbit is not located at the mid-plane, but rather north of the mid-plane. These individuals speculate that as our galaxy rushes toward the "great attractor" it generates cosmic rays which damage living organisms and the environment, and they argue that our planet is more exposed to this radiation when it is north of the mid-plane [43].

Beyond the potential dangers associated with our solar system's bobbing up and down in its elliptical orbit, there exists the possibility that some regions of that orbit are more hazardous than others. As I mentioned, it takes roughly 225 million years for our solar system to complete one trip around the Milky Way, and to date, our species has not seen much of the galactic superhighway that our planet is on. What awaits us on the other side of the galaxy is hard to predict, particularly since the galaxy is a dynamic place. As we circle the galaxy, new stars are coming into existence, old ones are exploding and imploding, and many cosmic structures such as stars and black holes are moving about in complex paths. There is always the possibility that we may end up in the wrong place at the wrong time. On a more optimistic note, it is also true that during the last 4.56 billion years, our solar system has made many trips around the galaxy, and overall, it appears that our Sun and planet have fared reasonably well. This suggests that the likelihood of our encountering existential threats during our sojourn around the perimeter of the galaxy is relatively small, and because of this, life on Earth has a significant advantage over many other planets.

IS THE EXISTENCE OF OUR SUN, WITH ALL ITS ANOMALIES, ANOTHER FORTUNATE EVENT FOR HUMANKIND, OR IS IT SIMPLY A STATISTICAL INEVITABILITY?

When a star is labeled "Sun-like," this often means that the star has a mass like that of our Sun. However, as we have seen, true "Sun-like" stars tend to lack stellar companions, and they have an unusual chemical composition. In addition, they have relatively little fluctuation in their luminosity, they have stable and serene galactic orbits, and they can generate and maintain a long-term habitable zone for their orbiting satellites. If one also considers all these factors, the number of true "Sun-like" stars drops precipitously.

So, how many truly "Sun-like" stars are there? Is our Sun one of an exceedingly small number of stars capable of supporting life, or are there many stars that could do the job? The answer to this question will depend, in part, on the number of stars that exist. Astronomers currently estimate that there are about 200 billion stars in our galaxy, and all totaled, there are about 300 sextillion (3×10^{23}) stars in the two to three trillion galaxies of our universe [17, 44]! This is a staggeringly large number and it has been said that there are about as many stars in the universe as there are cells in all of the humans on Earth [45]. Another way to try to begin to grasp the size of this number is to consider the fact that there are about 10 billion stars for every person who has lived [1].

With this many stars in the universe, it seems reasonable to speculate that there are many stars that outperform our Sun when it comes to supporting life. Included among this group may be some K dwarfs [46]. These stars are 45–80% as massive as our Sun and their moderate size may eliminate many of the limitations associated with the smaller M dwarf stars. K dwarfs are also about 60% more common than are the G dwarf stars such as our Sun, and because they are parsimonious when it comes to expending their nuclear fuel, they can outlive our Sun by billions of years. Life on planets that circle K dwarfs would have more time to evolve, and in the aftermath of a cataclysmic event, more time to recover.

Given that within our galaxy, close to 20% of the stars are either G or K dwarfs and assuming that there are 200 billion stars in the Milky Way, we can estimate that in our galaxy, there are about 40 billion stars that have masses similar to that of our Sun. Many of these will lack traits necessary for supporting life, but nevertheless, cosmologists have suggested that there are about 2.4 billion stars in the Milky Way alone that could be capable of doing so [47].

Even if these estimates are way off and only one in a billion G or K dwarf stars could support life, there would still be 40 such stars in our galaxy and perhaps 120 trillion of these entities in the observable universe. Therefore, the existence of other true Sun-like stars that are capable of supporting life is likely a near statistical certainty.

Knowing whether Sun-like stars likely exist and knowing why they likely exist are two very different questions, however. We know that Sun-like stars likely exist because there are so many stars in our universe, but to know why there are so many Sun-like stars, we also need to know why our universe contains so much matter and why this matter exhibits the traits that it does. To answer these questions, we need to unravel many of the mysteries discussed in the last several chapters. For example, we need to know why there is more matter than anti-matter in our universe, for had there been even slightly more anti-matter, we would live in a place with far fewer stars, and far fewer Sun-like entities capable of potentially supporting life.

Regardless of why Sun-like stars are here, we humans are clearly lucky that they are. We are fortunate that these stars can form from collections of hydrogen molecules, that they can use hydrogen to produce heavier elements, and that they can generate heat and light for eons. It is also good for us that these life-supporting stars can sometimes be found in a life-supporting region of a rare life-supporting galaxy. In addition, we are lucky that when the largest stars expire, they blow apart in a way that guarantees the widespread distribution of the heavy elements they manufactured, and we are fortunate that the remains of these forsaken stellar giants can sometimes form a black hole that helps maintain the integrity of the life-generating, habitable regions of a galaxy.

In the next chapter, we will look at the satellites that circle some potentially life-supporting stars, and we will see that these relatively small globes must have a litany of traits and characteristics for life as we know it to arise and thrive. We will also see that among all the true Sun-like stars in the observable universe, only a small fraction will have the exceptional planetary partners that they need to form a life-generating and life-supporting team.

REFERENCES

1. Scharf, C.A., *Gravity's engines: how bubble-blowing black holes rule galaxies, stars, and life in the cosmos.* First edition. ed. 2012, New York, NY: Scientific American/ Farrar, Straus and Giroux. ix, 252 p.
2. Mattila, S., et al., *A dust-enshrouded tidal disruption event with a resolved radio jet in a galaxy merger.* Science, 2018. **361**: p. 482–485.
3. B. Bromley, et al., *Hypervelocity stars: predicting the spectrum of ejection velocities.* Astrophysical Journal, 2006. **653**: p. 1194–1202.
4. Sergey E Koposov, et al., *Discovery of a nearby 1700 km/s star ejected from the Milky Way by Sgr A*.* Monthly Notices of the Royal Astronomical Society, 2019. arXiv:1907.11725 [astro-ph.GA]: p. 1–16.
5. A. Loeb and J. Guillochon, *Observational cosmology with semi-relativistic stars.* Physical Review Letters, 2015. arXiv:1411.5030 [astro-ph.CO].
6. Plotner, T., *Messier 15.* Universe Today, 2009.
7. Ellis, J. and D. Schramm, *Could a nearby supernova explosion have caused a mass extinction?* PNAS, 1995. **92**: p. 235–238.

8. Fraknoi, A., *How fast are you moving when you are sitting still?* The Universe in a Classroom, 2007.

9. Editor, S.-S., *Stars within 20 light-years.* Sol-Station, 2005. http://www.solstation.com/stars/s20ly.htm

10. Sahu, K.C., et al., *Microlensing events by Proxima Centauri in 2014 and 2016: opportunities for mass determination and possible planet detection.* Astrophysical Journal, 2014. **782**(2): p. 89.

11. NASA, *Sun-Earth Day—technology through time.* NASA, 2014.

12. Schiller, F. and N. Przybilla, *Quantitative spectroscopy of Deneb.* Astronomy and Astrophysics, 2008. **479**(3): p. 849–858.

13. Lineweaver, C., Y. Fenner, and B. Gibson, *The galactic habitable zone and the age distribution of complex life in the galaxy.* Science, 2004. **303**: p. 59–62.

14. Prantzos, N., *On the "galactic habitable zone."* Space Science Reviews, 2007. **135**(1–4): p. 313–322.

15. Ward, P.D. and D. Brownlee, *Rare Earth: why complex life is uncommon in the universe.* 2000, Göttingen: Copernicus Publications.

16. LeDrew, G., *The real starry sky.* Journal of the Royal Astronomical Society of Canada, 2001. **95**(1): p. 32–33.

17. van Dokkum, P.G. and C. Conroy, *A substantial population of low-mass stars in luminous elliptical galaxies.* Nature, 2010. **468**(7326): p. 940–942.

18. NASA, *M dwarfs: the search for life is on.* Astrobiology Magazine, 2005.

19. Phillips, T., *Solar variability and terrestrial climate.* NASA Science, 2013. https://science.nasa.gov/science-news/science-at-nasa/2013/08jan_sunclimate.

20. Wood, A., *Are tidally locked exoplanets more habitable than we once thought?* New Atlas, 2016. https://newatlas.com/tidally-locked-exoplanet-life-ku-leuven/44372/

21. Carone, L., R. Keppens, and L. Decin, *Connecting the dots—II. Phase changes in the climate dynamics of tidally locked terrestrial exoplanets.* Monthly Notices of the Royal Astronomical Society, 2015. **453**(3): p. 2412–2437.

22. Robles, J.A., et al., *A comprehensive comparison of the Sun to other stars: searching for self-selection effects.* Astrophysical Journal, 2008. **684**(1): p. 691–706.

23. Stevenson, D., *Planetary diversity.* Physics Today, 2004. **57**(4): p. 43–48.

24. R. Roskar, et al., *Riding the spiral waves: implications of stellar migration for the properties of galactic disks.* Astrophysical Journal, 2008. **684**: p. L79-L82.

25. Courtland, R., *Sun may have been thrown far from its birthplace.* New Scientist, 2008.

26. Courtland, R., *Titanium reveals explosive origins of the solar system.* New Scientist, 2009. **21**: p. 54.

27. Trinquier, A., et al., *Origin of nucleosynthetic isotope heterogeneity in the solar protoplanetary disk.* Science, 2009. **324**(5925): p. 374–376.

28. Radick, R., G.L.B. Skiff, and S. Baliunas, *Patterns of variation among Sun-like stars.* Astrophysical Journal, 1998. **118**: p. 239–258.

29. Editor, *Sol.* Sol-Station, 2011. http://www.solstation.com/stars/sol.htm

30. Grossman, L., *The Sun is less magnetically active than other similar stars.* Science News, 2020.

31. Brock, T.D., *Life at high temperatures.* 1994, Yellowstone National Park, Wyo.: Yellowstone Association. 31 p.

32. Clarke, A., et al., *A low temperature limit for life on Earth.* PLoS ONE, 2013. **8**(6): p. e66207.

33. Kogut, A., et al., *Dipole anisotrophy in the COBE differential microwave radiometers first-year sky maps.* Astrophysical Journal, 1993. **419**: p. 1–6.

34. Kocevski, D. and E. Harald, *On the origin of the local group's peculiar velocity.* Astrophysical Journal, 2006. **645**: p. 1043–1053.

35. Tully, R.B., et al., *The Laniakea supercluster of galaxies.* Nature, 2014. **513**(7516): p. 71–73.

36. Gonzalez, G., *Is the Sun anomalous.* Astronomy and Geophysics, 1999. **40**: p. 5.25–5.29.

37. Pavlov, A.A., *Passing through a giant molecular cloud: "snowball" glaciations produced by interstellar dust.* Geophysical Research Letters, 2005. **32**(3).

38. Randall, L., *Dark matter and the dinosaurs: the astounding interconnectedness of the universe.* First edition. ed. 2015, New York, NY: Ecco, an imprint of HarperCollins Publishers. xv, 412 p.

39. Horner, J. and B. Jones, *Jupiter—friend or foe? An answer.* Astronomy and Geophysics, 2010. **51**: p. 16–22.

40. Fan, J., et al., *Dark-disk universe.* Physical Review Letters, 2013. **110**(21).

41. Randall, L. and M. Reece, *Dark matter as a trigger for periodic comet impacts.* Physical Review Letters, 2014. **112**(16): p. 161301.

42. Gibney, E., *Did dark matter kill the dinosaurs?* Nature, 2014. https://doi.org/10.1038/nature.2014.14839.

43. Medvedev, Mikhail V. and A.L. Melott, *Do extragalactic cosmic rays induce cycles in fossil diversity?* Astrophysical Journal, 2007. **664**: p. 879–889.

44. Editor, *Number of galaxies in universe.* Nature, 2016. **537**: p. 453.

45. Borenstein, S., *Number of stars in the universe could be 300 sextillion, triple the amount scientists previously thought.* Huffington Post, 2010.

46. Heller, R. and J. Armstrong, *Superhabitable worlds.* Astrobiology, 2014. **14**(1): p. 50–66.

47. Gowanlock, M.G., D.R. Patton, and S.M. McConnell, *A model of habitability within the Milky Way galaxy.* Astrobiology, 2011. **11**(9): p. 855–873.

5 The Inimitable Earth

The Earth is the only world known so far to harbor life. There is nowhere else, at least in the near future, to which our species could migrate. Visit, yes. Settle, not yet. Like it or not, for the moment the Earth is where we make our stand.

Carl Sagan

Be grateful for the home you have, knowing that at this moment, all you have is all you need.

Sarah Ban Breathnach

I find it easier to believe that two Yankee professors would lie than that stones would fall from heaven.

Thomas Jefferson

Galaxies and stars capable of supporting life are atypical, and a life-supporting planet is even more unusual. To successfully nurture life over a period of billions of years, a planet must have many characteristics including a balanced mix of heavy and light elements, a stable atmosphere, a temperate climate, and an Earth-like mass. Planets with these traits take a long time to develop, and while it is possible that the chemistry that made life possible began as early as 10 million to 17 million years after the Big Bang [1, 2], there is good reason to think that billions of years are required before planets capable of supporting complex animal life can arise.

In this chapter, we will begin to investigate a few of the features that make Earth suitable for life. What will become apparent is that Earth is a very special place, and the emergence of complex life requires the genesis of a planet like the one that we are fortunate to have.

Over the last several centuries, many have criticized humankind's Earth-centric and anthropocentric orientation, and the move away from this mindset has been constant and steady since Copernicus demonstrated that Earth was not the center of the solar system. However, until another life-supporting planet is found, we may want to rethink this trend, for a strong argument can be made for considering Earth to be "the gem" of the universe. One day, other life-supporting planets may be discovered, and if that day comes, our planet may be downgraded from "the gem" to "a gem" of the cosmos. However, until our planet reaches its demise, perhaps 500 million years from now, it will be "a gem" at the very least, and it will likely forever be one of the relatively few outposts within our universe on which complex life can evolve.

THE SLOW PLANETARY CONSTRUCTION PROCESS

Planets like Earth are extraordinary in part because they contain moderate levels of life-supporting heavy elements such as carbon, nitrogen, oxygen, phosphate, and iron. However, as was mentioned in the previous chapters, it took an extended period

DOI: 10.1201/9781003270294-7

of time to produce these elements and another lengthy period to generate second- and third-generation stars that were enriched in these chemicals [3]. As a case in point, it took our universe 9.2 billion years to create our Sun.

Once stars like the Sun form, metal-rich, life-harboring planets can potentially appear relatively quickly thereafter. In the case of our own solar system, Earth, which is more than 96% iron, sulfur, silicon, oxygen, and magnesium [4], emerged within ten million years of Sun's genesis [5]. So, the universe generated the only metal-rich, life-supporting planet that we are currently aware of over a period of 9.2 billion years, and by anyone's standards, 9.2 billion years is indeed a very long time.

GETTING THE PLANETARY CHEMISTRY RIGHT

In addition to time, a life-harboring planet needs to be in a good location because where a planet is will influence its chemical composition. In our own solar system, the four outermost planets are giant, primarily gaseous structures, while the inner planets are metal-rich and rocky. The dramatic difference between the makeup of the outer and inner planets is due, in part, to the temperature differences that existed in the various regions of the developing solar system.

When the planets were taking shape in the outer sections of the solar system, the temperatures were in the neighborhood of -400°F, and molecules such as water and methane were plentiful. Over time, this material aggregated and formed the cores of the giant planets, and eventually, these proto-planets' gravity attracted the abundant hydrogen and helium atoms that were located around them. Ultimately, they grew to be enormous, with thick hydrogen and helium-based atmospheres. In our solar system, Jupiter is the largest of the gas giants, with a mass that is 318 times that of Earth.

While the outer planets were forming, another phenomenon was occurring in the inner solar system, where temperatures in some parts approached 3,200°F. The cosmic dust that entered this domain was often heated to high temperatures and many of the most plentiful elements in that dust were vaporized. Materials with high boiling points are less likely to be in the vapor phase, and so these substances became enriched within the sticky dust particles that survived. Eventually, these coalescing specks formed planetary embryos that had high concentrations of metals, and after a great deal of violence, the planetary proto structures merged to form the primitive early terrestrial planets.

As the inner section of the solar system cooled, volatile materials were delivered to the planet's surfaces by asteroids and comets that bombarded the region. Many of these icy and rocky bits of solar debris formed in the deeper and colder recesses of interstellar space, and consequently, they contained numerous compounds that the inner planets initially had in only sparse amounts [6, 7].

In the case of Earth, the iron and nickel that the planet accumulated from various sources gave rise to the inner and outer planetary cores, and the decaying unstable heavy radioactive elements that Earth gathered eventually produced most of the heat gradients that formed within the planet. The value of these materials is hard to overstate, and in their absence, Earth would lack a protective magnetosphere, plate tectonics, and life itself.

When the process of planet formation was over, Earth contained a balance of heavy elements that was conducive to the genesis of complex animal life. However, this was not guaranteed to be so. For example, Earth could have formed with too few radioactive isotopes or too many toxic heavy metals. Our planet could have also been produced with a vast amount of metals like iron or nickel, and as a result, our globe could have been nothing more than a large chunk of heavy metal orbiting a star. In a somewhat less dramatic scenario, Earth could have been endowed with so many heat-trapping, carbon-based molecules that the temperature on the planet's surface would have prohibited the evolution of metazoan life. There are indeed many scenarios in which Earth could have ended up with a composition ill-suited for animal life of the kind we are familiar with.

From our perspective, one substance that Earth was fortunate to receive is water. All life that we know of requires water, and indeed, most organisms are largely composed of this substance. We, for instance, are on average 65% water. In contrast to humans, Earth itself is comprised of only 0.023% water by mass [8], so as a fraction of our planet's mass, water is a rare molecule. But fortunately for us, a large amount of the water on our planet is where we need it to be: on the surface. In fact, our planet's exterior is literally flooded with water, and approximately 70% of the surface is covered with some form of this molecule. In some regions, such as our oceans, this seemingly ubiquitous substance is piled two miles deep on average, and in some sections, such as the Mariana Trench in the Pacific, the water is close to seven miles deep. In addition to all this surface water, there is a large amount of water within Earth's mantle, some of which makes its way to the planet's exterior as volatile material emitted from volcanoes. We are not sure exactly how much water lies below our feet, but by some estimates, it may be a volume equal to that of the Pacific Ocean [9], and according to others, it may be more than 100 times the volume of water found in the oceans [10].

The presence of all this surface water raises the question of how an inner terrestrial planet, which formed at high temperatures, managed to end up with so much of this life-sustaining molecule. The answer is still being debated, but it appears that some of Earth's water originated in the asteroid belt debris that formed the planetesimals [11, 12]. Apparently, despite the intense heat, some of the accreting planetesimals managed to trap water within their complex structures, and this water later ended up embedded within the mineral inclusion bodies that formed under high pressures within Earth's interior [13]. As a result of plate tectonic movements, a fraction of the water stored in the mantle eventually worked its way to the crust, and therefore Earth likely had a significant amount of water on its surface and in its atmosphere soon after it formed [8]. In addition to the water locked in planetesimals, Earth also acquired water from a large number of comets and asteroids that shelled the planet 4.1–3.8 billion years ago [14].

It is worth considering what our planet would be like now if the number of comets and asteroids that collided with Earth during the 300 million years of the late heavy bombardment (LHB) had been different. It is likely that fewer collisions would have generated a substantially drier world that lacked some of the elements required for the robust development of life. On the other hand, an increase in the intensity of bombardment may have left our planet completely submerged in a water-based, chemical-rich stew.

A LIFE-SUPPORTING PLANET MUST BE CLIMATE CONTROLLED

The chemical composition of a planet determines if there is any possibility of life developing, but for complex life to exist, a moderate climate is also essential. One of the crucial aspects of any planet's climate is its surface temperature and so the planet's distance from the star it orbits is of prime concern. If the planet is far away from its star, it will be an icy world, but if it is too close, it may end up being too hot or tidally locked.

As was mentioned previously, in the tidally locked scenario, one side of the globe will likely be perpetually boiling, while the other is permanently sealed in a deep freeze. In addition, the side of the planet that faces the star would be subjected to a substantially greater gravitational force than the side that faces away. This would deform the globe into an ellipsoidal structure, and it would cause internal friction increasing the probability of surface volcanism [15]. All of this decreases the probability that a tidally locked planet could support life.

As was also mentioned earlier, some have suggested that despite these potential scenarios, tidally locked planets may have borders between the two planetary faces that provide suitable environments for life. However, if this is so, the amenable conditions would only exist on a thin sliver of the planet, and the high-energy radiation emanating from the nearby star would likely make even those areas inhospitable to complex life forms.

In the case of our solar system, the inner rocky planets started out at extremely high temperatures, but they gradually cooled, and eventually, some of them acquired relatively moderate climates. In fact, some were cool enough that they were able to maintain liquid water on their surfaces, at least for some limited period. Such planets are often said to be within the habitable zone of their star.

The location of the habitable zone depends on many factors including the size and luminosity of the star, the planet's atmospheric pressure, the ability of the planet's atmosphere to trap heat (i.e., its "greenhouse effect"), the amount of incident light that is reflected by the planet (i.e., the planet's albedo), and the planet's axial orientation. In general, as the star ages, it becomes hotter and more luminous, which pushes the habitable zone further out from the star. Additionally, if the planet has a high greenhouse effect or a low albedo, the habitable zone for that planet will also be extended outward from its star. Conversely, if a star is young, or if a planet has a low greenhouse effect, or if it has a high albedo, the stellar habitable zone will tend to move inward, closer to the star.

Our planet obviously exists within our Sun's habitable zone, and this is part of the reason Earth has a relatively moderate temperature. It is worth noting, however, that our Sun's habitable zone is only about 20 million miles wide and Earth is not far from the inner edge of it. To be specific, the inner edge of the Sun's habitable zone begins at about 88 million miles out from the Sun, and on average, Earth orbits about 93 million miles from the Sun [16]. Consequently, if Earth had formed just a mere five million to ten million miles closer to the Sun, we would not be here.

It is also worth noting that the habitable zone, as it is being defined here, only means that liquid water is capable of existing on the surface of a planet. Such a planet is not necessarily habitable, at least by creatures like us. Under our current atmospheric conditions, liquid water will form on our globe at temperatures between 32°F and 212°F, but we wouldn't do well if the planet's average temperature was at either extreme. We and other complex organisms could not also tolerate significant or rapid fluctuations between these temperature extremes. In fact, if our orbit was reduced by just 0.3%, or 290,000 miles, the resulting planetary warming would make life on Earth difficult for us [17]. Two hundred ninety thousand miles is only about 21% of the average distance between Earth and the Moon, and it would take light just a mere 1.6 seconds to transit this gap. In cosmic terms, this is next to nothing. But, if Earth's orbit was just 290,000 miles closer to the Sun, Earth would be transformed from a sanctuary for complex multicellular life to just one more inhospitable rock in a universe that is likely dotted with them.

As it turns out, our Sun's relatively narrow habitable zone is not atypical, and most Sun-like stars likely create a narrow band in which life could potentially evolve. Due in part to this reality, it has been estimated that only about 11% of Sun-like star systems potentially have planets with liquid water [18], and an even smaller number of Sun-like star systems have planets that possess a surface temperature that remains stably ensconced within the moderate range of the habitable zone over long periods of geological time.

THE CONSTRUCTION OF EARTH LIKELY REQUIRED A LITTLE HELP FROM ITS CELESTIAL NEIGHBORS

Up to this point, I discussed the importance of the composition of the pre-solar nebula in determining the chemical characteristics of Earth, and I mentioned that the region of the solar system in which Earth formed also had a profound effect on our planet's physical traits and climate. However, a staggering number of additional events influenced the formation of our home planet including the amount of dust and debris that was in Earth's orbit while it formed, the frequency and nature of the collisions of the accreting material, the distance between the newly forming Earth and other galactic structures, and the activity of nearby planets [19]. In the case of Earth, the location and activity of Jupiter and Saturn in particular played a significant role in our planet's formation.

Jupiter is the oldest and largest planet in our solar system, and it likely took shape at about 3.5 AU from the Sun (1 AU is the average distance from Earth to the Sun, i.e., about 93 million miles), not far from the "snow line." The "snow line" refers to the place in our solar system where volatile compounds such as water, methane, and carbon dioxide can form ice grains even while under direct sunlight. However, while Jupiter may have formed at 3.5 AU, it didn't stay there for long. When the gas giant was about 70,000 years old, it became gravitationally attracted to the gas disks that were rotating around the Sun, and as a result, it began to move toward the inner solar system. As it did so, it scattered or acquired the material in its path [8] until it finally settled into an orbit at about 1.5 AU from the Sun [20]. Jupiter's stay at this location was also short, because soon after Saturn formed, it too began moving inward, and eventually, it entered an orbital resonance state with Jupiter. As a result of Saturn's migration, the two largest gas giants of our solar system engaged in periodic gravitational interactions, and eventually, this caused both planets to turn around and travel into the distant hinterlands, where they ultimately settled into their present-day orbits.

If this "roaming Jupiter" hypothesis is correct, Jupiter's sojourn into the inner realms of the solar system was a game changer, for as it moved inward, it removed material that would have formed "super-Earths," which are defined as being two to ten times more massive than our planet. While planets on the low end of this spectrum may have habitable characteristics, all of the planets studied to date with a radius more than double that of Earth are not rocky, and life as we know it appears to require the existence of a rocky planet [21]. Planets that are several times more massive than Earth are also likely to have suffocating, lethal atmospheres rich in hydrogen gas, and the mantle of these "super-Earths" would be highly pressurized and viscous. These conditions would impair the planet's ability to generate a magnetosphere and plate tectonic activity, and it would make the production of an oxygen-rich environment highly unlikely [22, 23]. As a result, "Super-Earths" would be less than super for intelligent life like us.

Viewed from our perspective, it was certainly good that Jupiter cleared out some of the material in Earth's orbit; however, it is also fortunate that Jupiter didn't remove too much. If it had, it is probable that a miniature version of Earth would now exist. A scaled-down Earth would also be far from ideal for complex life, as it would lack many of our planet's current traits, including the gravitational gravitas needed to hold onto an atmosphere. In what resembles yet another Goldilocks story, Jupiter appears to have cleared out just the right amount of material in Earth's orbit, and when the planetary giant's sojourn around the inner solar system was over, it had set the stage for the formation of a potentially habitable planet, i.e., it set the stage for the accretion of our Earth.

The "roaming Jupiter" hypothesis assigns considerable credit to Jupiter for the formation of current-day Earth; however, it is important to emphasize that the same hypothesis recognizes Saturn as another major player in our planet's formation, for without Saturn, Jupiter may have stayed in a location close to where Mars is now. In such a scenario, Earth's largest sibling might have flung much of the dust and debris that ultimately formed Earth into the Sun or it may have thrown that material clear out of the solar system. Alternatively, Jupiter could have captured and consigned what ended up being Earth-forming planetesimals to one of its many satellites, or it might have incorporated these structures into its own core. If any of these scenarios had occurred, Earth as we know it would not be here today [24].

Remarkably, even after returning to its current orbit, Jupiter continued to influence life on Earth, although exactly how it did so is a matter of debate. Some argue that our giant neighbor acts as a "cosmic bouncer" that regularly intercepts or alters the routes of comets and asteroids otherwise destined to hit Earth. Currently, a comet or an asteroid with a diameter of six miles or more smashes into our planet approximately once every 100 million years, and when this happens, life on Earth becomes exceedingly difficult. The last time such a collision occurred was 66 million years ago, and that cataclysmic event ushered in the demise of as much as 75% of Earth's plant and animal species, including the non-avian dinosaurs. Those who support the

"Jupiter as cosmic bouncer" hypothesis estimate that, in the absence of this planetary titan, the rate of extinction-level catastrophic disasters would increase 1,000 fold, and they posit that animal life would likely not have evolved on our globe under these conditions [25].

More recently, however, it has been suggested that rather than protecting Earth, Jupiter actually puts it in jeopardy. According to this view, Jupiter does not only intercept or alter the paths of some earthbound objects, but it also directs some comets and asteroids toward Earth, and on balance, the gas giant causes more cataclysmic disruptions than it prevents [26]. If this is true, then Jupiter may have triggered the K/Pg extinction that eliminated the dinosaurs, and as result, our giant neighbor may have created the conditions necessary for new niches to open and for creatures like us to emerge (see Chapter 10).

While these two scenarios differ dramatically, what they both suggest is that Jupiter had a profound effect on the evolution of complex life on Earth. Ultimately, it may not be necessary to have a solar system with a Jupiter-like planet to generate metazoans, but then again, perhaps having a Jupiter-like planet positioned relatively nearby does significantly alter the odds of intelligent life forming on a terrestrial globe. Of the star systems that have been studied to date, it appears that only about 3.3% have a Jupiter analog, but unfortunately we have no way of knowing what effect, if any, these enormous planets have on the evolution of life within their realms [27]. Nevertheless, despite our uncertainty about Jupiter's role in our evolution, what we do know is that a planet's location within its star system matters because it affects its interactions with its planetary siblings.

MARS—CELESTIAL EVIDENCE THAT CONSTRUCTING A LIFE-SUPPORTING TERRESTRIAL PLANET IS A TRICKY BUSINESS

While the movements of Jupiter may have portended a good outcome with regard to the formation of Earth, some of the potential negative consequences of a large gas giant's movements around the inner region of a solar system can be contemplated by analyzing the fate of Mars. As it turns out, Jupiter's temporary stay at 1.5 AU likely removed large amounts of material from the region of space that eventually gave rise to the red planet, and consequently, when Mars took shape, it was much smaller than astronomers expected, given its location [28]. Ultimately, Mars developed into a planet that has a diameter just slightly larger than half that of Earth and its gravitational field is only 38% as strong as our planet's.

At first, Mars may have seemed like a promising home for life. Despite its small size, about 4.3 billion to 4.1 billion years ago (i.e., about 300–500 million years after it formed), Mars likely had an atmosphere rich in oxygen, carbon dioxide, and methane [29, 30]. Carbon dioxide and methane are good at trapping heat, so even though the Sun's intensity was weaker four billion years ago than it is today, and notwithstanding Mars' distance from the Sun, these greenhouse gases likely made the surface of Mars warm enough to contain large amounts of liquid water. In fact, the Martian globe likely had an ocean that occupied more than half of its northern hemisphere, and on some parts of the planet, this body of water may have been more than a mile deep [31]. Within all that water, it is possible that life existed. However, approximately

4.1 billion to 3.8 billion years ago, Mars was hit by a number of comets and asteroids and the limitations associated with Mars' small mass became more evident. It now appears that, because of the extreme cosmic thrashing Mars endured, much of the Martian atmosphere and surface water was blasted into space. Earth was also a target for the shower of cosmic debris that was cascading into the inner solar system during this period, but because of its greater mass, it had a better chance of maintaining its atmospheric gases and surface water after the collisions.

Earth's mass also gave it another enormous advantage over Mars, namely, the ability to maintain its protective magnetosphere. Earth's magnetosphere depends on the existence of magnetic fields, and these are generated by our planet's moon-sized metal cores.

The inner region of our planet's core is an iron-rich, sphere-shaped area. Surrounding this is a nickel-rich outer core. Both regions formed as the accreted material that generated our planet cooled. During this process, the densest elements sank toward the center of our newly forming, molten globe, while the lighter elements floated upward.

Today, the inner core of Earth has a diameter of 760 miles, and it is under a pressure that is millions of times greater than that exerted by our atmosphere. Due to these surreal conditions, the inner core is a solid structure even at temperatures that exceed 10,000°F. The outer core, which has a diameter of about 1,400 miles, experiences less dramatic conditions, and it is liquefied. As a result, electron-rich metals in the outer core can circulate, and this movement is enhanced by the planet's internal heat gradients and its rotation on its axis.

As the charged metals within the outer core move, they generate electromagnetic fields. These fields emanate out from the planet for thousands of miles, and they form the planet's magnetosphere (see Figure 5.1). This magnetosphere functions like a giant shield that protects Earth from many of the ravages of the solar system, and without this protective dome, our planet would be blasted by cosmic particles. These particles would devastate biological molecules, destroy the atmosphere, and rip apart the planet's water supply and jettison it into space.

Before the massive late bombardment, the Martian world was protected by a magnetosphere that was likely produced by a process similar to the one that currently generates the magnetosphere that surrounds Earth [32]. However, there is evidence that near the start of the late bombardment period, one or more large comets or asteroids slammed into the red planet, superheated its surface, and eliminated its thermal gradients [33]. Without heat gradients to drive the convention cycles (see Figure 5.2), the Martian magnetic fields all but collapsed, and the planet's magnetosphere failed. This cataclysmic disaster exposed the planet's atmosphere to attacks from cosmic particles, and as the gases surrounding the Martian globe were slowly torn asunder, the planet's exosphere and surface oceans evaporated.

So, most of the atmosphere and surface water on Mars that was not blasted into space during the late massive bombardment eventually slowly floated away into the darkness of space after the Martian magnetosphere crumpled [34]. As a result, Mars now has a mostly dry surface [35, 36] and an atmospheric density equal to about 1% of Earth's. Recently, some liquid water was found on Mars [36], but that water contains large amounts of salts (i.e., perchlorates), and although life may be capable of living in this briny solution, that is far from certain. In any case, what we can

FIGURE 5.1   Depiction of Earth's magnetic field (shown in blue). The incoming waves of radiation from the Sun and other cosmic objects (shown in red orange and yellow) are deflected, in part, by our planet's magnetic field.

definitively state is that Mars will not be mistaken for Earth any time soon, and in the unlikely case that life exists on the red planet, it will probably be limited to simple, unicellular, prokaryotic-like creatures.

As tragic as the failure of the Marian magnetosphere was, the bad news for the red planet did not end there. As we will see later, the movement of surface crustal plates is also dependent on the formation of internal planetary heat gradients, and once the heat gradients collapsed, so too did the planet's plate tectonic activity. Without active plate tectonics, Mars lost its ability to modulate its surface temperatures, and it also lost the important geological recycling mechanism that was needed to keep essential life-supporting nutrients surface-bound.

As was mentioned earlier, large comets and asteroids also regularly slammed into Earth, but Earth's larger mass significantly decreased the possibility that these collisions would superheat its mantle and permanently stifle its internal temperature gradients. Consequently, unlike its smaller sibling, Earth maintained its magnetosphere and its plate tectonic activity.

Ultimately, Mars' diminutive size guaranteed that it would never be capable of supporting complex life [37, 38]. However, it was Jupiter that condemned Mars to this fate, for it was this menacing gas giant's journey through the inner solar system that ensured that Mars would never amount to more than a runt of a planet.

VENUS—YET ANOTHER EXAMPLE OF THE IMPROBABILITY OF FORMING A LIFE-SUPPORTING PLANET

The other planet that is in our immediate solar neighborhood is Earth's "twin sister," Venus. Named after the Roman goddess of desire, sex, and fertility, the Venus of today is far from desirable. With a surface temperature of about 900°F, a CO_2-laden atmosphere that is 90 times as thick as Earth's, and nearly no water, Venus is almost certainly devoid of life [39].

Like Mars, Venus may have once been a much nicer place. At first, this may seem surprising because the planet is located just 0.72 AU from the Sun, and therefore it receives almost twice as much thermal radiation as Earth. However, when the planets first formed, the Sun's energy output was only about 70% as intense as it is today [40], and the surface temperature of our neighbor may have been low enough for vast amounts of liquid water to pool. Early in its history, the Venusian atmosphere may have also been more Earth-like, and it is even possible that Venus was once home to microbial life. But, as the Sun intensified and the heat-trapping gases began to accumulate in the Venusian atmosphere, the planet warmed considerably. The warming effect was further potentiated by the Sun's gravitational pull on the surface elements of the Venusian globe, which slowed Venus's rotation to the point where its day is now equivalent to about 117 Earth days. Given that the Venusian year is only 225 Earth days long, the lengthening of the Venusian day ultimately created a near tidal lock on our nearest neighbor, and therefore, Venus now slowly roasts in much the same way as languorously rotating shrimp on a rotisserie.

To add to an already dire situation, the blistering heat on Venus also likely assured that the Venusian globe would not have plate tectonic activity. This turned out to be an additional impediment for the development of life on the planet, but before speculating as to why Venus lacks plate tectonics, it is worth outlining in more detail the benefits a planet like Earth accrues because of this unusual planetary activity.

At its most fundamental level, plate tectonic activity depends on the ability of crustal plates to slide over and then down into the hot, viscous mantle that underlies a planet's crust. On Earth, the energy needed to slide the relatively flexible plates over the more rigid mantle comes primarily from the heat gradients that generate convection currents in the outer core and mantle of the planet (see Figure 5.2).

Earth's internal heat gradients are powered primarily by decaying radioisotopes located deep within our planet. The heat released by these isotopes warms the molten metals of the outer core, and as these metals increase in temperature, their density decreases. This change triggers the heated metals' ascension within the outer molten liquid, and some of the metals' energy is transferred to the surrounding environment in the form of heat. Ultimately, this process cools the metals and increases their density, and as a result, they begin to sink downward again, back toward the inner region of the outer core from which they emanated. At this point, this "convection cycle" is ready to restart, and with each round, some of the thermal energy deep within the planet moves upward to the mantle and crustal regions of the globe.

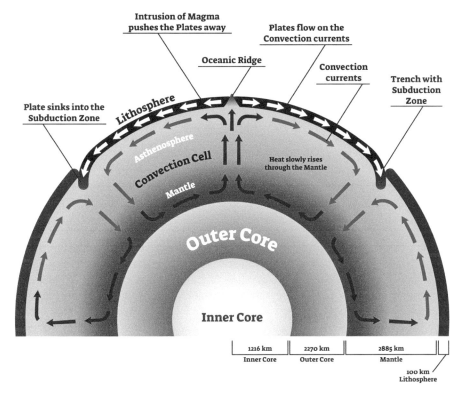

FIGURE 5.2 The heat-driven convection currents in the mantle of our planet. The subduction of the lithospheric plates that float above these currents is shown, as is the movement of magma to the surface.

Sometimes the energy generated by these currents can open a wedge in the crust, and when this happens, magma can flow to the surface. Occasionally escaping magma (which is referred to as lava) adds to the mass of the crustal plates and forces them to diverge and spread apart (see Figure 5.2), while in other situations, magma rushes up to the surface and belches out through volcanoes. Either way, when magma rises to the exterior of a planet, it can profoundly alter the geology, composition, and topology of the globe.

Some of Earth's most significant alterations occur in regions of the crust called subduction zones. In these areas, the spreading, dense oceanic plates tend to follow the descending convection currents that operate within the mantle, and as they do so, they slide under the less dense and relatively buoyant continental plates. Eventually, the subducted oceanic plate material merges with the underlying molten magma of the mantle, and at some point, later in time, the oceanic plate material re-emerges at the

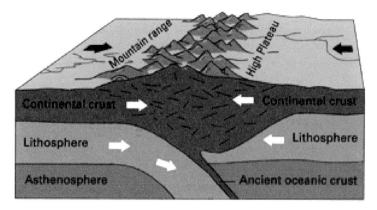

Continental-continental convergence

FIGURE 5.3 Depiction of the collision of continental crust and the subsequent emergence of mountain ranges.

planet's surface. This can occur when the mantle components percolate up to the crustal region through volcanoes and other weak spots on the exterior of the globe.

In addition to facilitating the recycling of Earth's crust, the oceanic plates can sometimes plow into and compress the continental plates. As this process unfolds, a continental landmass pushes upward (see Figure 5.3), which can lead to the formation of massive, geologically, and ecologically diverse, mountain ranges. Some of these mountain chains, such as the Andes, can span the entire length of a continent. Active plate tectonics also help to dissipate the thermal energy within the planet. In some cases, this is achieved by the expulsion of heated mantle material or through the physical movement of the massive crustal plates.

In the absence of plate tectonic activity, the thermal energy within the planet escalates and the temperature within the mantle increases. Eventually, as the upper layers of the planet warm, the thermal gradients move toward equilibrium, and the heat increases to the point where the globe's entire surface begins to melt. This can trigger a resurfacing of the planet, and that would be devastating to any complex life on the surface.

MAINTAINING A PLANET'S ATMOSPHERIC THERMOSTAT

Without plate tectonic activity, a planet also loses its ability to effectively modulate its atmospheric temperature. On Earth, the temperature of the atmosphere is regulated in part by the cycling of calcium and silicate. This process moves CO_2 back and forth between the planet's surface and the mantle, and in so doing, it influences the amount of CO_2 that is in the atmosphere. For complex life forms, CO_2 cycling is essential because if there is too much of this gas, intolerable global warming can result, while too little can cause a globe to freeze from one pole to the other.

On our planet, the atmospheric level of heat-trapping CO_2 is constantly increasing due to unremitting volcanic outgassing, but this situation can be modulated by binding CO_2 to calcium and silicate-containing mineral compounds such as wallastonite (i.e., $CaSiO_3$). The reaction produces limestone (i.e., $CaCO_3$) + silicon dioxide (i.e., SiO_2), and it effectively sequesters CO_2 within the limestone rock. If left unchecked, the sequestration of CO_2 by calcium silicates would eventually deplete a large volume of the heat-trapping CO_2 from the atmosphere, so from the perspective of a living organism, it is important to decrease the levels of CO_2 released by volcanic activity somewhat, but not to an extreme degree. In other words, it is vital that the calcium-silicate cycle be tightly regulated.

This brings us to the question of how this critical, temperature-modulating cycle is controlled. During the last several decades, we learned that the kinetics of the cycle are regulated by weathering and plate tectonic activity. During weathering, calcium-silicate rocks in Earth's crust are chemically altered, and as a result, the amount of calcium silicate that can form carbon-trapping limestone increases. Consequently, it is the weathering of rocks, such as granite, that generate the calcium silicate minerals needed to remove carbon dioxide from the atmosphere and lower the planet's surface temperature. However, the degree of geological weathering is proportional to the temperature. Therefore, as the temperature of the atmosphere drops, so does the rate of weathering. Eventually, when the temperature is low enough, the weathering process slows to the point where the atmospheric CO_2 levels increase faster than they are sequestered. This triggers a rise in the atmospheric temperature, which results in more weathering. Because of these processes, Earth effectively has a chemical thermostat that can modulate the temperature of its atmosphere over long periods of geological time.

Calcium silicates play a critical role in the maintenance of our planet's thermostat, but once these elements are bound in limestone and diatomaceous earth (which is a silicon-rich sedimentary rock), they tend to get buried under sediment and debris, and if this material is not recycled, the minerals needed to maintain Earth's surface temperature are eventually depleted from the crust. Fortunately, the plate tectonic-mediated recycling of the crust prevents rocks such as limestone and diatomaceous earth from permanently sequestering calcium-silicates (see Figure 5.4), and this allows our planet to maintain its climatic stability.

In addition to recycling carbon, plate tectonics also facilitates the reutilization of other nutrients such as nitrogen and phosphate. Carbon, nitrogen, and phosphate are the most prominent components of the fertilizers that we use to promote plant growth, and for billions of years, plate tectonics has made it possible for natural fertilizers to exist.

It is worth taking a moment to consider what our planet would be like if it lacked volcanism or plate tectonic activity. Without volcanic outgassing of heat-trapping gases, Earth's current average surface temperature would be about 0°F. However, with plate tectonics, a calcium-silicate cycle, and the release of volcanic greenhouse gases, we experience a much more pleasant average surface temperature of about 59°F [41]. In contrast, if we had just the outgassing but lacked the plate tectonic driven calcium-silicate cycle, the temperature of Earth's crust would increase, and

1. High atmospheric temperatures ➔ crustal rocks such as granite undergo weathering

➔ increase in free minerals such as $CaSiO_3$ (i.e., wallastonite)

2. $CaSiO_3 + CO_2$ ➔ $CaCO_3$ (i.e., limestone) $+ SiO_2$

➔   decrease in atmospheric CO_2
➔   decrease in atmospheric temperature

3. Decrease atmospheric temperature ➔ decrease weathering of crustal rocks

➔   increase in atmospheric CO_2 on planets with active volcanism
➔   increase in atmospheric temperature
➔   back to step 1

Notes:
a. Active plate tectonics drives the turnover of calcium-silicate rich rocks, and this results in an increase in the amounts of calcium-silicate rocks available on the surface to react with CO_2.

b. Without plate tectonics, the calcium-silicate rich rock would stay buried, and this would result in a decrease in the amount of calcium-silicate compounds available to react with CO_2. Ultimately, this would trigger potential runaway greenhouse effects.

FIGURE 5.4   An outline of a calcium-silicate-dependent geological cycle.

the atmosphere would accumulate large amounts of greenhouse gases. As a result, the planet would warm to the point where complex life could not survive. Therefore, from our perspective, the temperature of the planet's surface and the concentration of greenhouse gases currently in our atmosphere are at near ideal levels. This near ideal state has not always existed, and due to the burning of fossil fuels, it may not exist for much longer. However, for long periods of geological time, the plate tectonic activity of our planet and the greenhouse gases in its atmosphere have allowed complex life forms to endure.

Today, the primary greenhouse gas in our atmosphere is CO_2, and it currently makes up about 0.04% of the air we breathe. Mars and Venus have atmospheres very different from ours, and they are 96–97% CO_2. Without active plate tectonics, the heat-trapping gases on these planets will remain extremely high.

The ability of plate tectonics to promote temperature stability, crustal recycling, and geographical diversity suggests that this planetary feature is particularly important in establishing a life-supporting globe. The fact that the other planets in our solar system are devoid of plate tectonics is likely by itself a sufficient explanation for why these worlds are unlikely to host complex life. But, given the many similarities between Earth and its "twin sister" Venus, a question arises: "Why doesn't Venus have plate tectonic activity?"

Many investigators believe that Venus did indeed have active plate tectonics at one point in its history, but they suggest that, as the planet warmed, the crustal plates dried out and ultimately the Venusian surface became stiff, non-deformable, and immobile [42]. However, over the last few years, this idea has been challenged [43] and some now think that Venus never had active plate tectonics at all. According to this group, the geological plates that are normally generated as a result of the stress-induced fracturing of large masses of crust never formed on Venus because the extraordinary heat of the planet triggered the rapid repair of stress-induced crustal damage. Therefore, these geologists postulate that, instead of crustal plates, Venus ended up with a relatively intact and non-moveable crustal lid [44].

Regardless of the cause, the absence of plate tectonics eliminated Venus' ability to slide enormous blocks of crustal material on its surface, and in the absence of this process, there was no effective means to dissipate a large fraction of the thermal energy that was released within the Venusian globe. Consequently, the thermal gradients that stretch from the core to the surface of the planet were compromised, and over time, the heat increased throughout the interior of the planet. Eventually, the thermal energy under the crustal lid became so intense that the entire surface of the planet melted, and with that, some of the energy within the planet was released. This cycle of thermal energy accumulation and crustal lid melting continues to occur, and it is ostensibly causing the ongoing periodic resurfacing of Venus.

MORE BAD NEWS FOR VENUS

Venus' slow rate of rotation coupled with the attenuation of the planet's heat gradients ensured that Venus' inner molten metal core was unable to generate the strong electric current required to generate a robust intrinsic magnetic field. This left our nearest neighbor vulnerable to many of the vicissitudes of the cosmos, and without a magnetic dome, Venus' atmosphere was blasted by massive numbers of high-energy particles, including those produced by the relatively nearby Sun. As a result, much of the abundant water in the Venusian atmosphere was ripped into hydrogen and oxygen molecules, and the liberated hydrogen—and to a lesser extent, the oxygen—slowly bled off into the vastness of space. In the end, this devastation turned Venus into an arid world and today the Venusian atmosphere contains only about 0.0004% as much water as that found on the surface of Earth [39].

Interestingly, unlike Mars, Venus still manages to trap a large amount of atmospheric gases such as carbon dioxide despite its lack of a substantial intrinsic magnetosphere. This is due, in part, to the planet's possession of an extrinsic magnetosphere, which is generated by ionized gases in its upper atmosphere interacting with the intense solar winds that constantly pummel the planet. The resulting semi-protective dome is shaped differently than Earth's, and it has an intensity that is equal to about 10% of the intrinsic magnetosphere produced by our planet. Nevertheless, despite its different character and relative weakness, the Venusian extrinsic magnetosphere does offer some shielding, and consequently, there is still a massive amount of atmosphere around the surface of our drier-than-bone, sunward sister planet [45].

Not surprisingly, there are other proposed explanations for how Mars and Venus ended up becoming so miserably inhospitable [4], but what we do know with confidence is that Mars is unbearably cold and dry, while Venus, which resembles the Christian version of hell, sans the devil and his minions, is hot enough to melt lead and more than hot enough to sterilize medical equipment [39]. So, while Mars, Earth, and Venus are all rocky planets that formed from the same nebula, at approximately the same time, and in roughly the same zone of the developing solar system, they each ended up being very different from one another, and today Mars and Venus are nowhere near capable of hosting metazoan life.

WE ARE ENDLESSLY TRAVELING IN CIRCLES, OR NEARLY SO

In earlier chapters, I pointed out that our Milky Way galaxy is rushing toward the "Great Attractor" at 1.3 million mph, while the solar system is circling the center of the Milky Way at the dizzying speed of close to 500,000 mph. And as was noted earlier, these movements may affect life on Earth. However, in addition to engaging in sojourns within the Milky Way, our planet also moves within the solar system in ways that are familiar to most of us. Specifically, Earth circles the Sun at 67,000 mph, and the equator of our planet spins on its axis at slightly more than 1,000 mph. These actions give rise to the seasons of our year as well as our day/night cycle. The process through which our Earth completes these yearly and daily cycles is critical to the development of animal life on our globe.

While Earth has been rotating around the Sun since it was formed, over a period of hundreds of thousands of years, Earth's solar orbit has varied from nearly circular to slightly elliptical [46]. These changes are due to its gravitational interactions with the other planets in our solar system, particularly Jupiter, Saturn, and Venus, and although Earth's orbit is never more than 5% elliptical, even slight variations in our planet's path can have profound effects on our climate [47].

Since Earth's orbit is fairly circular, our planet doesn't take wide swings into and out of the habitable zone created by our Sun (see Figure 5.5), but many other planets have much more elliptical orbits [48]. These worlds are likely inimical to complex life because they lack the ability to maintain stable temperatures for extended periods. Recent work suggests that planets that are part of a multi-planet star system are more likely to have circular or near-circular orbits [49], and therefore, it is possible that being part of a multi-planet system is yet another characteristic that a life-supporting planet must possess.

While the circularity of a planet's orbit is important, so too are its precession (i.e., its wobble around its axial tilt and its elliptical orbital) and its obliquity (or axial tilt, which is equal to the angle between the planet's rotational axis and its orbital axis, see Figures 5.6 and 5.7. Shifts in these parameters occur more frequently than do changes in the planet's eccentricity (see Figure 5.8), and these fluctuations can also have large effects on the planet's climatic stability [50].

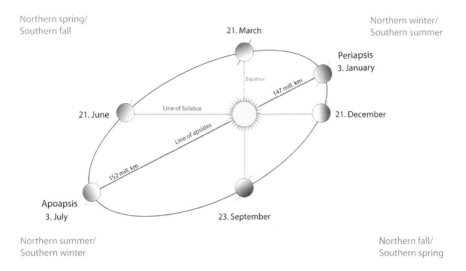

Northern spring/
Southern fall

21. March

Northern winter/
Southern summer

Periapsis
3. January

Equinox

147 mill. km

Line of Solstice

21. June

Line of apsides

21. December

152 mill. km

Apoapsis
3. July

23. September

Northern summer/
Southern winter

Northern fall/
Southern spring

FIGURE 5.5 Earth's orbit is slightly elliptical. The seasons of the year are influenced by the distance the planet is from the Sun as well as our globe's axial tilt. (From: Vector image: Gothika., CC BY-SA 3.0 <http://creativecommons.org/licenses/by-sa/3.0/>, via Wikimedia Commons.)

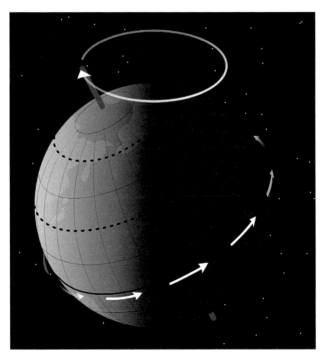

FIGURE 5.6 The precession of Earth. (From: NASA, Mysid, Public domain, via Wikimedia Commons.)

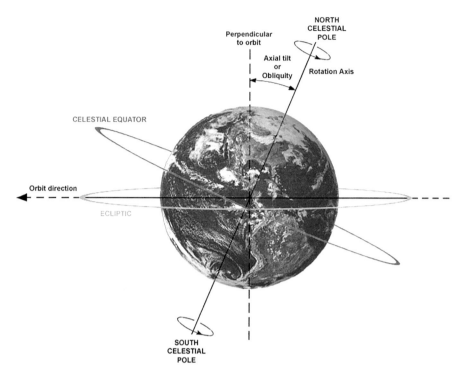

FIGURE 5.7 The axial tilt (or obliquity) of Earth. I, Dennis Nilsson, CC BY 3.0 (From: <https://creativecommons.org/licenses/by/3.0>, via Wikimedia Commons.)

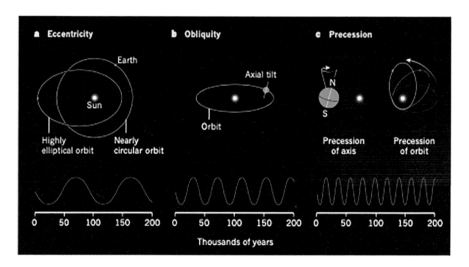

FIGURE 5.8 Earth's rotation around the Sun (i.e., its eccentricity and precession of orbit) and its tilt and rotation around its axis (i.e., its obliquity and precession of axis) varies over time. (This image is from Maslin [50].)

STABILIZING OBLIQUITY

For complex life of the type with which we are familiar, environmental predictability is good. In fact, it is more than good: it is necessary. One way to increase this vital factor is to increase the constancy of a planet's obliquity.

The obliquity of a globe affects its seasonal weather patterns. On our planet, the northern hemisphere is most inclined toward the Sun during June, July, and August, and it leans furthest from it during December, January, and February (see Figure 5.5). As it turns out, Earth's current slightly elliptical orbit places our planet furthest from the Sun around July 4, and closest to the Sun around January 3, but since the elliptical character of Earth's orbit is minimal, the slight differences in the distance to the Sun during July and January do not matter as much as the planet's 23.5°-or-so of axial tilt. Therefore, those of us in the northern hemisphere, experience the heat of summer in July and the cold of winter in January.

In the southern hemisphere, our planet's distance from the Sun and its angle of tilt both ensure that those in this part of the world experience their coldest temperatures in July and their warmest in January. Given that the southern hemisphere is closest to the Sun during a period when it is also most directly angled toward it, and that it is furthest from the Sun at the time when it is also leaning away from the Sun, one might expect the southern hemisphere to exhibit more extreme seasonal variation. However, the southern hemisphere's surface is 81% water (compare this to the northern hemisphere which is 61% water), and since water is slow to change temperature, the large aqueous surface area in the southern hemisphere minimizes extreme temperature fluctuations. Ultimately, our planet's current asymmetric distribution of land and water ensures that the northern and the southern hemispheres are presently both climatically stable.

THE IMPORTANCE OF KEEPING A PLANET IN KILTER

From our human perspective, we are fortunate that our planet has a moderate and stable axial tilt that slowly oscillates between 22.1° and 24.5° over a 41,000-year cycle (see Figure 5.8). Life on Earth would be very different if our globe had a less pronounced or a more pronounced angle of axial tilt. For example, if Earth had a very low angle of axial tilt our planet would have minimal seasonal variations, and as a result, the summers would be colder and the winters, warmer. At first, this may seem like a favorable situation, but the formation of ice sheets is particularly dependent on mild summers, and therefore if Earth had a low angle of axial tilt, large-scale glaciation would be more likely to occur [51]. On the other hand, if Earth had a much higher angle of axial tilt, the planet would exhibit extreme variations in seasonal temperatures [52], which would render large parts of the globe uninhabitable.

While Earth has a life-promoting and stable obliquity, our nearby neighbor, Mars, does not, and it can exhibit rapid chaotic obliquity shifts that range from 0° to 60°. It is interesting to note that Earth itself did not always have a stable obliquity, and at one point our planet's axial fluctuations may have been even more extreme than Mars, with a range spanning 0°–90° [4, 53]. Clearly, Earth's seasonal consistency and climatic stability are not something that necessarily had to be, and this raises the question of why Earth now has a moderate and relatively stable obliquity. In the 1990s, astronomer Jacques Lasker put forth a potential answer to this inquiry when he suggested that our planet's current stable axial tilt was generated and stabilized by Earth's unusually large moon [53].

EARTH AND THE MOON'S LONGTIME COSMIC PARTNERSHIP

The Moon often conjures up romantic images, but the beginning of Earth–Moon relationship was anything but quixotic and peaceful. In the opinion of many in the cosmology community, the story of the Moon's creation began between 4.53 billion and 4.48 billion years ago, not long after the end of Earth's accretion process. During this period, the newly formed Earth sideswiped a Mars-sized, would-be planet, called "Theia" [54–56], and the impact released 100 million times more energy than the dinosaur-killing asteroid that hit the Yucatan Peninsula 66 million years ago. To put this into perspective, consider that the impact of the rock that eliminated the non-avian dinosaurs was itself 100,000 times more powerful than the force that would be generated if both the United States and Russia decided to release all of their nuclear weapons at one time [57].

The collision between Earth and Theia completely annihilated the smaller sphere, and some have speculated that if the collision had been head-on instead of off-center, Earth itself may have been destroyed. This calamitous pile-up was the worst disaster our planet has endured, and it initiated the single most destructive and violent day in our planet's history. When the event was over, our globe was left in a deformed, highly disrupted, molten state.

Those that support the giant impactor hypothesis suggest that soon after Theia disintegrated, its iron core melted and merged with the molten iron center of our planet [58]. They also argue that, because of the collision, a vast amount of rocky mantle material from both worlds was jettisoned into interplanetary space where it was lost forever. However, some of the debris from the collision remained relatively nearby the marred Earth and was eventually captured by our planet's gravity. If this did indeed happen, the remnants trapped by Earth would have accreted, and within a short period (perhaps as little as one year), they would have generated our Moon [58].

There are some problems inherent in the giant impactor model. In particular, this hypothesis suggests that about 20% of the Moon's mass should be composed of Earth material and 80% should be derived from Theia. However, analysis of moon rocks demonstrates that the Moon has the same composition as Earth, and since it is unlikely that Theia and Earth had the same composition, this raises the question of what happened to the material from Theia.

To explain these incongruities, Raluca Rufu and colleagues have proposed that the Moon did not arise because of one impactor, but rather it was generated as a result of 20 or so smaller impacts between Earth and objects with a mass equal to 0.01 and 0.1 times that of the current Moon. Each of these impactors would have disintegrated upon striking Earth, but as they did, they would have ripped up part of Earth's mantle and blasted parts of it into orbit around the planet. This debris would have formed "moonlets," and according to this hypothesis, these "moonlets" eventually accreted to form our Moon [59].

Regardless of whether the Moon formed because of one giant collision or multiple smaller ones, it appears that Earth's large satellite is the product of stochastic events. Why then, do the other inner terrestrial planets in our solar system lack similar large moons? All these bodies formed at about the same time and in roughly the same area; however, Mercury and Venus are moonless, and Mars, with its two minuscule moons both less than 14 miles in diameter, is nearly so. Perhaps the answer is that Earth is simply an atypical planet that just happened to end up with a large moon.

What we can say with confidence is that the impact of the Moon on Earth was significant. Soon after the Moon formed, the obliquity of Earth was about 70°; however, over time, the gravitational interactions between Earth and its large moon resulted in our planet settling into its presently stable obliquity of about 23° [60].

In addition to altering and stabilizing our planet's obliquity, the collision between Earth and the Moon-generating celestial object (or objects) likely increased our globe's rate of rotation. Ultimately, the Moon itself attenuated this enhanced rotational speed by exerting tidal friction on our planet, and in so doing, facilitated the transfer of angular momentum from Earth to itself.

Over geological time, the Moon's drag on Earth increased the length of the day from six hours to its current 24 hours. As a result, our planet rotates more slowly into and out of the Sun's rays, and this has helped the globe maintain a moderate and stable temperature conducive to the evolution of complex life. In addition, this slow increase in the length of a day likely decreased the average wind speed on our planet as well as the intensity and frequency of violent windstorms. Without the Moon, some have estimated that the typical wind speed on Earth would be approximately 125 mph [61]. So, it is worth keeping in mind that in the absence of the Moon, the worst windstorms of today would seem downright mild, and without the Moon, the winds on Earth would have made it almost impossible for land plants and surface-dwelling animals to evolve.

Over the last several years, some have suggested the Moon's importance in maintaining the stability of our planet's axial tilt has been overstated. These investigators have computer models that indicate that if Earth had been spinning faster (i.e., if a day lasted less than 10 hours), or if it had been rotating in the reverse orientation (i.e., if Earth had a retrograde instead of prograde rotation), Earth's obliquity would be relatively stable due to gravitational interactions with Jupiter and other relatively nearby planets. They concede that the axial stability would not be as robust as it now is, but they state that significant changes in Earth's axial tilt would manifest slowly over a period of billions of years, and they suggest that this slow rate of change would not necessarily have impaired the evolution of complex life [62, 63]. In the end, however, what is most noteworthy is that, unlike many other planets, Earth has a relatively stable obliquity, and it has a rate of rotation that is conducive to the evolution of complex, surface-dwelling life forms. Earth's moon helped make climate-stabilizing events possible, and as a plant and land-dependent life form, we can appreciate the good fortune of being on a globe that is distinguished by its atypical, long-term, cosmic partner.

SIZING UP OUR LIVING QUARTERS

Earth enjoys a favorable intergalactic location within the life-supporting Milky Way, and it is also obvious that, within our solar system, Earth is a far superior venue for complex life than Mars, Venus, or any of the other planets or moons that orbit our Sun. This makes Earth unique within our star system. Given this reality, if you were tasked with selling cosmic real estate to alien humanoids and you were trying to sell planet Earth, you would have an easy time of it. As part of your sales pitch, you would no doubt discuss Earth's favorable chemical composition and its life-supporting atmosphere. You would also talk up the fact that the planet has a moderate gravitational field, a protective, intrinsic magnetic field, active plate tectonics, a stable and almost circular orbital, and a modest and stable axial tilt. And you would certainly point out that, because of these traits, our world also has a favorable distribution of liquid water and dry land, impressive geological and ecological diversity, and a mild climate that can maintain its temperature stability over long periods of geological time. But assuming your alien customers could travel the galaxy at the speed of light, would you have competition, i.e., would your customers be tempted to buy cosmic real estate in some nearby star system? To date, the evidence indicates that they would not be, because there are likely few, if any, properties like Earth in our galactic neighborhood, and even if some do exist, getting there from Earth would be a major undertaking, even for an alien species capable of traveling at the speed of light.

The aforementioned statement may seem surprising given that some have argued that there are billions of "Earth-like" planets circling "Sun-like" stars within our galaxy. However, those who make these claims often employ rather elastic definitions when characterizing a planet as "Earth-like" or a star as "Sun-like." For example, in one study, Earth-like planets were defined as those receiving 0.25–4 times as much stellar energy as Earth and whose mass is one to two times that of Earth [64]. By this definition, Venus is clearly in the "Earth-like" category, however, while Venus is similar in size to Earth, the fact that it receives 1.9 times as much insolation as our planet is significant. After a visit to Venus, you would not call it "Earth-like," and you certainly wouldn't call it Earth's twin. In addition, in many studies "Sun-like stars" are defined as being similar in size, color, and age to our Sun. However, as was discussed in the previous chapter, our Sun has many other traits that distinguish it from that crowd.

Still, it is exciting to speculate that there might be another Earth somewhere "nearby," and when astronomers found that Proxima Centauri may have a "potentially life-harboring planet" orbiting it [65], the imaginations of scientists and non-scientists alike lit up [66, 67]. Proxima Centauri is the closest star to us, and at "just 4.25 light years away," some have speculated that we may eventually settle on this potentially "Earth-like" world. However, this newly discovered planet, which was named Proxima b, is likely bombarded by 100 times more X-rays and other forms of high energy radiation than Earth, and it is probably tidally locked [68, 69]. So, even without seeing Proxima b up close, it is fair to say that it will never be our home away from home.

But let's assume for the sake of argument that Proxima b is just like Earth. If that were true, and if we wanted to visit this planet, we would have to take on the non-trivial issue of getting there. Four and a quarter light years is an extraordinary distance, and in fact, it is equal to about 25 trillion miles. Even if we could adapt our current technology so that we could undertake this trip, it would take many millennia to cover that distance.

Currently, NASA is exploring the idea of building a Nuclear Thermal Propulsion (NTP) engine that would use nuclear energy to heat liquid hydrogen. This process would produce plasma consisting of ionized hydrogen, and in theory, this plasma could be used to propel a rocket over unlimited distances at speeds far greater than those we can currently achieve. But, if such a rocket were available, it would still take about 1,000 years to make a one-way trip to Proxima b [70]. To complicate matters, we now know that even relatively short trips in space would expose travelers to radiation that would kill approximately 5% of their cells and significantly increase their risk of cancer [71]. In addition, this radiation would trigger irreversible, behavior-altering, and potentially lethal, brain damage [72, 73]. Consequently, undertaking a 1,000-year voyage is not in the cards anytime soon.

Of course, there is always the possibility that technology will evolve well past what we can currently realistically contemplate. For example, we may one day be able to significantly increase how fast we can travel, and perhaps we will find a way to generate an artificial magnetosphere that would protect space travelers from cancer-inducing and brain-damaging radiation. However, these advances, if they come, are part of the distant future, and it is sobering to realize that, even with such prodigious and near-miraculous innovations, we still will be highly constrained by the realities of our universe. This is particularly true when it comes to our rate of travel, because no matter how wonderful our technology, all the available evidence suggests that we will never be able to go faster than the speed of light. So, even if we could one day safely travel at the universe's ultimate speed limit, how many of us would want to embark on a 4.25-year, non-stop, one-way journey to a desolate, rocky world that probably lacks an atmosphere, and which is likely unbearably hot on one side and intolerably cold on the other? And 4.25 years is the least amount of time it would take to arrive at a planet outside our solar system. Even if we could travel at the speed of light, most other planets with Earth-like characteristics would take much, much longer to get to.

This begs the question of why our culture gets so excited by the prospect of finding another "Earth-like planet." In an attempt to answer this riddle, *New York Times* writer Lisa Messeri suggests that some of us hope to find "Earth-like planets" because their presence may indicate that celestial refuges are available if we destroy our own planet. She also suggests that some within our society believe that finding another Earth means that we will one day meet beings like ourselves, and as a result, our "cosmic loneliness" will come to an end [74].

These ideas are interesting, but the excitement is not warranted. The fact is that, even if there are tens of billions of planets outside of our solar system that have some traits in common with Earth [75], their sheer distance nearly guarantees that we will never visit any of them. If our species does survive into the far distant future, and if we find that we must leave our planet, we will likely head off as small groups of nomads, assuming we have the technology to do so. These groups will move through space for numerous generations, and if they are lucky, they will one day reach another

"Earth-like" planet. However, when they do, what they will almost certainly discover is that their newly found land is not like the one they left many generations ago, and they will also likely find that their evolutionary heritage is ill-suited for life on other planets.

Regarding other intelligent life forms, I would heavily wager that somewhere within our inconceivably vast universe, there are colonies of intelligent and sentient beings that can appreciate the significance of their existence. However, as I mentioned earlier, I don't think we will ever travel to these outposts, and I doubt that we will ever meet our cosmic brethren. Instead, we will have to make our way through this universe alone, and for the foreseeable future, we will do so inextricably bound to our planet.

THE EMERGENCE OF A PLACE WE CALL HOME

Over the last five chapters, we have considered some of the conditions that made it possible for a planet like Earth to come into existence, and we saw that many aspects of the universe that produced our home are inscrutable, improbable, and truly amazing. However, despite the enigmatic nature of our universe, we do know that we live on an exceedingly remarkable planet that orbits an unusual star, and we know that our star system is located within a special section of an extraordinary galaxy. As magnificent as our physical milieu is, however, perhaps what is most astonishing is what we see when we glance in a mirror: a conscious, relatively hairless, perceptive ape that is part of a group that somehow managed to come forth on this tiny spec of a planet.

In the next section of the text, we will explore how creatures like us came to be. We will begin by investigating how the early Earth, which was initially nothing more than a molten rock hurling through space, managed to give rise to prokaryotic life, and then we will embark on a journey to learn how metazoans in general, and human primates in particular, arose. As we consider this information, we will see that the genesis of complex life on Earth was a long, circuitous, and tenuous process. We will also discover that the emergence of primates like us was never a certainty of nature, and unfortunately, neither is our long-term survival.

REFERENCES

1. Dreifus, C., *Avi Loeb ponders the early universe, nature and life*. New York Times, 2014.
2. Loeb, A., *The habitable epoch of the early universe*. International Journal of Astrobiology, 2014. arXiv:1312.0613 [astro-ph.CO].
3. Larson, R.B. and V. Bromm, *The first stars in the universe*. Scientific American, 2001. **285**(6): p. 64–71.
4. Ward, P.D. and D. Brownlee, *Rare Earth: why complex life is uncommon in the universe*. 2000, Göttingen: Copernicus Publications.
5. Elkins-Tanton, L., *Solar system smash up*. Scientific American, 2016. **315**: p. 42–49.
6. Lin, D., *The genesis of planets*. Scientific American, 2008: p. 50–59.
7. Editors, *Planets and how they formed*. Las Cumbres Observatory Global Telescope Network, 2016.

8. Crockett, C., *How did Earth get its water?* Science News, 2015. **187**(10): p. 18.

9. Gorder, P.F., *Study hints that ancient earth made its own water—geologically | news room.* The Ohio State University News Room, 2014.

10. Hazen, R.M., *The story of Earth: the first 4.5 billion years, from stardust to living planet.* 2012, New York, NY: Viking. 306 p.

11. Mason, B., *The carbonaceous chondrites.* Space Science Reviews, 1963. **1**(4): p. 621–646.

12. Alexander, C.M., et al., *The provenances of asteroids, and their contributions to the volatile inventories of the terrestrial planets.* Science, 2012. **337**(6095): p. 721–723.

13. Lydia J. Hallis, et al., *Evidence for primordial water in Earth's deep mantle.* Science, 2015. **350**(6262): p. 795–797.

14. Daly, R.T. and P.H. Schultz, *The delivery of water by impacts from planetary accretion to present.* Science Advances, 2018. **4**(4): p. eaar2632.

15. Barnes, R., et al., *Tidal Venuses: triggering a climate catastrophe via tidal heating.* Astrobiology, 2013. **13**(3): p. 225–250.

16. Perkins, S., *Earth is only just within the Sun's habitable zone.* Nature, 2013. **504**.

17. Ray, C.C., *Even tiny changes in Earth's orbit would yield global catastrophe.* New York Times, 2017.

18. Leconte, J., et al., *Increased insolation threshold for runaway greenhouse processes on Earth-like planets.* Nature, 2013. **504**(7479): p. 268–271.

19. Morbidelli, A., et al., *Building terrestrial planets.* Annual Review of Earth and Planetary Sciences, 2012. **40**(1): p. 251–275.

20. Walsh, K.J., et al., *A low mass for Mars from Jupiter's early gas-driven migration.* Nature, 2011. **475**(7355): p. 206–209.

21. Lissauer, J.J., R.I. Dawson, and S. Tremaine, *Advances in exoplanet science from Kepler.* Nature, 2014. **513**(7518): p. 336–344.

22. Heller, R. and J. Armstrong, *Superhabitable worlds.* Astrobiology, 2014. **14**(1): p. 50–66.

23. Lenton, T. and A.J. Watson, *Revolutions that made the Earth.* 2011, Oxford: Oxford University Press.

24. Billings, L., *Jupiter, destroyer of worlds, may have paved the way for Earth.* Scientific American, 2015. https://www.scientificamerican.com/article/jupiter-destroyer-of-worlds-may-have-paved-the-way-for-earth/.

25. Wetherill, G., *Possible consequences of absence of "Jupiters" in planetary systems.* Astrophysics and Space Science, 1994. **212**: p. 23–32.

26. Horner, J. and B. Jones, *Jupiter—friend or foe? An answer.* Astronomy and Geophysics, 2010. **51**: p. 16–22.

27. Wittenmyer, R.A., et al., *On the frequency of Jupiter analogs.* Astrophysical Journal, 2011. **727**(2): p. 102.

28. Atkinson, N., *Solving the puzzle of why Mars is so small.* Astrobiology Magazine, 2011.

29. Wordsworth, R., et al., *Transient reducing greenhouse warming on early Mars.* Geophysical Research Letters, 2017. **44**(2): p. 665–671.

30. Editors, *Mars atmosphere was oxygen-rich 4 billion years ago.* Astrobiology Magazine, 2013.

31. Editors, *NASA research suggests Mars once had more water than Arctic Ocean.* NASA, 2015. https://www.nasa.gov/press/2015/march/nasa-research-suggests-mars-once-had-more-water-than-earth-s-arctic-ocean

32. Benjamin P. Weissa, et al., *Records of an ancient Martian magnetic field in ALH84001.* Earth and Planetary Science Letters, 2002. **201**: p. 449–463.

33. Roberts, J.H., R.J. Lillis, and M. Manga, *Giant impacts on early Mars and the cessation of the Martian dynamo.* Journal of Geophysical Research, 2009. **114**(E4).

34. B.M. Jakosky, et al., *MAVEN observations of the response of Mars to an interplanetary coronal mass ejection.* Science, 2015. **350**(6261). DOI: 10.1126/science.aad0210.

35. Dantonio, M., *The case for standing water on Mars.* Huffington Post, 2015.

36. Lujendra Ojha, et al., *Spectral evidence for hydrated salts in recurring slope lineae on Mars.* Nature Geoscience, 2015. **8**: p. 829–832.

37. Aguilar, D.A., *Earth: a borderline planet for life?* Harvard Smithsonian Center for Astrophysics (News Bulletin), 2008.

38. R. Orosei, et al., *Radar evidence of subglacial liquid water on Mars.* Science, 2018. **361**: p. 490–493.

39. Way, M.J., et al., *Was Venus the first habitable world of our solar system?* Geophysical Research Letters, 2016. **43**(16). https://doi.org/10.1002/2016GL069790.

40. Goldblatt, C. and K.J. Zahnle, *Clouds and the faint young Sun paradox.* Climate of the Past, 2011. **7**(1): p. 203–220.

41. Ma, Q., *Greenhouse Gases: Refining the Role of Carbon Dioxide, NASA, 1998,* https://www.giss.nasa.gov/research//briefs/1998_ma_01/

42. Solomatov, V.S. and L.N. Moresi, *Three regimes of mantle convection with non-Newtonian viscosity and stagnant lid convection on the terrestrial planets.* Geophysical Research Letters, 1997. **24**(15): p. 1907–1910.

43. Hongzhan Fei, et al., *Small effect of water on upper-mantle rheology based on silicon self-diffusion coefficients.* Nature, 2013. **498**(7453): p. 213–215.

44. Bercovici, D. and Y. Ricard, *Plate tectonics, damage and inheritance.* Nature, 2014. **508**(7497): p. 513–516.

45. T.L. Zhang, et al., *Magnetic reconnection in the near Venusian magnetotail.* Science, 2012. **336**: p. 567–570.

46. Crampton, J.S., et al., *Pacing of Paleozoic macroevolutionary rates by Milankovitch grand cycles.* Proceedings of the National Academy of Sciences of the United States of America, 2018. **115**: p. 5686—5691.

47. Hays, J.D., J. Imbrie, and N.J. Shackleton, *Variations in the Earth's orbit: pacemaker of the ice ages.* Science, 1976. **194**(4270): p. 1121–1132.

48. Editors, *Elusive Earths.* Astrobiology Magazine, 2005.

49. Limbach, M.A. and E.L. Turner, *Exoplanet orbital eccentricity: multiplicity relation and the solar system.* Proceedings of the National Academy of Sciences of the United States of America, 2015. **112**(1): p. 20–24.

50. Maslin, M., *Forty years of linking orbits to ice ages.* Nature, 2016. **540**: p. 208–210.

51. Heller, R., J. Leconte, and R. Barnes, *Tidal obliquity evolution of potentially habitable planets.* Astronomy & Astrophysics, 2011. **528**: p. A27.

52. Williams D.M. and K.J.F., *Habitable planets with high obliquities.* Icarus, 1997. **129**: p. 254–267.

53. Laskar, J., F. Joutel, and P. Robutel, *Stabilization of the Earth's obliquity by the Moon.* Nature, 1993. **361**(6413): p. 615–617.

54. Young, E.D., et al., *Oxygen isotopic evidence for vigorous mixing during the Moon-forming giant impact.* Science, 2016. **351**(6272): p. 493–496.

55. Canup, R.M. and E. Asphaug, *Origin of the Moon in a giant impact near the end of the Earth's formation.* Nature, 2001. **412**(6848): p. 708–712.

56. Halliday, A.N., *A young Moon-forming giant impact at 70–110 million years accompanied by late-stage mixing, core formation and degassing of the Earth.* Philosophical Transactions of the Royal Society A, 2008. **366**(1883): p. 4163–4181.

57. Brannen, P., *The ends of the world: volcanic apocalypses, lethal oceans, and our quest to understand Earth's past mass extinctions.* First edition. ed. 2017, New York, NY: Ecco, an imprint of HarperCollins Publishers. x, 322 p.

58. Ida, S., R.M. Canup, and G.R. Stewart, *Lunar accretion from an impact-generated disk.* Nature, 1997. **389**(6649): p. 353–357.

59. Rufu, R., O. Aharonson, and H.B. Perets, *A multiple-impact origin for the Moon.* Nature Geoscience, 2017. **10**: p. 89–94.

60. Williams, G., *History of the Earth's obliquity.* Earth-Science Reviews, 1993. **34**(1): p. 1–45.
61. Conway Morris, S. *Life's solution: inevitable humans in a lonely universe.* 2003, Cambridge: Cambridge University Press.
62. Schilling, G., *Who needs a Moon?* Science, 2011(May 27). https://www.science.org/content/article/who-needs-moon.
63. Cooper, K., *Earth's Moon may not be critical to life.* Astrobiology Magazine, 2015.
64. Petigura, E., A.H. Howard, and G. Marcy, *Prevalence of Earth-size planets orbiting Sun-like stars.* Proceedings of the National Academy of Sciences, 2013. **110**(48): p. 19273–19278.
65. Anglada-Escude, G., et al., *A terrestrial planet candidate in a temperate orbit around Proxima Centauri.* Nature, 2016. **536**(7617): p. 437–440.
66. Chang, K., *One star over, a planet that might be another Earth.* New York Times, 2016.
67. Greenfield Boyce, N., *This planet just outside our solar system is 'potentially habitable'.* NPR, 2016, https://www.npr.org/sections/thetwo-way/2016/08/24/490947403/this-planet-just-outside-our-solar-system-is-potentially-habitable
68. Hatzes, A.P., *Earth-like planet around Sun's neighbour.* Nature, 2016. **536**: p. 408–409.
69. Vladimir S. Airapetian, et al., *How hospitable are space weather affected habitable zones? The role of ion escape.* Astrophysical Journal Letters, 2017. **836**(1): p. L3.
70. Williams, M., *How long would it take to travel to the nearest star?* Universe Today, 2016. https://www.universetoday.com/15403/how-long-would-it-take-to-travel-to-the-nearest-star/
71. Mack, E., *Discovery could reverse aging and help us live on Mars.* CNET Magazine, 2017(Spring 2017).
72. Zimmer, C., *Scott Kelly spent a year in orbit. His body is not quite the same.* New York Times, 2019.
73. Limoli, C.L., *Deep-space deal breaker.* Scientific American, 2017. **316**(2): p. 54–59.
74. Messeri, L., *What's so special about another Earth?* New York Times, 2016.
75. Borenstein, S., *8.8 billion habitable Earth-size planets exist in Milky Way alone.* NBC News, 2013.

Part II

The Emergence of Life
From a Lifeless World

6 Let There Be Life
The Rise of the Prokaryotes

It is mere rubbish thinking at present of the origin of life; one might as well think of the origin of matter.

Charles Darwin

The fact that life evolved out of nearly nothing, some 10 billion years after the universe evolved out of literally nothing, is a fact so staggering that I would be mad to attempt words to do it justice.

Richard Dawkins

I think that science suggests to us, tentatively of course, a picture of a universe that is inventive or even creative; of a universe in which new things emerge on new levels.

Karl Popper

THE SIGNIFICANCE OF LIFE

There has been a great deal of interest of late in critical transitions within the universe, which are, by definition, periods during which the nature of the universe fundamentally changes. Depending on to whom you speak, the number of these events can vary significantly [1], but I will argue that there are at least four, namely, the genesis of the universe, the production of element-generating stars, the origin of life, and the emergence of sentient, conscious beings like us.

The first of these transitions was initiated by the Big Bang. This event defined our universe as we know it, and it is hard to imagine a more significant transition than one that moves from the absence of a universe to the existence of one. The events that characterize the formation of the universe appear to be complex in that the laws of physics and the physical constants of the universe somehow came into being during this period. However, in terms of its composition, the early cosmos was relatively simple, consisting primarily of energy and a few light elements.

When the first stars formed about 200 million years after the Big Bang, the structural complexity of the universe began to increase. Suddenly, heavy elements appeared, and with these elements present in abundance, Sun-like stars and rocky planets began to materialize. This set of events set the stage for the next great transition: The emergence of cellular life.

Life, in the form of bacteria-like creatures, imparted new properties onto the universe, because unlike abiotic atoms, living organisms grow, independently reproduce, metabolize, maintain homeostasis, react to stimuli, and evolve. A collection of abiotic atoms may do some of these things. For example, they may react in some way to a stimulus, but they cannot carry out all the remarkable functions of something

DOI: 10.1201/9781003270294-9

that is "alive." The genesis of living organisms redefined what could occur in the cosmos.

Bacteria, and their single-celled cousins, the Archaea, are extraordinary, but even more remarkable are sentient, conscious beings that are capable of sophisticated, abstract thought, complex language, and collective learning. Here the creatures I am speaking of are humans, and our appearance also changed the properties of the universe itself. With our genesis, the universe was able to reflect on its own nature, and because of the emergence of complex minds, advanced cultures of immense complexity arose.

As was mentioned earlier, elevating the appearance of humans to this level of significance does go against the current trend of decentralizing the importance of Earth and humanity, but we are part of the universe, and our appearance fundamentally changed what the universe can do. Complex thought, self-reflection, and the formation of sophisticated cultures are not trivial, and as far as we know, our universe did not possess these attributes before we came to be.

Many theists will no doubt argue that the existence of God bestowed upon the universe a far more impressive set of characteristics than those that I ascribe to humans but, according to most theists, God is not of the universe. Rather, God created the universe, and He stands apart from it. This is not true of humans. We are certainly part of this universe.

Others suggest that, since every creature has evolved to survive in its environment, each is special in its own way, and we are therefore no more exceptional than any other organism. To some extent, this is true. We do not represent the pinnacle of performance in every domain. We can't launch into an airborne, aerodynamic tuck and hit speeds in excess of 200 mph like a peregrine falcon, which is the world's fastest bird. We also can't survive under extreme conditions, as can a tardigrade. But, when it comes to complexity, the human brain has no known rival, and in that regard, we humans appear to be unique. If indeed there are no other creatures like us, and if humans were to vanish, then the complexity of the universe would decrease, and the nature of the universe would be more limited and circumscribed.

It is important to note that this argument does not necessarily imply that we are the ultimate apogee of intellectual development in our universe. Additional levels of neurological complexity may arise in the future, and entities with traits beyond those that we can imagine may yet appear. Bacteria existing 3.5 billion years ago obviously did not anticipate our appearance, and despite our complex minds, it is likely that we cannot appreciate what may exist 3.5 billion years from now.

It is also worth noting that the universe is an old place, and if life initiated long ago in some distant part of the cosmos, our abilities may have already been eclipsed by beings that far exceed us in mental capacity. If so, then our disappearance from the universe would not decrease its complexity, but to date, we know of no such entities. Therefore, to the best of our knowledge, we are the most complex systems in our local area, and we could well be one of the most, or even the most, complex systems within the cosmos.

So, how did the complexity associated with life arise? The simple answer is that we don't know. But there are many fascinating hypotheses, and we are beginning to generate empirically based theories about how life on Earth came to be.

THE MYSTERIOUS APPEARANCE OF LIFE

In the middle of the 19th century, many people believed that life arose spontaneously, and in the 1600s, some of humanity's leading intellectuals argued that even mammalian life could come into being in such a way. For those who have studied biology, the thought that an intricate living system complete with highly integrated, homeostatic, and mind-bogglingly sophisticated organ systems could somehow appear out of thin air is hard to fathom. Making a fully programmed computer materialize in front of your eyes would be child's play by comparison. But, without an appreciation for the complexity of life, and an understanding of the sheer improbability of life's very existence, many strange thoughts seem plausible, and within the collection of humanity's odd ideas about how life began, the Flemish physician and chemist, Jan Baptista van Helmont, generated some of the strangest. Operating in the 16th century, van Helmont believed that a mouse could be produced by placing wheat and soiled underwear together in an open mouth jar for three weeks. During this time, van Helmont argued, "the reaction of the leaven in the shirt with fumes from the wheat" will "transform the wheat into mice" [2]. Who would have guessed that dirty underwear possessed such life-generating powers?

Of course, today very few educated people hold these beliefs, in large part because of improvements in our understanding of biology. Nevertheless, even in the absence of our current knowledge, one wonders why the veracity of van Helmont's claim went unchallenged by so many. Surely, in the 16th century, many individuals considered the possibility that mice didn't actually spontaneously form in a jar, and they must have realized that the mice could have simply sauntered into the container. Many people must also have recognized that organizing a continuous three-week observational vigil during the experiment would be one way to test the validity of van Helmont's hypothesis. Granted three weeks is a lot of time to stare at a jar, but presumably, if you believed van Helmont was correct, the effort would have been worthwhile, because as part of the team holding the vigil, you could have been the first person in history to witness the spontaneous generation of a complex life form!

But let's assume a three-week vigil was too much to organize. Investigators still could have eliminated the possibility that a mouse would meander into the jar simply by placing a lid on it. van Helmont most likely didn't take this precaution because he believed that spontaneous generation required air. However, had he placed a large number of small holes in the lid, he would have provided air to the system and simultaneously prevented any neighborhood mice from walking into the container.

While it is easy for those of us living in the 21st century to criticize the experimental design of van Helmont, it is harder to conceive of a strategy to test whether bacteria arise spontaneously in a nutrient soup, and up until the mid-1800s, many within the scientific community firmly believed that bacteria did indeed spontaneously appear within such broths. However, the validity of this belief was debunked less than 200 years ago by Louis Pasteur. Pasteur suspected that airborne organisms were entering the broth, but since the scientific community of the 1800s also believed that spontaneous generation required air, he had to find a way to filter the gases that were mixing with the liquid broth solution. Today, this task could be easily

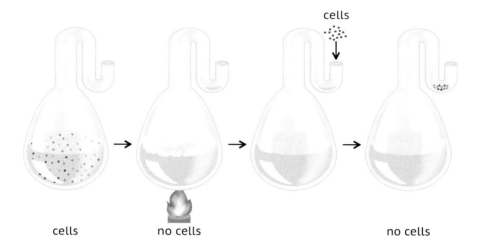

FIGURE 6.1 Pasteur's swan neck flask experiment. The contaminated media in the flask (depicted on the far left) was sterilized by heating (second flask from the left). After this, contaminants from the air surrounding the flask are not able to enter the media because they are retained in the crook of the S-shaped glassware (third and fourth flasks).

accomplished with a microfilter, but in 1859, there were no such filters, so Pasteur needed to find an acceptable substitute. Ultimately, he very cleverly managed to filter the air by placing the liquid into a swan-neck flask that he constructed (see Figure 6.1). This invention allowed air to mix with the broth, but the many twists and turns in the flagon's neck impeded the movement of airborne contaminants and successfully prevented microbes from entering the nutrient soup. With his new piece of glassware at hand, Pasteur boiled the broth in his flask and then exposed its contents to the air. In what, no doubt, was a great surprise to many, he found that the nutrient broth remained sterile for extended periods [3]. So, by 1859, most of the scientific community accepted the idea that even a relatively simple bacterium couldn't arise spontaneously. But, if bacteria could not simply pop into existence, how did they come to inhabit the early Earth?

BACTERIA: VISITORS FROM AFAR?

One potential answer to this question is that life migrated to Earth from some other world. In the 1800s, Svante Arrhenius and William Thomson generated considerable excitement when they argued for this hypothesis, and in his presidential address to the British Science Association in 1871, Thomson, (better known as Lord Kelvin), stated, "we must regard it as probable in the highest degree that there are countless seed-bearing meteoric stones moving about through space" [4]. Clearly, Arrhenius and Thomson had no difficulty envisioning how life may have hitched a ride to our planet.

Almost a century later, Francis Crick, of DNA structure fame, and his associate Leslie Orgel, revived this old panspermia, or "life is everywhere" paradigm, when they published a paper suggesting that extraterrestrials seeded primitive life forms on Earth in the distant past [5]. But, although Crick and Orgel's suggestion was the subject of a very entertaining *Star Trek* episode [6], there is—to date—no real evidence that ancient space aliens purposefully initiated life on Earth during a layover on their journey elsewhere.

In addition to the extremely imaginative ideas of Crick and Orgel, the 20th-century origins of life community were treated to the creative thoughts of Fred Hoyle, who had earlier in his career proposed a mechanism for the nuclear synthesis of carbon (see Chapter 2). Together with his former student and long-time colleague Chandra Wickramasinghe, Hoyle argued that viruses and cellular life routinely arrive at Earth on asteroids [7], and he and Wickramasinghe suggested these extraterrestrial viruses were responsible for many of the large-scale pandemics that have plagued humankind [8]. Again, due to a lack of evidentiary support, the majority within the scientific community rejected this concept.

Despite the lack of support for the pangenesis hypothesis, efforts to examine meteors and asteroids for signs of exobiological cellular life continue to this day [9], and although extraterrestrial viruses or bacteria have still not turned up in meteors, complex molecules have [10]. In addition, recent investigations have shown that some bacteria can potentially survive the extreme acceleration required to escape Earth's gravity [11], and some can remain viable under conditions that simulate space travel [12, 13]. These findings leave open the possibility that we may yet discover that bacteria arrived on our planet from another world, and it is possible that, one day, we may all trace our ancestry back to some other yet-to-be-identified globe. However, even if we did find extraterrestrial bacteria in space debris, it would remain likely that life on Earth did not arise elsewhere. This is true because molecular analysis of cellular evolution suggests that we arose from protocells that were less advanced than a modern-day bacterium, and the panspermia hypothesis typically suggests that an organism similar to a current-day bacterium was necessary to initiate life on Earth [14]. So, while it is not possible to definitively rule out the panspermists' idea that life began in another world, the currently available evidence suggests otherwise. Therefore, given this reality, what can we say about life's potential origination on Earth?

NAILING DOWN LIFE'S BIRTHDATE

To start, it is worth asking when life on Earth first appeared. This question is intrinsically interesting, and knowing the answer would provide insight into the conditions that prevailed when life first appeared. With this information, we could then consider the types of chemical and geophysical processes that may have occurred on Earth during the period of life's debut.

To state the obvious, life on Earth cannot be older than Earth itself, so we can bracket life's origin by estimating the age of our planet. In 1650, the Anglo-Irish archbishop James Ussher attempted to do exactly this, and with the help of biblical scripture, he concluded that Earth was about 6,000 years old. In fact, Ussher argued that he could be even more precise, and he stated that our planet was created at noon on October 23, 4004 BC [15]. Up until the start of the 19th century, a large fraction of humanity agreed with Ussher's estimation, and this led to the conclusion that life could not be more than 6,000 years old. However, during the 17th and 18th centuries, some natural philosophers began to make geological observations that led them to question the veracity of Ussher's calculations, and by 1778, the Jesuit-educated Georges-Louis Leclerc, Comte de Buffon, publicly stated that Ussher's conclusion was wrong. Unlike most of his predecessors, Buffon was of the opinion that Earth formed as a result of colliding astronomical bodies, and using Earth's temperature and his estimate of the rate at which colliding bodies cooled, he calculated that our planet was 74,832 years old [4]. For his troubles, Buffon earned the animosity of the Vatican, but his work helped pave the way for an empirically based analysis of the age of Earth and the life that it supports.

By 1859, Charles Darwin had employed his understanding of the geological processes that operate in southern England to estimate that Earth was 306,662,400 years old. This evaluation allowed Darwin to argue that life could have evolved over a much longer period than possible if Ussher or Buffon were correct, but Darwin's assessment fell under intense criticism, and even his son, the astronomer George Darwin, argued that Earth was at most 56 million years old.

At the beginning of the 20th century, new strategies for determining the age of Earth were developed. By 1899, Henri Becquerel and Marie Curie had discovered the existence of radioisotopes, and soon after, geologists were able to measure the rate of radioactive decay of isotopes in rocks. With this information, more accurate assessments of the age of Earth were produced, and if there were fossils in the rocks being studied, it was also possible to say when the life forms that generated these fossils lived. Consequently, by the middle of the 20th century, the scientific community was able to state that Earth was 4.54 billion years old, and the available evidence now indicates that life has existed on our planet for most of that time. However, despite advances in radiometric dating, it has been difficult to precisely determine when life first appeared. This is due, in part, to the fact that it is inherently difficult to determine if the object being studied in an ancient rock is a fossil of a primordial organism or simply an artifact of a geological process. Furthermore, it is hard to find primordial cellular fossils because bacteria and other single-celled organisms do not fossilize well, and most of Earth's oldest rocks have been destroyed by the LHB or because of plate tectonic activity. Nevertheless, despite these limitations, the scientific community has concluded that cellular life was present on Earth 3.56 billion years ago. There are still open questions about whether life was present even earlier than that, and it is worth noting that some recent estimates suggest that life may have first appeared on our planet 3.77–4.28 billion years ago [16].

If life did begin 4.28 billion years ago, it would have emerged just 260 million years after Earth formed. This would be surprisingly fast given that, at end of Earth's accretion, the surface of the planet was a roiling tempest of molten magma with a temperature of 3,680°F. But, after 50–100 million years of dispensing heat into the vast and frigid space that surrounds our globe, the temperature of Earth's surface dropped, and the red-hot magma at the exterior of the planet began to crystallize and form rocks such as peridotite.

Ultimately, the peridotite layer that formed on Earth's surface didn't last long, because peridotite is denser than the magma over which it crystalized. Soon after it formed, the peridotite layer sank down into the planet, only to be replaced by more red-hot magma from below. As this cyclical process of peridotite formation continued, some of the layers of submerged peridotite began to partially melt (i.e., the elements with the lowest melting temperature in the peridotite liquefied), and the magma produced as a result was rich in calcium, silicon, and aluminum. This initial melt material was less dense than the peridotite from which it was derived, and once it cooled, it formed a rock called basalt, which floated above the peridotite slush. Ultimately, basalt encased the globe [17], but the basalt rock that initially lined our planet was far too hot to have liquid water floating above it, and without liquid water, there could be no life.

We don't know exactly how long it took for Earth to cool to the point where water vapor in the atmosphere could condense and the oceans could begin to cover the basalt, but some have estimated that these great bodies of water may have formed about 130 million years after the planet accreted. If true, it is possible that the first cellular life appeared on Earth within 130–850 million years of the formation of the oceans, and this would imply that cellular life arose on Earth relatively soon after it was possible for it to do so.

THE PATH TO PROTOCELLULAR LIFE

It is important to note that the aforementioned estimates are referring to cellular life, which technically may not have been the first life on the planet. The definition of life varies depending on the authority consulted, but according to NASA, life is "a self-sustaining chemical system capable of incorporating novelty and undergoing Darwinian evolution." By this definition, it is likely that the first living organism was a protocell that formed on the planet soon after the globe was cool enough to harbor liquid water. However, to date, we have not found any extant fossils of these protocells.

Although the first "living" protocells may not have bequeathed humanity with fossils, they did give rise to something much more important, namely, a cellular bacterium that lies at the base of every organism's phylogenetic tree [18]. This organism is life's last universal common ancestor (also known as LUCA), and it is an ancestor that we share with every organism that has ever inhabited the planet. How long it took for LUCA to evolve from the protocell that marked the beginning of life on our planet is not known, but since some have estimated that life appeared on Earth in the neighborhood of 4.28 billion years ago [16, 19], and LUCA arrived about 3.5–3.8 billion years ago, there may have been more than 700 million years separating the first protocellular life from LUCA.

It is also worth noting that, if living organisms were present on Earth 4.28 billion years ago, their descendants had to endure the LHB of 3.8–4.1 billion years ago. To live through this time, the descendants of the first protocells needed to be exceptionally resilient and adaptable. During the extraordinarily violent period of the LHB, cosmic assaults superheated the planet, and some of the oceans literally vaporized [20]. Given the hellish conditions that characterized this epoch, the most successful organisms of this era may well have been single-cell thermophiles that derived their energy through the inorganic oxidation of reduced hydrothermal vent compounds [21]. Interestingly, some of the oldest putative cellular-like fossils found to date were likely produced by thermophiles embedded in seafloor hydrothermal vent precipitates [16].

THE QUEST TO UNDERSTAND THE GENESIS OF THE FIRST PROTOCELL

There is a great interest in the ontology of the first protocell, but everything about this creature is mysterious, including how and where it arose. In 1871, Darwin stated that he didn't think we could determine how the first living creature came to be, but he nevertheless speculated that it may have appeared in a "warm little pond, with all sorts of ammonia and phosphoric salts, light, heat, electricity, etc." [22]. Others have postulated that the first life on the planet may have originated in tide pools, or hot springs, or deep underground, or near a deep-sea hydrothermal vent, and as was mentioned previously, some have argued that the first living organisms were transported to Earth from afar.

While Darwin held out little hope that we could determine how life arose, today there is reason to be more optimistic. Though we lack details, we are now able to generate plausible scenarios regarding life's origins, and if we can establish the environmental and geological state of the early Earth, we may soon be able to describe with a reasonable degree of confidence a plausible pathway that gave rise to Earth's first life forms. If we manage to do this, humanity will have taken a significant step forward in understanding itself and the complexity that exists within our universe.

There are many ideas about how life might have begun, and more than a few books have explored this topic. I will not attempt to discuss all the possible mechanisms that have been proposed; instead, I will focus on one of the most strongly supported ideas, namely, those that constitute one version of the "RNA first" hypothesis. For other related versions of how the first RNA-based protocells may have arisen, I suggest reading some of the recent fascinating articles by Roy Black and his colleagues [23–25].

AN RNA FIRST MODEL

As I mentioned earlier, in the 1960s, Francis Crick and Leslie Orgel suggested that space aliens seeded life on Earth, but perhaps to hedge their bets, Crick and Orgel joined Carl Woese, and together they suggested that it was also possible that life initiated with the formation of RNA-directed protocells. Crick, Orgel, Woese, and other early "RNA first" advocates were particularly impressed by the structural, biochemical, and genetic properties of a type of RNA called transfer RNA, or tRNA for short. This molecule was discovered in 1956, and based in part on its functional capabilities, Crick, Orgel, and Woese envisioned tRNA, and by extension RNA in general, as having many potential roles in an early protocell. But, with no convincing experimental evidence to support their idea, the early adopters of this "RNA first" hypothesis saw their thinking consigned to the same intellectual bin that the pansper- mia concept had been relegated to.

In the mid-1980s, however, the status of the "RNA first" concept began to improve. By then, David Baltimore and Howard Temin had demonstrated that RNA was used to store genetic information in retroviruses [26], and Sidney Altman and Thomas Cech established that some RNA molecules possess catalytic activities similar to those of proteins [27, 28]. By 1986, the excitement generated by the study of the multifaceted biochemical capacities of RNA inspired the Nobel Laureate Walter Gilbert to outline a mechanism in which RNA may have generated a pro- tocell [29]. But, despite the enthusiasm for Gilbert's so-called "RNA world" and the mounting evidence for the "RNA first" hypothesis, many in the origin-of-life research community rejected the idea that RNA was key to the initiation of life on our planet. This was due in large part to the belief that RNA could not be efficiently synthesized under the conditions that characterized the primordial Earth of four billion years ago. However, since the start of the 21st century, some of these con- cerns have abated, and we now have reason to believe that RNA may have formed on the prebiotic early Earth.

Over the next several pages, I will outline, in a step-by-step fashion, one possible mechanism for the synthesis of RNA-based protocells. For those who are not par- ticularly interested in how RNA and protocells may have been synthesized and rep- licated on the ancient Earth, feel free to skip over this material and continue reading at the section entitled, "Something New Under the Sun." But, having said this, I hope that you will consider embarking on this somewhat technical and detailed section of the journey in order to learn more about potential RNA-based protocells, which may have been the first living creatures on our planet. So, without further ado, let's look at how these primitive forms of life may have arisen.

THE FORMATION OF AUTONOMOUS, RNA-BASED PROTOCELLS

STEP 1—THE CONSTRUCTION, CONCENTRATION, AND ACTIVATION OF BIOLOGICALLY IMPORTANT RNA PRECURSOR MOLECULES

RNA is a long, single-stranded polymer that is composed of four nucleotide subunits (see Figure 6.2), so if RNA was made on the primordial Earth, it must have been possible to generate nucleotide precursors using the energy sources and molecules that were present on the planet at the time. To the surprise of many, in 2015, John Sutherland demonstrated that RNA nucleotide precursors could be produced using just UV light, hydrogen cyanide, hydrogen sulfide, and inorganic metal catalysts [30–32]. Sutherland also revealed that other biologically important molecules such as lipids, which are an essential part of cellular membranes, and amino acids, which form the basic building blocks of proteins, could also be produced using these start- ing substances.

Importantly, Sutherland's starting materials were likely very common on the ancient Earth. Four billion years ago, the planet lacked an ozone layer, so it would have been flooded with UV light, and on an oxygen-poor, heavily irradiated globe like Earth of four billion years ago, the all-important hydrogen cyanide molecule could have formed in abundance from the planet's plentiful sources of hydrogen, carbon, and nitrogen. Furthermore, hydrogen sulfide would have been easy to come by during this period because it was quite literally belching out of Earth through the tops of the planet's many active volcanoes. But, while Sutherland's chemical reactions outlined a pathway by which the precursors of nucleotides and other biologically important molecules could have been produced, he initially did not explain how these nucleotide precursors were stabilized and concentrated in a manner that would allow them to be used to construct nucleotides and RNA molecules.

To address this concern, Sutherland considered the likely geophysical processes at work on Earth's surface four billion years ago. Based on his understanding of these events, he described a plausible mechanism for stabilizing and concentrating the RNA precursors. Specifically, Sutherland argued that the environment may have affected the precursors' concentration and reactivity, and he suggested that on some parts of the early Earth, the environment may have been similar to that of modern-day Yellowstone Lake, located in Wyoming's Yellowstone National Park.

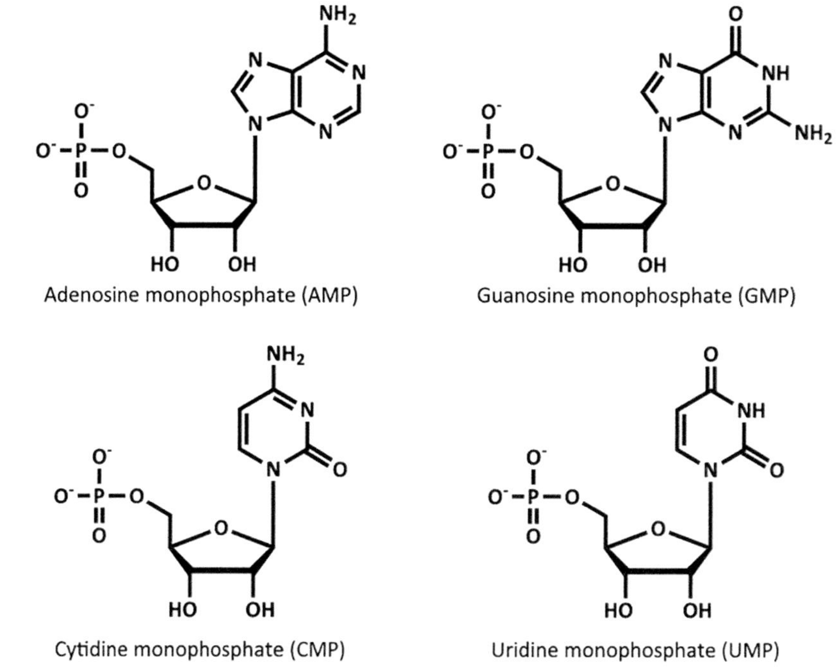

PART A: The four ribonucleotides that compose RNA.

FIGURE 6.2   RNA subunits and RNA structure.

RNA Structure

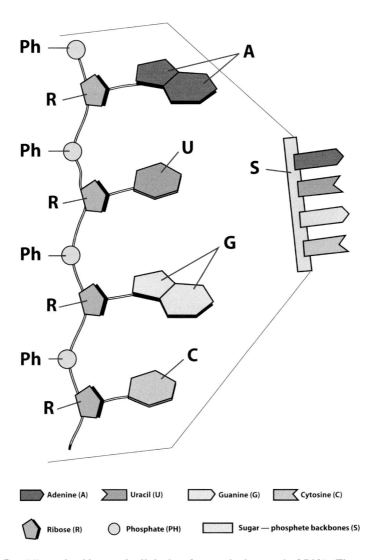

Adenine (A)   Uracil (U)   Guanine (G)   Cytosine (C)

Ribose (R)   Phosphate (PH)   Sugar — phosphete backbones (S)

PART B: Ribonucleotides can be linked to form a single strand of RNA. The sugar (i.e., ribose) and phosphate molecules form the backbone of the RNA molecule, and the nitrogenous bases are attached to this backbone.

FIGURE 6.2 (Continued)

In this volcanically active area, magma heats the groundwater, causing it to circulate over the local rocks. As this water moves about, it leaches iron from stones, and some of this aqueous, metal-laced solution mixes with rainwater that eventually flows into the lake. Four billion years ago, the iron entering a body of water like Yellowstone Lake would have immediately interacted with hydrogen cyanide, which, as I mentioned earlier, was literally raining down onto Earth's surface during this period. As a result, the stable and insoluble compound ferrocyanide would have formed, and over the millennia, this substance would have precipitated to the bottom of the lake where it likely formed a thick, cyanide-rich sediment.

Sutherland speculates that, due to the geological instability of the primordial Earth, and because asteroids were constantly pummeling the planet, most lakes would have dried up at some point, and he suggests that the heat that desiccated a lake environment would have also transformed the ferrocyanide salts into more reactive compounds via well-document chemical pathways. Eventually, these molecules would have rehydrated, and in this state, they could have produced the stable, concentrated, and activated molecules necessary to produce the nucleotides found in RNA, as well as many of the amino acids and lipids found in modern cells [33].

If Sutherland's speculations are correct, RNA nucleotide precursor molecules may have been present in abundance on some parts of ancient Earth. However, if this is true, the question that arises is how did these precursor molecules get assembled into the actual RNA nucleotides, and given that RNA is a chain of nucleotides, how did the RNA nucleotides get linked to form nucleotide chains?

THE FORMATION OF AUTONOMOUS RNA-BASED PROTOCELLS

Step 2—The Formation of Nucleotides From Precursor Molecules, and the Linking of Nucleotides to Form Single-Stranded RNA

Attempting to address the aforementioned questions, Sutherland managed an impressive feat. In 2009, he demonstrated how pyrimidine ribonucleotides could be synthesized under the conditions thought to exist on the primordial Earth [34]. The pyrimidines cytidine monophosphate (sometimes referred to as CMP or C) and uridine monophosphate (also called UMP or U) constitute two of the four nucleotide subunits of RNA, and the production of these nucleotides under putative early Earth conditions was a major advancement for the "RNA first" hypothesis.

Seven years after Sutherland published his seminal paper, Thomas Carell generated additional support for the "RNA first" hypothesis when he established that purines could also be produced under putative prebiotic conditions [35, 36]. The purines adenosine monophosphate (also known as AMP or A) and guanosine monophosphate (commonly abbreviated GMP or G) constitute the remaining two fundamental building blocks of RNA. So, by 2016, the work of Sutherland, Carell, and their colleagues had coalesced into a demonstration of how the four basic building blocks of RNA could have formed billions of years ago.

That it may have been possible to synthesize RNA precursor molecules on the primordial Earth was encouraging, but synthetic pathways are not always compatible, and consequently, the fact that all four RNA nucleotides could have been synthesized in the past does not mean that they were. As it turns out, the reactions that Sutherland and Carell employed in the laboratory to construct the pyrimidine and purines were indeed incompatible, and consequently, it was difficult to imagine how these seemingly incongruous processes could have generated an "early Earth RNA world." To the relief of the "RNA first" supporters, this concern was attenuated in 2017 when Jack Szostak demonstrated that compatible, prebiotically plausible routes for the synthesis of all four RNA subunits may have indeed existed on the primordial Earth [37]. But, if the four ribonucleotide building blocks of RNA were all present in the same area and at the same moment in time four billion years ago, how could they be linked to form a single strand of RNA?

In a modern cell, the joining of nucleotides is an enzyme-mediated process, but in the absence of protein catalysts, it is not immediately obvious how long polymers of RNA nucleotides could have efficiently assembled in a prebiotic world. However, RNA nucleotides can polymerize into chains in the presence of some types of clay minerals [38] which form as a result of rock weathering. Due to their charged surfaces, Na^+-clays, for example, are particularly good at absorbing, concentrating, and aligning nucleotides. So, it is possible that these processes facilitated the formation of a bond between nucleotides which in turn led to the formation of nucleotide chains [39]. It is also possible that nucleotide polymerization occurred in ice that contains magnesium and lead because like the clay surfaces, these materials also facilitate the linkage of nucleotides into nucleotide chains [40].

Taken together, these facts suggest that four billion years ago, it may have been feasible for abiotic processes to generate the four ribonucleotide building blocks needed to produce RNA, and furthermore, these data argue that it may have also been possible to link these ribonucleotide subunits to form a strand of RNA. But RNA strands are not protocells. So, given a potentially plausible route for RNA synthesis, we are now left with the question of how this RNA could be used to generate a protocell.

THE FORMATION OF AUTONOMOUS RNA-BASED PROTOCELLS

STEP 3—USING SINGLE-STRANDED RNA TO GENERATE PROTOCELLS

For a protocell to form, some of the newly formed RNA molecules needed to isolate themselves so that they could establish their own metabolically active microenvironments. So, how might this process have occurred on the ancient Earth? In 2017, David Zwicker put forth a potential answer. Zwicker demonstrated that, when RNA and peptides are synthesized in an aqueous environment, they form liquid-like aggregates that he termed "liquid droplets." These spherically shaped structures can co-exist with the fluid that surrounds them, while remaining separate from it. As distinct entities, the well-defined surfaces of the liquid droplets allow many substances to diffuse in and out, and within the microenvironment of the droplet itself, specific chemical reactions can occur. Amazingly, if these chemical reactions are driven by an external energy source, and if they result in the production of some substances and the destruction of others, the droplets grow and ultimately spontaneously divide into two smaller droplets of equal size [41, 42].

Zwicker's work essentially suggests that, with energy that could have been derived from convection currents, tides, hydrothermal vents, or chemical compounds, liquid droplets may have acquired a primitive form of metabolism. And most importantly, with their simple metabolism and distinct boundaries, these humble entities may have organized the chemistry of RNA replication, and in so doing, they may have formed the first primitive RNA-based, self-reproducing protocells!

Zwicker also speculated that the liquid droplets could have acquired a membrane because amphiphilic molecules such as fatty acids tend to gather around a liquid droplet's surface [41]. If this occurred, then the protocells would have taken on an additional degree of sophistication. With this primitive lipid membrane, the free RNA nucleotides required to drive RNA synthesis would have been able to permeate the early protocells, and therefore these pre-cellular structures would not need the more complex protein-based transportation machinery found in modern cellular membranes in order to produce additional RNA [38]. Furthermore, the work of Jack Szostak's team indicates that once a fatty acid membrane forms, simply adding energy and more fatty acids to the spherical structure is enough to trigger its division [33]. Therefore, if a fatty acid-encased, protocellular membrane formed, it could also potentially be replicated.

THE FORMATION OF AUTONOMOUS RNA-BASED PROTOCELLS

STEP 4—NON-CATALYTIC REPLICATION OF PROTOCELLULAR RNA

Getting the liquid droplet or fatty acid membrane of a protocell to replicate is significant, but how could the RNA molecule within the protocell replicate? In a series of papers, Szostak, Orgel, and a number of other investigators took on this question.

It has been known for some time that, under some conditions, RNA can form a double-stranded structure in which adenosine (A) can chemically bond with uridine (U), and guanosine (G) can bond with cytidine (C), and given this, it was suggested that an RNA molecule in a protocell could have served as a template for the synthesis of a new strand of RNA [43]. Specifically, under certain conditions, purines and pyrimidine in the original single-stranded RNA molecule could pair up with their complementary partner nucleotides, and a double-stranded RNA molecule could form [33] (see Figure 6.3). Therefore, it is possible for a single-stranded RNA template to potentially generate a double-stranded RNA molecule.

However, once these double-stranded RNA structures appear within a protocell, they could potentially remain in the double-stranded state until they degrade. If this were to happen, the production of new RNA strands would be compromised. But, in thermally graded bodies of water such as Yellowstone Lake, which are often cold at one end and near boiling at the other, the double-stranded state of the RNA molecule need not be permanent. In this setting, the protocell could generate

```
G ---- C
A ---- U
C ---- G
U ---- A
C ---- G
U ---- A
A ---- U
A ---- U
G ---- C
```

PART A: The pairing of RNA nucleotides in an original strand of RNA (displayed in red) with a complementary RNA nucleotides (depicted in blue), followed by the linking of the newly assembled RNA nucleotides generates a double-stranded RNA molecule.

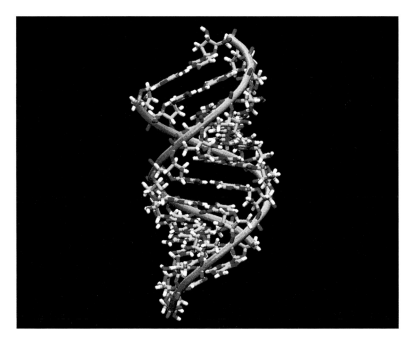

PART B: The three-dimensional image of the double-stranded RNA (From: Supyyyy, CC BY-SA 4.0 <https://creativecommons.org/licenses/by-sa/4.0>, via Wikimedia Commons.)

FIGURE 6.3   The formation of a double-stranded RNA molecule.

FIGURE 6.4   Under high temperatures, the double-stranded RNA molecules depicted in Figure 6.3 can melt and give rise to single-stranded RNA strands. Consequently, as a result of the processes described in Figure 6.3 and and in this figure, the number of RNA molecules in a protocell doubles.

double-stranded RNA in a cool environment, and if this RNA eventually drifted into a warmer section of the lake, the double-stranded structure would melt and give rise to two single strands of RNA (see Figure 6.4). As a result, under the right conditions, when a protocell does divide, each of the two daughter proto-cells could end up with its own copy of a single-stranded RNA molecule [33, 38]. Presumably, after repeated cycles of this process, the protocell population would increase exponentially.

THE FORMATION OF AUTONOMOUS RNA-BASED PROTOCELLS

STEP 5—ENHANCING THE RATE OF RNA REPLICATION DURING PROTOCELLULAR DIVISION

Having been among those who speculated as to how RNA could be replicated non-enzymatically, Jack Szostak and his team began to test aspects of this hypothesis. They found that they were able to generate new RNA molecules in the absence of an enzyme, but the process occurred very slowly. This was a potential problem because RNA is not a highly stable molecule, and a very slow rate of RNA rep-lication would limit the protocell's ability to maintain its complement of RNA. Interestingly, however, Szostak's group found that by making a small chemical modification to the RNA nucleotides, they could increase the rate of RNA synthe-sis by over two orders of magnitude [38, 44, 45], and based on this finding, they suggested that the RNA in the first protocells may have been slightly different from the RNA found in modern cells.

However, Szostak's non-enzymatic RNA replication model is also complicated by the fact that it requires magnesium to be present during RNA synthesis, which is a surprising necessity, given that magnesium degrades single-stranded RNA and destroys fatty acid membranes. So, this leaves the question of how a protocell with self-replicating RNA could exist if the metal ion needed to replicate RNA destroys the original RNA template and the lipid casing that delineates the protocell.

One potential explanation is that the activity of the magnesium ion was constrained by a "chelating agent." Such an agent would allow magnesium to facilitate its role in RNA replication, but it could also inhibit the metal ion from degrading single-stranded RNA and lipids. Szostak sought a substance that could do this, and remarkably, he found that a chemical called citrate could carry out the task. Although citrate was likely not present on the prebiotic Earth, Szostak believes that a short acidic peptide (which is similar to the magnesium-binding peptide that forms part of the modern cell's RNA synthesis machinery) may have substituted for citrate [46].

Ultimately, even with chemically modified nucleotides and magnesium chelating agents, non-enzymatic RNA replication would have occurred inefficiently and at a very slow rate in the prebiotic world. However, this problem may not be as significant as it first appears, because even if it took an RNA molecule a week to replicate, given sufficient precursor molecules, there could be a million cubic kilometers of RNA produced within two years of the appearance of the first self-replicating RNA [17]. Nevertheless, to significantly improve their speed of replication, the protocells needed an enzymatically catalyzed system, and presumably, because of chance copying errors that occurred in a non-enzymatically replicating RNA molecule, some protocellular RNA molecules eventually acquired the ability to catalyze RNA replication.

In effect to gain insight into how catalytic, self-replicating RNA molecules potentially arose a number of investigators are attempting to recreate the conditions necessary to produce these molecules. Ultimately, the goal of their work is to generate a self-replicating RNA molecule. To date, no one has succeeded in this task, but some investigators have found RNA molecules that are capable of synthesizing other long RNA sequences [47]. This has led to the speculation that Darwinian evolution may have produced groups of RNA molecules that cooperated and replicated one another [48, 49]. If so, teams of cooperating RNA molecules may have ensured that catalytically facilitated RNA replication occurred in some primordial protocells.

SOMETHING NEW UNDER THE SUN

The appearance of protocells during the Hadean eon represents a major milestone in the history of Earth, because unlike other structures that had developed on the planet, protocells were capable of mutating and self-replicating, and they constituted "a self-sustaining chemical system capable of incorporating novelty and undergoing Darwinian evolution." In other words, these primitive protocells were remarkable, because, by NASA's definition, they were "alive."

Having come into existence, protocells began competing for the resources they needed to survive and reproduce. Once this competition began, the protocellular entities crossed the biological equivalent of the Rubicon. From that moment forward, life on our planet began its inexorable march into the future, and to quote Darwin, "from so simple a beginning endless forms most beautiful and most wonderful have been, and are being, evolved."

FROM PROTOCELL TO TRUE BACTERIAL CELL—A FUNDAMENTAL TRANSITION IN THE HISTORY OF LIFE

On our quest to understand the origin of life, the next way station is the appearance of the first "true bacterial cell". Like many topics related to life's beginnings, there is much we still do not know. However, we are confident that during the transition from protocell to true bacterial cells, genetic information began to be stored in the form of DNA.

As an information storage system DNA has many advantages over RNA. For starters DNA is chemically more stable than RNA and therefore less likely to degrade. In addition, during the process of replication, DNA in modern cells is proof-read and many of the errors missed by the proofreaders are corrected by a group of DNA repair enzymes. As a result, the rate of error incorporation into a newly replicated DNA molecule is decreased by 10,000-fold or more [50], and this makes DNA a far more reliable vehicle than RNA for the high fidelity transmission of genetic information.

We also know that DNA in modern cells can encode various forms of RNA, and these RNA molecules can team up with proteins to generate the complex biological machinery necessary to produce new proteins. In modern cells, proteins do the lion's share of the work, and without them, chromosomes cannot be replicated, RNA molecules cannot be produced, and new proteins cannot be made. So, to produce true cells, mechanisms for generating DNA and proteins had to evolve. However, true cells need existing DNA and proteins to make new DNA and proteins, and therefore, before we can contemplate how true bacterial cells came to be, it is necessary to determine how protocells first produced DNA and genetically encoded proteins.

THE APPEARANCE OF THE FIRST GENETICALLY SPECIFIED PROTEINS AND DNA

Some have speculated that the RNA of protocells may have produced the first proteins and that this may have occurred with the help of a specialized molecule that was similar to modern transfer RNA (tRNA). Transfer RNA is an adaptor that binds messenger RNA (which has information for making proteins) and amino acids (which are the basic building blocks of proteins). As a result of its actions, tRNA can use the genetic information encoded in an mRNA molecule to assist in the assembly of a specific protein.

John Maynard Smith suggests that a group of ribozymes, which are RNA molecules with catalytic properties, may have been converted to structures like tRNAs, when they began to recruit amino acids as cofactors. Cofactors increase the range and specificity of enzyme-catalyzed reactions, and therefore ribozymes that interacted with amino acids may have imparted a metabolic advantage on the protocell they were in.

Maynard Smith also speculates that at some point, a subset of these tRNA precursor molecules may have evolved the ability to bind more than one amino acid, and yet another set of ribozymes may have acquired the ability to link amino acids that are bound to other RNAs. If this occurred, and if these latter two types of RNA were present in the same protocell, then they could have worked together to produce chains of amino acids. One or more chains of amino acids can form a protein, so this hypothesis suggests that teams of protocellular ribozymes may have formed the first proteins by generating protein-synthesizing machinery similar in function to the protein-generating ribosomes of modern cells.

In addition to explaining how tRNAs, the first proteins, and ribosomes may have initially come into being, Maynard Smith's hypothesis is fascinating because it implies that the first protein-based enzymes did not somehow evolve independently of the catalytic ribozymes found in the first protocells. Instead, this schema proposes that an RNA-based ribozyme was slowly converted into a protein-based enzyme as the number of amino acids associated with a specific RNA molecule gradually increased over time.

Still another notable feature of Maynard Smith's hypothesis is that some of the first precursor tRNAs may have interacted with other cellular RNAs. This would occur because single-stranded RNA molecules within the tRNA precursor would have been capable of interacting with other cellular RNAs through complementary base pairing. If this occurred, then the interactions between precursors of tRNAs and protocellular RNA may have given rise to the genetic code that forms the foundation of modern cellular heredity.

If the first genetically encoded proteins did evolve via a process similar to the one described, it is likely that some of the first proteins produced by the precursor of the modern cell were enzymes that converted RNA into DNA. Such enzymes can be found today in retroviruses such as HIV, which is the agent that causes acquired immune deficiency syndrome (i.e., AIDS). As was mentioned earlier, DNA is more stable than RNA, so once DNA was present in protocells, it would have eventually usurped RNA's role as the primary data storage molecule. DNA's place in the hereditary system of the cell would have been further entrenched once the cell evolved the ability to proofread and repair the DNA molecule.

This sketch of the protocell to true cell transition is obviously a highly speculative outline of a process that may have occurred, and it will be some time before experimentalists can generate convincing detailed evidence that supports or refutes its basic tenets. Furthermore, even if the experimental evidence does eventually suggest that these ideas are correct, there are still many additional questions about the transition from protocells to true cells that need to be addressed. For example, to understand this critical event in the history of life, we still need to determine how protocellular DNA formed a chromosome, how materials moved into and out of membrane-bound protocells, how protocells developed DNA proofreading and repair systems, and how the first cells partitioned DNA and other subcellular material to their descendants. The evolution of other important subcellular structures, the development of specialized roles for RNA, the development of gene regulatory mechanisms, and the nature and development of cellular metabolic systems are also among the many additional questions that need to be explored. The existence of all these questions indicates that we are a long way from understanding the details of how protocells gave rise to true cells, but the investigation into this fundamentally important transition has begun, and we are beginning to glean insights into an aspect of biology that seemed completely inscrutable just 100 years ago.

ALL IN GOOD TIME

As for how long it took protocells to give rise to true cells, that too is unclear, but the process likely took a very long time. As I mentioned earlier, most scientists agree that true bacterial cells were present on the planet by 3.56 billion years ago, and it is possible that protocells first formed on small volcanic islands during a period that preceded the LHB of 3.8–4.1 billion years ago. If these dates are correct, the transition from protocells to true cells may have taken more than half a billion years. During this long interval, some of the descendants of the original protocells may have evolved the ability to survive in diverse settings, including those found around deep-sea hydrothermal vents.

The ability to live in various environments may have been particularly important when life was first evolving, and as it turned out, protocells that could live in deep-sea waters may have had a great advantage during the Hadean and early Achaean eons of 4.6–3.6 billion years ago. This would have been especially true during much of the LHB period during which time the surface of Earth was superheated and significant fractions of the oceans were vaporized [20, 51]. It may have been that during this period, the safest place for a protocell was at the bottom of the ocean near a deep-sea vent.

Ultimately, over hundreds of millions of years, protocells may have generated many different types of more complex cells. However, if they did, only one (i.e., LUCA) was successful in producing extant descendants.

GETTING TO KNOW LUCA (THE LAST UNIVERSAL COMMON ANCESTOR)

The evidence that strongly suggests that all life originated from a single common ancestral cell comes from many sources. Included among these are genetic and biochemical data that indicate all living organisms use the same set of nucleotides to construct nucleic acids (i.e., DNA and RNA), and they all use the same set of amino acids to synthesize proteins. Given that there are a large number of nucleotides and amino acids that could have been used by living cells, the odds that all living creatures just happened to independently evolve biochemical pathways that use these same precursor molecules are extremely small. It is also worth noting that the DNA within living systems makes use of a nearly universal code to produce proteins, and once again, the likelihood of this happening by chance is exceedingly remote.

In an attempt to quantitate just how probable it was that life arose from a universal common ancestor, Douglas Theobald used model selection theory to analyze a group of proteins that were found in the three major domains of life, namely, bacteria, archaea, and eukaryotes. Theobald's analysis, which was published in 2010 in the journal *Nature*, indicates that the model that suggests life arose from a universal common ancestor was $10^{2,860}$ times more likely to be correct than the next most probable model, which assumed that the various domains of living creatures arose independently from genetically unrelated organisms [52]. I am not a betting man, but I like those odds, and consequently, I feel comfortable in concluding that the idea that all organisms share a common ancestor is a near certainty, or at least as close to a certainty as one can get in science.

To gain a better understanding of this LUCA, a German team headed up by William Martin analyzed more than six million different genes found in bacteria and Achaea. They found a group of 355 that were likely present in LUCA, and based on the analysis of these genes, Martin and his colleagues postulated that LUCA probably lived near deep-sea hydrothermal vents and that it was a thermophilic, anaerobic, autotroph (i.e., LUCA was a heat-loving organism that could harness energy independently of other life forms and in the absence of oxygen). They also argued that LUCA used the metals, hydrogen-rich compounds, carbon dioxide, and various inorganic compounds that were released from the depths of the planet to synthesize the carbon-based molecules it needed to survive [53, 54]. If these interpretations of the data are correct, we, and every other organism that has appeared on Earth, can trace our ancestry back to a cell that lived near a larva-belching opening on the ocean floor some 3.56 billion years ago.

WAS THE EVOLUTION OF SELF-REPLICATING RNA REALLY THE KEY TO GENERATING LIFE ON EARTH?

Much of what I have discussed assumes that we can accurately describe and recreate the geochemical conditions that existed on the early prebiotic Earth, but the truth is that we are not certain what Earth was like four billion years ago [32, 55]. Furthermore, even if we were confident that we could accurately recreate the conditions on the early Earth, and even if a protocell with a lipid boundary, a primitive metabolism, and a self-replicating RNA-based information system could have arisen as outlined earlier, we still would not know with certainty that a protocell actually did arise this way. The best we can do is sketch one potential pathway by which protocellular life on our planet *might have* arisen.

Since we cannot go back in time and witness the genesis of life, it is very difficult to rule out the possibility that protocellular life arose by another yet-to-be-described pathway, or that it arose multiple times via numerous pathways, or that it was the result of a merger of structures that began by various yet-to-be-described, independent pathways. In the end, the detailed and definitive story of how life came to exist on Earth will likely remain a mystery. But, despite this uncertainty, we are nevertheless beginning to get a glimpse of a well-reasoned and potentially plausible picture of how life may have come forth on our planet.

WHAT ARE THE ODDS?

If the "RNA first" advocates are correct, life likely began under a reducing atmosphere (such as one rich in hydrogen sulfide, methane, ammonia, and other hydrogen-rich gases), and it probably did so in a heavily UV-irradiated, thermally graded, mineral-rich, hydrogen cyanide-polluted, geologically unstable, lake environment [56]. Given this, and assuming life would have to begin via a similar pathway elsewhere, how likely is it that simple protocellular life exists on other planets or moons within our universe?

Before we can address this question, we first must determine if the laws of physics are the same everywhere in the universe. Ultimately, it is the laws of physics that direct the emergence of complexity, and consequently, if these laws vary in different sections of the universe, the probability that life could form in other parts of the universe will also vary. To date, we do not have compelling evidence that the laws of physics do fluctuate within our universe, but as was mentioned in Chapter 1, John Webb does have controversial data that suggest that they may. In particular, he argues that the electromagnetic force that governs chemical interactions is different in various regions of space, and if he is correct, it is possible that life as we know it may only be possible in some sectors of the cosmos [57, 58]. But let's assume for now that the laws of physics are constant throughout the universe. What other factors affected the likelihood of life emerging on Earth, and which factors would also likely affect the emergence of life elsewhere in the cosmos?

Clearly, one critical feature of a life-supporting world is its energy state. For life to form on Earth, liquid water needed to be present on the planet's surface, and that can only occur if the temperature on the planet falls within a narrow range. Furthermore, according to at least some renditions of the "RNA first" hypothesis, the planet's surface temperature had to vary in local regions in such a way that, in some relatively small bodies of water, it was possible to find areas that were cold, as well as areas that were near boiling.

The "RNA first" hypothesis also suggests that, on Earth, a moderate amount of energy in the form of UV light was needed to produce RNA, protein, and lipid precursors. It is worth noting that an important modifier in the previous sentence is the word "moderate," for if our planet had been immersed in enormous amounts of UV light, the surface temperature of Earth would have been too high for liquid water to exist, and since UV light damages DNA and proteins, intense amounts of UV light would have also irreparably damaged important biological molecules that did somehow manage to form. While too much UV light would be problematic, too little would also cause problems since UV light appears to have been necessary for the synthesis of RNA.

In addition to being on a planet that has the proper type and distribution of energy sources, for life to emerge, a planet also needs to have a precise chemical composition. Liquid water is essential for life as we know it, but a life-harboring world like ours also needs the right complement of numerous specific chemical elements such as carbon, nitrogen, oxygen, and hydrogen, and some of these elements likely need to be arranged in the form of catalytically important minerals. As was stated in Chapter 4, in many parts of our galaxy, the composition of the stars and planets is not conducive to the emergence of life. But, even when a planet has the chemical composition it needs to potentially support life, the chemical elements on that planet must be in the right place at the right time and in the right concentrations for life to plausibly emerge.

The implication of the last sentence is that, when it comes to forming a planet suitable for hosting life, there is much that can go wrong. In the case of Earth, carbon was essential for the emergence of life but had there been too much carbon in the atmosphere, the surface temperature of our planet would have been too high to support life. Conversely, had the carbon levels been very low, our planet would have been frigid and inhospitable to protocellular life. In addition, recent work on the "RNA first" hypothesis suggests that, for molecules such as purines and pyrimidine to form prebiotically, a reducing atmosphere may have been necessary, so in addition to having an optimal amount of carbon in its atmosphere, it may also be important for a life-hosting planet to sequester oxygen, at least during the early stages of the genesis of protocellular life.

In the case of our own planet, most atmospheric scientists believe that the chemical composition of the early Earth's atmosphere was determined by impact degassing (i.e., the release of gases after asteroids and comets collided with the planet) and by the volcanic outgassing of volatiles in magma. Up until recently, most geologists believed that the gases billowing out of early Earth's volcanoes were responsible for generating the reducing atmosphere needed for RNA synthesis, but to the surprise of many, a 2011 study published in the journal *Nature* suggests that the volcanic outgassing process on the early Earth may have generated an oxidizing atmosphere (i.e., one with molecules such as CO_2, H_2O, and SO_2) [55] and not a reducing one. If true, the RNA hypothesis may be in jeopardy, and because of the seeming incongruity of this study and the "RNA first" hypothesis, one of the *Nature* paper's authors argues that his data are most consistent with the panspermia hypothesis [59].

Nevertheless, despite evidence suggesting that volcanoes may have generated an oxidizing atmosphere on the early Earth, the "RNA first" hypothesis is not necessarily defunct. The authors of *Nature* article acknowledge that the oxidation state of the planet immediately after it formed is unknown, and therefore, even if Earth's atmosphere was oxidized very early in its history, life still could have emerged before or during a period when the atmosphere was transitioning from a reductive state to more oxidized one. However, if this did indeed happen, the emergence of life on Earth may have been constrained to a relatively small window of time. Consequently, if these types of atmospheric changes were to occur on some other life-hosting planet, the first RNA-producing protocells would have to appear after the planet is cool enough for liquid water to accumulate on its surface, but before the planet's volcanic emissions have significantly oxidized its atmosphere.

TRYING TO ESTIMATE THE INESTIMABLE

In addition to needing an ideal energy state and the correct distribution of a litany of chemical elements, the emergence of life on Earth may have required the existence of many other specific conditions, including the presence of specific geophysical features such as volcanic basalt islands dotted with chemically rich, thermally graded, volcanic lakes. Ultimately, all these factors may have determined which steps in the emergence of life were rate-limiting and which were low probability events.

At this time, we do not know how many low probability events there were in the process that produced the first protocells, nor do we know how probable it was that a protocell would evolve into a true cell. It may be that life and bacterial-like cells were certain to emerge on our planet, given the conditions that prevailed four billion years ago, but then again, it is also possible that, even under ideal conditions, the emergence of life on Earth was extremely unlikely. And as I have suggested earlier, it is even possible that life never did emerge on Earth, and the reason we are here is that bacterial life was transported to our planet on an asteroid that by chance was caught in Earth's gravitational field. In other words, it is possible that the emergence of life on Earth, or the colonization of Earth by life forms from elsewhere, was not a biochemical and biophysical certainty. Instead, it may have been a matter of luck. Until we know more about the detailed conditions that gave rise to the first protocells and the first true cells, we cannot comment with any confidence about the probability of life arising on Earth or elsewhere in the universe.

THE ENTHUSIASTIC SEARCH FOR EXTRATERRESTRIAL LIFE

Despite our uncertainty about how life arises, many are optimistic that life has emerged many times throughout our galaxy. The reason for this optimism varies, but in some instances, it is encouraged by the observation that some prokaryotes can exist in incredibly hostile and challenging environments. However, the robust single-celled organisms that currently live in some of the most arduous places on the planet did not arise fully formed, and it is likely that the protocells that gave rise to these modern-day survivalists were not as durable or resilient as their distant descendants.

Others are optimistic that life exists elsewhere because they believe that life on Earth arose relatively quickly (i.e., within the first 260–1,000 million years after the planet formed), and to them, this suggests that life may routinely appear as soon as the conditions are right. This may indeed be the case, but then again, it may not be true. As I mentioned earlier, we are largely ignorant of the rate-limiting steps that occurred during the formation of life, and we don't know how probable it is for life to arise on a planet that is like that of the primordial Earth of four billion years ago. Furthermore, given that we only know of one planet with life on it, a rigorous statistical evaluation of the probability that life will emerge in any given world is not possible, and it will likely not be possible to generate this type of statistical analysis any time soon.

Of course, life could be fundamentally different depending on where it evolved, and we may one day find organisms whose biochemistry and genetics are radically different than ours. Such creatures may use dramatically different genetic templates, chemical building blocks, solvents, and energy sources than those employed by life forms on our planet. If so, life within the universe may be much more diverse than we imagine, and there may be many novel pathways that can give rise to various chemically and genetically distinct forms of life. If there are many fundamentally different forms of life, the odds that life exists elsewhere would significantly increase, but we currently have no reason to believe (i.e., we have no data to suggest) that such profoundly different life forms can be found within our cosmos.

As we begin to search Mars, the moons of Jupiter, and some of the other exotic worlds of our solar system and beyond, the hypothesis that life is common throughout the Milky Way will be tested. But let's move on and explore the question of how bacterial cells gave rise to structurally complex, multicellular creatures. Regardless of how life began, we know that on Earth, the journey from a bacterial cell to a multicellular mammalian primate like us was complex. However, by studying some of the events that occurred during this protracted trip, we can develop insights into the nature of life and the origin of our species. So, let's investigate the next chapter of life's journey, and let's examine a few of the many critical forces and events that shaped the formation of complex multicellular life.

REFERENCES

1. Christian, D., *Big history the Big Bang, life on Earth, and the rise of humanity*. 2008, Chantilly, VA: Teaching Co.
2. Christy A. Tremonti, et al., *The origin of the mass-metallicity relation: insights from 53,000 star-forming galaxies in the Sloan Digital Sky Survey*. Astrophysical Journal, 2014. **613**: p. 898–913.
3. Pasteur, L., *On spontaneous generation: an address delivered by Louis Pasteur at the 'Sorbonne Scientific Soiree' on April 7, 1864*. Revue des cours scientifiques, 1864: p. 257–264.
4. Mesler, B. and H.J. Cleeves, *A brief history of creation: science and the search for the origin of life*. First edition. ed. 2016, New York, NY: W.W. Norton & Company, Inc. xvii, 312 p.
5. Crick, F. and L. Orgel, *Directed panspermia*. Icarus, 1973. **19**: p. 341–346.
6. J. Mensoky and R. Moore, *The chase (Star Trek: The Next Generation)*. Wikipedia, 1993.
7. Hoyle, F. and N.C. Wickramasinghe, *The case for life as a cosmic phenomenon*. Nature, 1986. **322**: p. 509–511.
8. Hoyle, F. and N.C. Wickramasinghe, *Influenza-evidence against contagion: discussion paper*. Journal of the Royal Society of Medicine, 1990. **83**: p. 258–261.
9. Crenson, M., *After 10 years, few believe life on Mars*. USA TODAY, 2006(August 6).
10. Schmitt-Kopplin, P., et al., *High molecular diversity of extraterrestrial organic matter in Murchison meteorite revealed 40 years after its fall*. Proceedings of the National Academy of Sciences of the United States of America, 2010. **107**(7): p. 2763–2768.
11. R.M.E. Mastrapa, et al., *Survival of bacteria exposed to extreme acceleration: implications for panspermia*. Earth and Planetary Science Letters, 2001. **189**: p. 1–8.
12. Rothschild, Lynn J. and R.L. Mancinelli, *Life in extreme environments*. Nature, 2001. **409**: p. 1092–1101.
13. Editor, *Bacteria can survive space travel, ISS research shows*. Nature World News, 2014. www.natureworldnews.com/articles/6877/20140503/bacteria-survive-space-travel-iss-research-shows.htm.
14. Di Giulio, M., *Biological evidence against the panspermia theory*. Journal of Theoretical Biology, 2010. **266**(4): p. 569–572.
15. Gould, S.J., *Fall in the house of Ussher*. Natural History, 1991: p. 12–21.
16. Dodd, M.S., et al., *Evidence for early life in Earth's oldest hydrothermal vent precipitates*. Nature, 2017. **543**(7643): p. 60–64.

17. Hazen, R.M., *The story of Earth: the first 4.5 billion years, from stardust to living planet.* 2012, New York, NY: Viking. 306 p.

18. Fischetti, M., *The circle of life.* Scientific American, 2016. **314**(3): p. 76–76.

19. Pearce, B.K.D., et al., *Origin of the RNA world: the fate of nucleobases in warm little ponds.* Proceedings of the National Academy of Sciences, 2017: **114**(43): p. 11327–11332.

20. Abramov, O. and S.J. Mojzsis, *Microbial habitability of the Hadean Earth during the late heavy bombardment.* Nature, 2009. **459**(7245): p. 419–422.

21. Sleep, N., et al., *Annihilation of ecosystems by large asteroid impacts on the early Earth.* Nature, 1989. **342**: p. 139–142.

22. Darwin, C., *Darwin correspondence project "letter no. 7471,"* 1871. www.darwinproject.ac.uk/DCP-LETT-7471.

23. Cornell, C.E., et al., *Prebiotic amino acids bind to and stabilize prebiotic fatty acid membranes.* Proceedings of the National Academy of Sciences of the United States of America, 2019. **116**(35): p. 17239–17244.

24. Blacka, Roy A., et al., *Nucleobases bind to and stabilize aggregates of a prebiotic amphiphile, providing a viable mechanism for the emergence of protocells.* Proceedings of the National Academy of Sciences, 2013. **110**(33): p. 13272–13276.

25. Black, R.A. and M.C. Blosser, *A self-assembled aggregate composed of a fatty acid membrane and the building blocks of biological polymers provides a first step in the emergence of protocells.* Life (Basel), 2016. **6**(3).

26. Baltimore, D., *Viral RNA-dependent DNA polymerase.* Nature, 1970. **226**: p. 1209–1211.

27. C. Guerrier-Takada, et al., *The RNA moiety of ribonuclease P is the catalytic subunit of the enzyme.* Cell, 1983. **35**: p. 849–857.

28. Zaug, Arthur and T. Cech, *The intervening sequence RNA of tetrahymena is an enzyme.* Science, 1986. **231**: p. 470–475.

29. Gilbert, W., *The RNA world.* Nature, 1986. **319**: p. 618.

30. Service, R., *Researchers may have solved origin-of-life conundrum.* Science, 2015. DOI: 10.1126/science.aab0325.

31. Bhavesh H. Patel, et al., *Common origins of RNA, protein and lipid precursors in a cyanosulfidic protometabolism.* Nature Chemistry, 2015. **7**: p. 301–307.

32. Sutherland, J., *Studies on the origin of life—the end of the beginning.* Nature Reviews. Chemistry, 2017. **1**: p. 1–7.

33. Szostak, J., *On the origin of life.* Medicina, 2016. **76**: p. 199–203.

34. Powner, M.W., B. Gerland, and J.D. Sutherland, *Synthesis of activated pyrimidine ribonucleotides in prebiotically plausible conditions.* Nature, 2009. **459**(7244): p. 239–242.

35. Sidney Becker, et al., *A high-yielding, strictly regioselective prebiotic purine nucleoside formation pathway.* Science, 2016. **352**(6287): p. 833–836.

36. Service, R., *'RNA world' inches closer to explaining origins of life.* Science, 2016. DOI: 10.1126/science.aaf5719.

37. Stairs, S., et al., *Divergent prebiotic synthesis of pyrimidine and 8-oxo-purine ribonucleotides.* Nature Communications, 2017. **8**: p. 15270.

38. Alonso Ricardo, J.W.S., *Life on Earth.* Scientific American, 2009. **301**(3): p. 54–61.

39. Ferris J.P., Ertem G., and A.V., *Mineral catalysis of the formation of dimers of 5'-AMP in aqueous solution: the possible role of montmorillonite clays in the prebiotic synthesis of RNA.* Origins of Life and Evolution of Biospheres, 1989. **19**(2): p. 165–178.

40. Pierre-Alain Monnard, K. Anastassia, and D.W. Deamer, *Eutectic phase polymerization of activated ribonucleotide mixtures yields quasi-equimolar incorporation of purine and pyrimidine nucleobases.* Journal of the American Chemical Society, 2017. **203**(125): p. 13734–13740.

41. David Zwicker, et al., *Growth and division of active droplets provides a model for protocells.* Nature Physics, 2017. **13**: p. 408–414.

42. Golestanian, R., *Division for multiplication*. Nature Physics, 2017. **13**: p. 323–324.

43. Mansy, S.S. and J.W. Szostak, *Thermostability of model protocell membranes*. Proceedings of the National Academy of Sciences of the United States of America, 2008. **105**(36): p. 13351–13355.

44. Szostak, J.W., *The eightfold path to non-enzymatic RNA replication*. Journal of Systems Chemistry, 2012. **3**(1): p. 2.

45. Li, L., et al., *Enhanced nonenzymatic RNA copying with 2-aminoimidazole activated nucleotides*. Journal of the American Chemical Society, 2017. **139**(5): p. 1810–1813.

46. Adamala, Katarzyna and J.W. Szostak, *Nonenzymatic template-directed RNA synthesis inside model protocells*. Science, 2013. **342**: p. 1098.

47. Attwater, J., A. Wochner, and P. Holliger, *In-ice evolution of RNA polymerase ribozyme activity*. Nature Chemistry, 2013. **5**: p. 1011–1018.

48. Higgs, P.G. and N. Lehman, *The RNA world: molecular cooperation at the origins of life*. Nature Reviews Genetics, 2015. **16**(1): p. 7–17.

49. Dawkins, R., *The ancestor's tale: a pilgrimage to the dawn of evolution*. 2004, Boston, MA: Houghton Mifflin.

50. Maynard Smith, J. and E. Szathmry, *The origins of life: from the birth of life to the origin of language*. 1999, Oxford: Oxford University Press.

51. Daley, J., Behold LUCA, *The last universal common ancestor of life on Earth*. Smithsonian Smart News, 2016.

52. Theobald, D.L., *A formal test of the theory of universal common ancestry*. Nature, 2010. **465**(7295): p. 219–222.

53. Weiss, M.C., et al., *The physiology and habitat of the last universal common ancestor*. Nature Microbiology, 2016. **1**(9): p. 16116.

54. Wade, N., Meet Luca, *The ancestor of all living things*. New York Times, 2016.

55. Trail, D., E.B. Watson, and N.D. Tailby, *The oxidation state of Hadean magmas and implications for early Earth's atmosphere*. Nature, 2011. **480**(7375): p. 79–82.

56. Van Kranendonk, M., D. Deamer, and T. Djokic, *Life springs*. Scientific American, 2017. **317**(1): p. 2835.

57. Webb, J.K., et al., *Indications of a spatial variation of the fine structure constant*. Physical Review Letters, 2011. **107**(19): p. 191101.

58. Webb, J., *Is life on Earth due to a quirk in the laws of physics?* The Conversation, 2011.

59. Editors, *Setting the stage for life—scientists make key discovery about the atmosphere of early Earth*. Rensselaer University Bulletin, 2011.

7 Evolution
The Generator of Life's Diversity

Nothing in biology makes sense except in the light of evolution.

Theodosius Dobzhansky

Our world is not an optimal place, fine-tuned by omnipotent forces of selection. It is a quirky mass of imperfections, working well enough (often admirably); a jury-rigged set of adaptations built of curious parts made available by past histories in different contexts.

Stephen J. Gould

Life can only be understood backwards; but it must be lived forwards.

Soren Kierkegaard

The journey from prokaryotes to humans covers a span of almost four billion years, and during that time, life adapted, diverged, converged, grew in complexity, and radiated across the globe. Since its formation, it has been estimated that our planet has hosted trillions of species [1]. Most of these creatures were single-celled organisms, but others were large and complex—complete with organ systems composed of billions of cells with trillions of connections. When viewed as a whole, the menagerie of life on our planet is overwhelming, and pondering this diversity gives rise to two deceptively simple and important questions, namely:

- How did all these various life forms arise?
- Was it inevitable that creatures like us would come to be?

In the first section of this chapter, we will explore answers to the first question. Then, at the end of this chapter, and in those that follow, we will examine whether our appearance was inevitable.

THE GENESIS OF EARTH'S VARIED LIFE FORMS

How did all of life's diversity arise? It is a straightforward question, but many different answers have been posited. One common belief is that God created each species as it currently appears, and in so doing, He created a hierarchy of life that was rigid, unchanging, and enduring. Some argued that these life forms composed part of a "Great Chain of Being," which encompassed everything from inanimate objects to angelic and satanic supernatural entities (see Figure 7.1).

DOI: 10.1201/9781003270294-10

FIGURE 7.1   Didacus Valades' 1579 depiction of the "Great Chain of Being." Valades' illustration, which was constructed at the dawn of modern science, depicts connections among living creatures and supernatural entities. Approximately 300 years after this image was constructed, the thinking of Charles Darwin upended many of the beliefs depicted here. (From: Didacus Valades (Diego Valades), Public domain, via Wikimedia Commons.)

The concept of the "Great Chain of Being" is based on the thinking of the Greek philosophers Aristotle, Plato, and Plotinus. Some modern-day creationists continue to hold many of the beliefs, but there is no scientific evidence to support the idea that all beings were created at one time and in one place, nor is there evidence that species are forever unchanging and enduring [2–4]. Indeed, the fluidity of morphological, physiological, and heritable change within and between species has been well documented, and there is compelling support for the view that countless numbers of species that once roamed Earth are now extinct. In some instances, species were around for millions of years before they vanished, and in other cases, a species died off after a brief and ephemeral period.

EVOLUTION—THE MIRACLE OF LIFE

If species were not created in one time and in one place, then how did they come to be? This question puzzled many natural philosophers during the early part of the 19th century. At this time, geologists were just starting to actively identify and characterize evidence of long-extinct creatures, and the study of the fossils generated by these organisms was quickly becoming a specialized discipline. By 1822, the term paleontology (which means the study of prehistoric beings) had entered the lexicon, and by then, more than a few individuals were trying to explain how some of the species enshrined as fossils arose. In many cases, these same individuals were struggling to explain why some allegedly unchanging and enduring fossilized organisms were apparently no longer in existence, while other species, which are alive now, have no trace in the ancient fossil record.

At around this time, European naturalists also documented the existence of similar species on different continents, and this led some to ponder how these seemingly related organisms ended up so far apart and so distantly removed from the putative center of creation. To add to the mysteries of this age, comparative morphologists documented the existence of apparently functionless appendages such as vestigial leg bones in pythons and tailbones in humans. Given that all creatures within the "Great Chain of Being" were presumably created in a perfect state, the presence of these structures was puzzling.

Presented with this evidence, some naturalists began to question the tenets of the "Great Chain of Being." Among them was the prominent French Academy of Science member, Jean-Baptiste Lamarck. Lamarck suggested that organisms changed because they had a drive toward perfection, and as a result, they induced heritable morphological and physiological alterations through their own efforts. Still other notable biologists of the time, such as Georges Cuvier, argued that the "Great Chain of Being" was not stable, and he proposed that some species did indeed go extinct. Cuvier also left open the possibility that some organisms may have come into existence relatively recently [5–7].

In the midst of this intellectual tumult, the English naturalist, Charles Darwin, appeared on the world stage. As a young man, Darwin had set out to circumnavigate the world as part of the crew of the exploratory ship, *The Beagle*. During the *Beagle's* nearly five-year journey, Darwin pondered many of the issues that had engrossed his contemporaries, and in 1858 and 1859, he published his own answer to the question of how life's diversity arose. As we will see, in many ways, Darwin's thinking on this topic was straightforward, extraordinarily visionary, and remarkably accurate [8, 9], but despite this, his theory was vigorously opposed by many. Today, more than 98% of biologists and geologists believe that the evidence gathered since 1858 strongly validates Darwin's ideas, and yet more than a third of American adults reject evolutionary theory in its entirety [10]. So, what were Darwin's ideas, and why do modern scientists overwhelmingly accept them? And what are the implications of Darwin's thoughts for our understanding of how we came to be?

DARWIN IN A NUTSHELL

In essence, Darwin believed that the diversity of life could be explained by five basic hypotheses:

1. All organisms descend from a common ancestor.
2. Offspring differ from their parents, which implies that during reproduction, there is "descent with modification."
3. In any given environment, some offspring are better able to survive and reproduce than others, and consequently, there is a form of natural selection that operates in the living world.
4. Due primarily (although not exclusively) to natural selection, over long periods of time, the genetic composition of a population changes, and
5. As a result of these genetic changes in populations, new species gradually arise.

It should be noted that Darwin did not actually express his thinking as a set of five separate but related hypotheses, but his ideas can be viewed in this way.

Interestingly, many of these hypotheses did not actually originate with Darwin. For example, long before Darwin, others had noted that parents differ from their offspring. In addition, in the 18th century, several individuals, including Darwin's grandfather, Erasmus Darwin, suggested that populations change over time [11]. However, what distinguished young Charles' thinking from that of his grandfather and his many intellectual predecessors was his belief that nature preferentially selected the offspring that were best able to survive and reproduce.

From a 21st-century perspective, it is easy to present a strong case for the rationality of Darwin's ideas. To start with, it makes sense to suggest that life may have started with a particular organism. This doesn't have to be the case, but it is quite reasonable to postulate that there may have been "a first living creature" or a "universal common ancestor." It also seems eminently reasonable to speculate that offspring differ from their parents. You only need to look at yourself and your own parents to know this second assumption is true. Furthermore, the idea that some individuals may be better at surviving and reproducing than others is also obvious. In my own family, I had a brother who was afflicted with paranoid schizophrenia, which is a particularly vile genetic illness. It didn't take long for me to realize that my brother would have a more difficult time surviving and reproducing than I or my other two, non-afflicted brothers. Nature had dealt my stricken brother a nasty hand,

and without proper treatment, he was at an immense disadvantage. Unfortunately, my sick brother did not have medical insurance, and it was therefore hard for him to find competent medical care. Many of the hospitals that "treated" him were focused primarily on getting him out the door as soon as possible, and sadly, due to a series of rushed and inaccurate diagnoses, or no diagnosis at all, he was forced to endure more than 25 years of horrifying voices and delusions. During this period, he learned to sedate himself with alcohol, but this drug worked synergistically with his disease, and it further eroded his ability to reason and function. Ultimately, the effects of schizophrenia and alcoholism made him a particularly easy mark for the villainous and vicious people with whom we share this planet, and for decades, he struggled to just stay alive.

After finally getting an extensive evaluation by a group of caring and competent psychiatrists affiliated with Brown University, my brother received the medical care he needed, but by then, significant damage had been done. Ultimately, he never married or had children, and he died at the age of 44. Nature and my brother's environment limited his ability to survive and reproduce, and he lived the most difficult and unhappy life I have personally witnessed. So, the idea that we exist in a world with natural selection is not hard for me to comprehend.

Sadly, the types of inequalities that nature can impart upon its living creatures are varied and enormous in number, and so too are the ways in which nature can select one organism over another. As a result, more than a few of us know from personal experience that genes are not distributed fairly, and we are aware that our ability to survive and reproduce is influenced by many factors that are completely outside our control. For better or worse, "Nature" has much to say about our fate, and it often has the last word about who will win the contest of survival and reproduction.

CHANGING POPULATIONS

Given the reality, harsh though it may sometimes be, of natural selection, Darwin reasoned that, if members of a population were to end up separated in different environments, the heritable material of the individuals in each subgroup would slowly diverge. He further speculated that, over time, variations exhibited by sexually reproducing organisms would accumulate, and eventually, the day would come when members of the different subgroups could not, or would not, mate and reproduce if they were reunited. At that point, Darwin argued that they would have become separate species. Furthermore, if their environments remained different, Darwin reasoned that they would likely continue to diverge and become increasingly different over time. Consequently, Darwin concluded that the modification and proliferation of species occurs within the natural world because of descent with modification, natural selection, and the slow accumulation of heritable changes within a population.

Over most of the last two centuries, legions of scientists have investigated the veracity of Darwin's claims, and more than a few would have liked to have established their academic reputations by creating an intellectual revolution that overturned Darwin's ideas. But alas, to date, no one has succeeded in negating Darwin's ideas, and after prolonged and intense scrutiny, the fundamental tenets of Darwin's theory have held up remarkably well.

While challenges to the essence of Darwin's theory have been unsuccessful, the extensive examination of his ideas was nevertheless fruitful, and we now have a much more detailed and nuanced understanding of the processes he described. As an example, one can review the evidence suggesting that all living creatures share a common ancestor, which we discussed in chapter six. During Darwin's time, the idea that there was a common ancestor was simply speculation, but today, this concept rests on a much stronger foundation of empirical and statistical data.

Studies of Darwin's ideas, and the independent work of geneticists such as Gregor Mendel, have also provided a clearer and more accurate view of how descent with modification occurs. When Darwin was doing his work, it was obvious that offspring did not always resemble their parents, but it was not at all apparent how the variation that exists between parents and offspring was generated. To generate an explanation, Darwin suggested that an organism's traits were transmitted from various parts of the body to the gametes (i.e., sperm and eggs) via hypothetical particles, which he called "gemmules." He further speculated that the gemmule-sculpted gametes of the parents blended to give rise to offspring. In short, using what is now called the "panagenesis hypothesis," Darwin argued that it was the blending of characteristics that produced the variation necessary for descent with modification [12].

It didn't take long before Darwin started to question his panagenesis hypothesis, however, and upon reflection, he realized that there were many examples in which his views on how variation arose within a population did not comport with the available data. In one case, Darwin noted that the circumcision of males within the Jewish community did not result in future generations of males being born without foreskin, and this, along with other observations, led him to conclude that his hypothesis that variation was attained by the blending of acquired traits was likely flawed. Furthermore, Fleeming Jenkin, a contemporary of Darwin, and an engineering professor at the University of Edinburgh, amplified Darwin's doubts when he asserted that an atypical trait that is blended with more common ones generally disappears over time, and consequently, the population tends to revert toward the mean [12, 13].

Darwin's inability to explain how variation arose undermined his concept of natural selection because natural selection depends on there being variation to select from. Hence, although it was clear to Darwin that variation arose and descent with modification occurred, he could not comprehend how or why these processes took place.

MENDEL TO THE RESCUE?

While Darwin was struggling to explain the mechanistic basis of his set of hypotheses, Gregor Mendel, the other great biologist of the 19th century, was busily opening new vistas into the nature of heredity. Working in what is now Brno in the Czech Republic, Mendel was living out his vocation as a Roman Catholic Augustinian monk. Trained as a theologian and mathematician, Mendel was also curious about the natural world, and like Darwin, he sought to understand how heritable variation arose within populations. Initially, he wanted to answer this question by studying animal crosses, but when his supervising bishop learned that he was conducting what he perceived to be questionable mating experiments, he ordered Mendel to stop his work. Mendel initially persevered, but eventually, under duress, he stopped his animal studies, and he switched to analyzing genetic variation and inheritance in pea plants [14]. In response to the bishop's actions, Mendel wrote, "I turned from animal breeding to plant breeding. You see the bishop did not understand that plants also have sex."

Mendel's deft tactical maneuvering proved that a little creativity and ingenuity could go a long way in the face of ideological opposition. But Mendel soon realized that the bishop's disapproval might have been propitious since the pea plant ended up being a very good source of answers for questions regarding heredity in sexually reproducing organisms. Using data he collected from his seven-year study of pea plant hybridizations, Mendel published a seminal paper in 1868, which demonstrated that heritable traits were not blended as they passed from one generation to another. Instead, Mendel found that genetic information maintained its integrity, and it could endure within a population, even when not expressed for generations. Darwin's and Mendel's publications constitute the two greatest documents produced by 19th-century biologists, and today, we understand that these documents complement each other, and together, they help us understand the processes that generate life's diversity. Unfortunately, when these papers were first published, this truth was not at all apparent, and up until the end of the 19th century, Mendel's paper was apparently cited only three times.

While the greater scientific community did not initially understand the significance of his work, Mendel did seem to have realized that it provided insight into the processes of descent with modification and natural selection. Soon after his paper was published he wrote, "This seems to be the one correct way of finally reaching a solution to a question whose significance for the evolutionary history of organic forms cannot be underestimated" [15], and in an attempt to convince others of the value of his findings, he paid a handsome price to distribute 40 reprints of his 44-page paper. It is likely that one target of Mendel's advertising campaign was Charles Darwin. However, if Darwin did receive Mendel's manuscript, he likely did not read it carefully, if at all, and he never realized how Mendel's findings could help provide the mechanistic basis for his descent with modification hypothesis.

Ultimately, it is unfortunate that Darwin did not appreciate the significance of Mendel's discoveries, and it is also unfortunate that Mendel was unable to draw significant attention to the implications of his findings. Had the situation been different, progress in evolutionary biology and genetics would have been greatly accelerated. But as it turned out, the unification of Darwin's and Mendel's conceptual insights would have to wait until decades after both men passed away.

AND WHERE'S THE EVIDENCE FOR NATURAL SELECTION?

In addition to not being able to explain how descent with modification occurred, Darwin was only able to produce anecdotal evidence for his natural selection hypothesis, and he was never able to collect compelling empirical data in support of the theoretical cornerstone within his set of hypotheses. Furthermore, even if one did accept Darwin's contention that natural selection influenced traits such as the size of a bird's beak, it was not at all obvious that the accumulation of small adaptations could lead to speciation. On this topic Fleeming Jenkin, Darwin's contemporary, once again rendered forceful criticism when he wrote in 1868, "the mind cannot conceive a multiplier vast enough to convert this trifling change by accumulation into differences commensurate with those between a butterfly and an elephant" [13]. Decades later, Thomas Morgan, one of the 20th-century's greatest geneticists, was among the many prominent scientists who were still arguing against the idea that new species could arise because of descent with modification and natural selection, and as the new century approached, it was obvious that Darwin's seminal contributions needed more evidentiary support before the scientific community would fully accept his set of ideas.

Among the first to provide some of the much-needed experimental data was a young biologist by the name of Hermon Bumpus. On February 1, 1898, the 36-year-old professor was walking up College Hill in Providence, Rhode Island, on his way to his laboratory at Brown University, when he spotted a group of 136 grounded and enfeebled English swallows. The birds had been viciously battered by a brutal New England winter storm, and they were fighting to survive. Seeing an opportunity to learn more about natural selection, Bumpus gathered up the distraught animals, and he carried them back to his lab to see which would live. In total, 72 of the birds persevered, and after carefully measuring a number of morphological features of the survivors, he concluded that nature had selected those with intermediate characteristics, and it eliminated the "throng of degenerates" whose features demonstrated a "disregard for structural qualifications" [16]. Since Bumpus' documentation of natural selection on the streets of Providence, Rhode Island scores of others have experimentally confirmed that natural selection occurs both in the laboratory and in the wild [17, 18]. Consequently, over the last 125 years, the biologists' verdict on the veracity of Darwin's natural selection hypothesis morphed from incredulity to near-universal acceptance.

Soon after Bumpus had completed his studies, the aforementioned Darwinian and Mendelian critic, Thomas Morgan was hard at work trying to understand if mutated genes behaved like Mendel's pea genes. By exposing the fruit fly *Drosophila melanogaster* to various chemicals and radiation, Morgan and his students were able to generate heritable alterations that were transmitted from parent to offspring in a manner consistent with hypotheses Mendel generated. This suggested that gene variants could arise because of environmental insults, and it also indicated that mutated fly genes, like Mendel's pea genes, maintained their integrity as they passed from parent to offspring. Ultimately, these studies revealed that:

1. Mutations can contribute to the heritable variation needed for descent with modification, and
2. Because mutated genes are not blended, and because they maintain their integrity, it is possible for nature to select individuals with variations that facilitate survival and reproduction.

In the end, Morgan's findings lent further credence to the ideas generated by Darwin and Mendel, and to his credit, Morgan's open-mindedness allowed him to ultimately morph into a supporter of his great scientific predecessors.

TIME IS OF THE ESSENCE

Despite the progress mentioned earlier, many still questioned how small changes in a group of organisms (i.e., microevolution) could lead to large-scale alternations (i.e., macroevolution) and speciation. For some, the transition from microevolution to macroevolution seemed to require something more than what Darwin proposed, perhaps something akin to divine intervention. However, an analogy used by Richard Dawkins in his text *"The Ancestor's Tale"* helps show how small modifications, such as those that occur during childhood, can compound, and ultimately produce large-scale alterations over time. In the case of a child, the continuous changes occurring over an 18-year period can convert a non-verbal, immobile, small, and helpless baby into a fully functioning adult. At any given moment during the short 18-year interval between birth and adulthood, the changes occurring may appear trivial, but collectively, they make a significant difference in the capabilities of the individual. Dawkins argues that in the case of Darwinian evolution, the macro changes that lead to speciation and the eventual radical divergence of one species from another can clearly be viewed as emerging from microevolutionary changes that occur over extended periods [19].

The evidence suggests that Darwin, and those who supported his incipient theory, intuitively understood the argument recently put forth by Dawkins. However, Darwin argued that the process of acquiring microevolutionary changes was very slow, and given the estimated age of Earth in the 1800s, many wondered if life had been in existence long enough to generate all its diversity.

As was mentioned in the last chapter, we now know that Earth is 4.54 billion years old, which makes the planet at least 81 times older than Darwin's contemporaries thought it was. Furthermore, current evidence indicates that life first appeared on Earth 3.5–4.3 billion years ago, and consequently, we know that life did indeed have a substantial amount of time to change. But one could still ask, "just how much can happen in four billion years," and "is four billion years long enough to generate the diversity of life that we see around us"?

WRESTLING WITH THE CONCEPT OF GEOLOGICAL TIME

Commenting on the vastness of four billion years is easy to do; however, it is much harder to truly grasp how long this period is. We have difficulty comprehending geological time because our brains did not evolve to do so, and therefore, we struggle to understand the magnitude and significance of it.

The fact that large numbers are difficult to comprehend was made particularly clear to me when I attended the "Women's March" in Boston, Massachusetts, in January of 2017. The gathering took place on a cold winter morning, and after a few hours, I found myself ensconced in an extraordinarily large crowd. Looking around at the sea of humanity, I guessed there were over one million individuals present, but my estimate wasn't even close: There were "only" about 175,000 souls on Boston Common that day. This experience helped me appreciate the fact that I do not have a good sense of what a million is, never mind what a billion is. Therefore, to begin to appreciate geological time, I find it helpful to reflect on variations of time analogies that are used by paleontologists and geologists. Paleontologists and geologists routinely contemplate vast periods, and consequently, unlike most of us, they must work hard to get a sense of extremely large scales of time.

One of these time analogies presents the evolution of life in the form of a trip across a landscape. In this thought experiment, one step (which is about 2.5 feet) is equivalent to 100 years. After moving just one pace on this journey, one would travel back to the days before cell phones, televisions, the Internet, and antibiotics. Take another two steps, and one would discover a world where landline telephones, radios, electrical grids, and the United States of America did not exist. After a mere 20 paces, one could encounter Jesus of Nazareth, and after walking the length of a football field you would be in a world with no cities, wheels, writing, or agriculture. In fact, when you reached the end of the football field, you would be 12,000 years in the past, and at that point, all your ancestors would be living as hunters and gathers.

So, in this analogy, all recorded human history, and much more, could be covered in 100 yards. However, to get to the time when prokaryotes first appeared, a stroll across a football field or two would not do. To get to that moment, you would have to walk the length of a marathon every day, seven days a week, for more than 2.5 years! Even if one could move through time at this incredible pace, getting to the moment when life first appeared would require a truly Herculean effort.

In the second analogy, deep time can be envisioned by letting the length of time that prokaryotes have existed on the planet equal the height of the Liberty Tower in New York City. In comparison, the length of time since anatomically modern humans first appeared would be equal in height to a pile of about 26 quarters, and the length of time behaviorally modern humans (i.e., humans whose brains likely functioned like ours) have been on Earth would only be equal in height to a stack of about four quarters.

When I last visited the Liberty Tower, I placed four quarters on the ground in front of me, and then I stared up at the full height of the building. I found this exercise to be an awesome display of just how briefly we behaviorally modern humans have shared the planet with the prokaryotes. To extend the analogy, I pondered just how long I had to enjoy my own "moment in the Sun," and I calculated that, given an average life expectancy, the length of my life in this analogy would be equal to the height of about 0.0068 quarters, which is approximately 0.00048 inches. In other words, in comparison to the length of the time prokaryotic life has been on Earth, a human lifespan is barely perceptible, and it is as ephemeral as a camera flash. Some find this depressing, but I find contemplating this reality particularly comforting because it helps me put my worries in perspective.

SLOW AND STEADY WORKS

So, four billion years is truly a very long time, and during this period, even minuscule changes can accumulate to generate colossal differences. Those studying plate tectonics know this to be true because they revealed that the rearrangement of the continents was due to continuous, but barely perceptible, movements of the continental plates. Specifically, geologists have demonstrated that the continental plates move about 1–20 millimeters per year, and over a period of a year, or even a lifetime, these changes are not noticeable. But, over a period of hundreds of thousands of years, these languorous movements can cause continental plates to come together and break apart. Remarkably, by moving slower than your fingernails grow, the continental plates may have merged and formed a supercontinent, and then fractured and moved apart, as many as six times during the last three billion years [20, 21]!

At their current rate of movement, the Americas are predicted to join Africa and Eurasia in another 250 million years. I must admit that it is hard for me to imagine that these tiny alterations will cause my home state of Massachusetts to abut the continent of Africa in a quarter of a billion years, which is less than half the amount of time since the Cambrian period's explosive proliferation of multicellular life forms. So, if continents can move at an imperceptible pace across the thousands of miles in a quarter of a billion years, imagine how life might change during an interval that is 16 times as long.

To address the idea that there has not been enough time for evolution to have generated all life's variation, the University of Pennsylvania mathematicians Herbert Wilf and Warren Ewens took a less intuitive approach and they estimated how much time was needed using a model of evolution that assumes that favorable alleles are selected by nature in parallel. After working up their data and analyzing the implications of their model, they proclaimed the essence of their conclusions in the title of their *Proceedings of the National Academy* paper, namely, "There has been plenty of time for evolution" [22].

WAS DARWIN RIGHT ON ALL COUNTS?

As indicated earlier, recent empirical analysis has strongly supported Darwin's idea that "all organisms share a common ancestor" and it has also buttressed Darwin's descent with modification and natural selection hypotheses. However, Darwin's belief that evolutionary adaptation occurred gradually may be true in some situations, but not in others [17, 23, 24]. Recently, biologists have shown that the environment and a population's size heavily influence the rate of evolutionary adaptation in any given group of organisms within a species. Typically, species under environmental stress are subject to more robust forms of natural selection than those not experiencing stress, and the rate at which genetic adaptations spread tends to be higher in small populations [17, 23, 25]. Consequently, environmental stress and small group size tend to accelerate the evolutionary process. So, not all heritable change and adaptation within a population appears to occur as slowly as Darwin envisioned.

Related to the issue of adaptation is the process of new species formation. It was the process of speciation that ultimately generated the diversity of life that we see on the planet, and Darwin noted that this could occur because of anagenesis or cladogenesis.

Anagenesis (also known as phyletic evolution) results from the slow, gradual accumulation of heritable alterations within a population, and it can occur because of changing environmental conditions over time. As a result of this activity, a descendant population may eventually become so different from an ancestral group that the two can no longer be considered members of the same species. When this occurs, evolutionary biologists refer to these temporally separated groups as chronospecies. It is important to note, however, that while anagenesis increases the number of species that have appeared through time, it does not increase the number of taxa (i.e., groups of species) on the planet at any particular moment. Consequently, the multiplicity of species that we currently see, and thus the diversity of life that our planet enjoys, is primarily due to cladogenesis. During this process, a reproducing lineage splits into two subpopulations, both of which share a common ancestor. This form of speciation can be triggered by different events, but most often, it occurs because of geological isolation.

As a result of geological isolation, populations adapt to their unique conditions, and over time, they will likely develop one or more reproductive isolation mechanisms. As a result, if the barrier that separates the two subgroups is removed, the individuals in one population may find that they cannot, or will not, mate and reproduce with the members of the other group. When this occurs, most biologists consider the two sub-populations to have speciated (i.e., the two have become separate species).

During cladogenesis, the mechanisms that reproductively isolate the sexually reproducing members of the two subgroups can differ substantially. In some cases, individuals in one population may eliminate individuals in the other group from their collection of potential mates, because members of each group find the behaviors or appearance of the members of the other group unacceptable. In this case, the groups are said to have become behaviorally isolated. Reproductive isolation can also be achieved when subpopulations mate at different times (i.e., when they are temporally isolated), or because gametes from one group cannot fertilize those of the other (i.e., when the groups become gametically isolated). Furthermore, reproductive isolation may also occur when members of the two subgroups try to copulate but fail to do so because their reproductive structures are physically incompatible (i.e., the groups have become mechanically isolated), and it can occur when two groups are no longer able to coexist in the same environment (i.e., the groups have become ecologically isolated). While these processes differ in how they impose segregation, they all effectively achieve the same end, because they all prevent the interacting members of each subgroup from producing a zygote (i.e., a fertilized egg).

In addition to preventing the formation of a fertilized egg, the process of speciation can also occur when two formerly separated populations form fertilized eggs that fail to develop (i.e., when the mating between members of the different groups produce an organism that is spontaneously aborted *in utero*), and speciation can be achieved when two populations produce a viable offspring, but that offspring is sterile. An example of this latter scenario occurs when a horse and a donkey produce a sterile mule.

Given the many possible processes that can lead to speciation, it is not hard to envision how organisms may change in ways that prevent them from sharing their genes. These speciation mechanisms are useful from the perspective of each population because they help each maintain the harmonious functioning of the set of genes that it acquired over long periods of adaptation [26]. In the absence of speciation, teams of genes that work well together would be disrupted, and the resulting hybrid organisms would be less fit, or to put it in other words, they would be less likely to survive and reproduce. Ultimately, speciation, at least in sexually reproducing organisms, protects the integrity of a set of genes that produce reproductively successful organisms in a particular environment.

DARWIN'S INSIGHTS ON THE NATURE OF SPECIATION

Of the two types of species-producing processes, cladogenesis occurs more often, and as was mentioned earlier, it produces the biological diversity that we see. However, the evolutionary biologists Stephen J. Gould and Ernst Mayr indicate that while Darwin was aware that both forms of speciation occurred, he often "muddled" the two, and he usually focused his discussions primarily on anagenesis [27, 28].

In addition to being vague about the dynamics of speciation, Darwin was also unsure of what a species actually is. In the 1830s he asserted, "my definition of species . . . is simply, an instinctive impulse to keep separate," and he went on to say that "the dislike of two species for each other is evidently an instinct—and this prevents (inter) breeding [28]." Based on these statements, Mayr argues that, early in his career, Darwin was essentially employing what is now called the biological species concept (BSC), and as a result, he recognized species as true biological entities that maintain their integrity by forming reproductively isolated groups. However, Mayr also argues that, later in his life, Darwin identified a species using a nominalist typological perspective, and this is borne out by documentation in which Darwin states that species are "merely artificial combinations made for convenience."

As a result of this confusion in his own thinking, Darwin struggled to define a species for many years, and ultimately, he concluded that "no definition has satisfied all naturalists; yet every naturalist knows vaguely what he means when he speaks of a species." In many ways, Darwin's comments about species foreshadowed Potter Stewart's attempts to define "obscenity" while he was serving as a Supreme Court Justice in 1964. Unable to adequately do so, Stewart made a retort that essentially encapsulates Darwin's views about a species, namely, "I know it when I see it" [29].

Given Darwin's confusion, it is worth pondering why he elected to entitle his paradigm-shifting book, "*On the Origin of Species*." By making this choice, Darwin emphasized that it was the proliferation of species that generated life's diversity, but his nomenclature problem placed him in the position of trying to explain the origin of entities he could not explicitly define.

Surprisingly, 160 years later, the scientific definition of a "species" still remains nebulous. Currently, most biologists do as I did earlier in the text, and they employ the BSC when defining the term, "species." According to this idea, species are populations of reproductively isolated, sexually reproducing organisms with harmonious genotypes. However, the BSC definition of a species has its limitations, including the inability to distinguish species within populations of asexually reproducing organisms. Given that asexually reproducing prokaryotes make up the bulk of life on the planet, this is a significant issue, and in an attempt to resolve this and other concerns, modern scientists have proposed more than two-dozen definitions for the word "species" [30]. So, even today, the scientific community cannot agree as to what exactly can emerge when descent with modification and natural selection occurs within populations.

REFLECTING UPON OUR UNDERSTANDING OF HOW LIFE'S DIVERSITY AROSE

For most of the two million years humans have been in existence, humankind did not have an accurate, empirically based understanding of how life's diversity arose. However, most of Darwin's perceptions and intuitions have now been vindicated by empirical evidence from numerous disciplines including molecular biology, paleontology, anatomy, biochemistry, developmental biology, and geology, to name just a few, and thanks to Darwin's insights, and the work of many scientists who came after him, we can now generate a detailed description of how life evolved. As a result, we can now make nuanced comparisons of the evolutionary relationships between populations of organisms (see Figure 7.2).

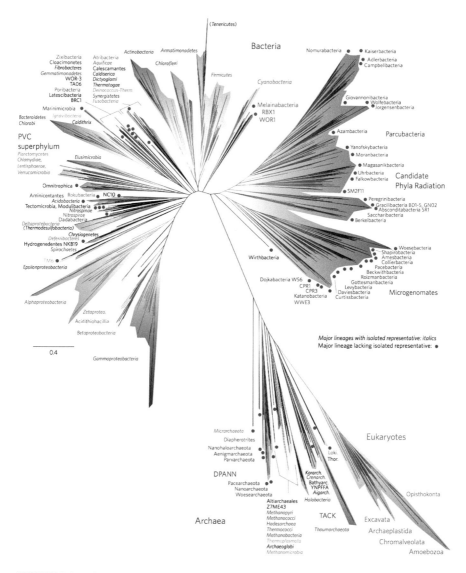

FIGURE 7.2 The phylogenetic tree of life. In this diagram, which is based on the molecular analysis of rRNA genes, the putative relationships between phyla within the three domains of life (i.e., Bacteria, Achaea, and Eukaryota) are depicted. We belong to the group of eukaryotes called "opisthokonta," which is located on the lower right-hand portion of the diagram. (From: This figure was generated in 2016 by Hug et al., [31], Laura A. Hug, Brett J. Baker, Karthik Anantharaman, Christopher T. Brown, Alexander J. Probst, Cindy J. Castelle, Cristina N. Butterfield, Alex W. Hernsdorf, Yuki Amano, Kotaro Ise, Yohey Suzuki, Natasha Dudek, David A. Relman, Kari M. Finstad, Ronald Amundson, Brian C. Thomas and Jillian F. Banfield, CC BY-SA 4.0 <https://creativecommons.org/licenses/by-sa/4.0>, via Wikimedia Commons.)

In the end, Darwin's insights allowed us to begin to appreciate how a simple pro-karyote could eventually give rise to a neurologically and behaviorally complex crea-ture that can ponder its own existence. So, ultimately, the ideas of Charles Darwin, a quiet, unassuming 19th-century biologist, helped us better understand who and what we are, as well as how we came to be.

THE MODERN HUMAN—DESTINED TO BE OR HERE BY CHANCE?

Now that we have briefly examined Darwin's thinking and some of the mechanisms that produced the variety of life that we see around us, it is time to explore whether our branch within the great tree of diversity was destined to be. In any attempt to answer this question, the first issue that must be addressed is whether there is physi-cal (or nomological) determinism. Those who argue for physical determinism believe that current events and actions are completely determined by past events. If this is true, and if the universe were to begin again in the way it did 13.8 billion years ago, we would be here today, for there simply could be no other possibility. Therefore, to know whether we were destined to be, it is necessary to understand whether the universe is physically deterministic.

One of the best-known advocates for physical determinism was the brilliant French mathematician and philosopher, Pierre-Simmon Laplace. In 1814, Laplace wrote,

> We ought . . . to regard the present state of the universe as the effect of its anterior state and as the cause of the one which is to follow. Given for one instant an intelligence which could comprehend all the forces by which nature is animated and the respec-tive situation of the beings who compose it—an intelligence sufficiently vast to submit these data to analysis—it would embrace in the same formula the movements of the greatest bodies of the universe and those of the lightest atom; for it, nothing would be uncertain and the future, as the past, would be present to its eyes [32].

Laplace was convinced that Newtonian physics made it possible for an entity with complete knowledge of the past, often referred to as Laplace's demon, to predict the future with certainty. However, during the early part of the 20th century, the scientific community witnessed the birth of quantum mechanics, and investigations carried out by particle physicists during this period led many to question the validity of Laplace's conclusion. One of the clearest threats to the Laplacian worldview came as the result of the work of Werner Heisenberg in 1927. In his study of subatomic particles, Heisenberg learned that there was a limit to the precision one could use in describing pairs of complementary physical characteristics. In the case of the loca-tion and velocity of an electron, Heisenberg found that the more he could say about one of these physical parameters at any moment, the less he could specify about the other [33]. Importantly, Heisenberg noted that these restraints were not due to the limitations of his measuring ability; rather, the uncertainty was an intrinsic property of the particle itself [34]. In short, Heisenberg revealed that the universe was a fun-damentally nondeterministic place.

But alas, subatomic particles are, by definition, very small entities, and while their properties may be nondeterministic, the question remains, does the behavior of these subatomic particles have any relevance to large biological systems like you and me? Some scientists and philosophers, such as Stephen Hawking and Daniel Dennett, have argued that they do not. They contend that subatomic uncertainties get averaged out in large systems, and consequently, structures such as cells and organisms are deterministic [35, 36]. This seems like a reasonable supposition, but the nondeterministic properties of some subatomic particles are in fact not averaged out. In 1935, Erwin Schrodinger suggested that this may be the case, and he postulated that it might even be possible to amplify the effects of a subatomic uncertainty, at least in principle. To explain how this would work, Schrodinger described his now-famous thought experiment in which he envisions a cat penned up in a hypothetical poisoning apparatus. In this contraption, the life of the animal depends on whether an atom of a radioactive substance decays while it is in the poisoning apparatus. If the cat is unlucky, the radioactive atom disintegrates, and it stimulates a Geiger counter, which in turn triggers the release of a hammer strategically positioned over a flask of gaseous poison [37]. This thought experiment clearly amplifies the effects of the nondeterministic decay of a radioactive atom, and the life of the cat hangs in the balance. However, Schrodinger's story describes an imaginary world, and as fun as it may be to ponder, the question remains, are there any real-life biological mechanisms that prevent the averaging out of the effects of nondeterministic events, and could these mechanisms affect the process of evolution?

EVIDENCE OF A BIOLOGICALLY NONDETERMINISTIC WORLD

The short answer to the aforementioned question is yes there are. To begin with, there is a class of spontaneous mutations called tautomerizations that arise as a result of the nondeterministic movement of electrons within a DNA molecule, [38–40]. The environment can affect the probability that a tautomeric mutation will occur, but the movement of the electrons and protons that trigger these mutations can transpire without cause [40, 41]. According to the philosopher of science Carl Popper, movements such as these are,

> elementary physical processes which are not further analyzable in terms of causal chains, but which consist of so-called "quantum jumps"; and a quantum jump is supposed to be an absolutely unpredictable event which is controlled by neither causal laws nor by the coincidence of causal laws, but by probabilistic laws alone [41].

Since a mutation changes the structure of a DNA molecule, the quantum movements that trigger tautomerizations are not averaged out, but rather, these quantum mechanical alterations become fixed within the DNA molecule in the form of a mutation, and ultimately this can affect the fate of the organism and its biosphere.

At first, the aforementioned claim may seem to be an overstatement because nucleotide tautomerizations alter a single nucleotide pair among the billions that may exist within an organism's genome. However, there are many ways in which

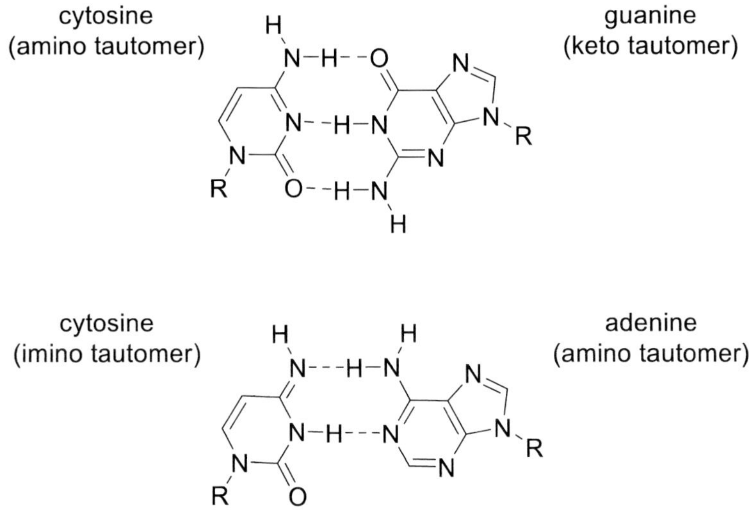

FIGURE 7.3 Cytosine tautomerization and subsequent DNA mutation. Several nucleotide tautomers exist. However, this figure focuses on the nitrogenous base found in the cytosine nucleotide tautomers. In the image, the DNA nitrogenous base, cytosine, is shown in the amino configuration in the top set of base pairs, and in the bottom set, cytosine's tautomeric imino form is displayed. When cytosine is in the amino configuration, it base pairs with guanine. However, due to the nondeterministic movements of subatomic particles, cytosine can occasionally assume an imino configuration, and in that state, it base pairs with adenine. When the cytosine is part of a stable, non-replicating DNA molecule, the transition of cytosine from th e amino to the imino configuration has little effect on the structure of the DNA molecule. However, during DNA replication, each strand of the original DNA molecule is used as a template to replicate a new DNA molecule (see Figure 7.4 for a depiction of the semi-conservative process of DNA replication), and if a cytosine in the template strand switches from the amino to imino configuration at the moment the cytosine containing nucleotide is being used as a template for the synthesis of the new DNA, the cytosine will base pair with adenine. If this occurs, the newly replicated strand of DNA will have acquired a mutation. If the mutation in the newly synthesized DNA strand is not repaired, the alteration will become fixed in the new DNA strand. Eventually, this mutated DNA strand will be used as a template for the synthesis of yet another strand of DNA, and as a result, the incorrectly incorporated adenine nucleotide will likely base pair with a thymine-containing nucleotide. When this happens, the DNA molecule is said to have undergone a "CG to AT transition." This image was constructed by Elizabeth Colby Davie.

a single nucleotide alteration can affect the biology of the organism, and within the human population, many pathologies are caused by single-nucleotide mutations. A few of the more common examples are some forms of sickle cell anemia [42], cystic fibrosis [43], spinal muscular atrophy [44], and hemophilia [45]. Interestingly, the mutated alleles that cause pathologies such as sickle cell anemia and cystic fibrosis can sometimes be helpful to an organism's quest for survival, if they are present in the heterozygous state (i.e., if the organism has one normal copy of the gene and one mutated form of this gene). So, not all single nucleotide mutations are detrimental to

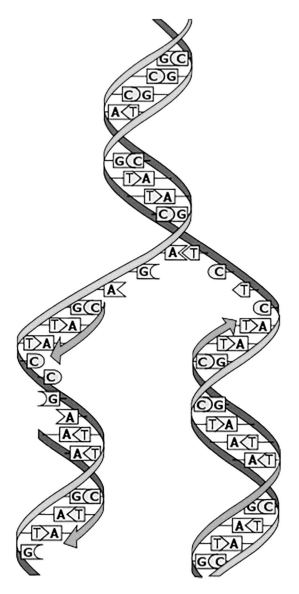

FIGURE 7.4 Semiconservative DNA replication. During the process of semiconservative DNA replication, DNA strands are separated and each strand of the original DNA molecule (shown in turquoise) is used as a template for the synthesis of a new strand of DNA (depicted in green). (From: I, Madprime, CC0, via Wikimedia Commons.)

an organism, and spontaneous tautomeric mutations have very likely produced genes that have helped organisms survive and reproduce.

If an altered gene is beneficial, it can spread throughout a population and it can contribute to the success of an entire group. In some cases, a genetic alteration may even help the affected organisms drive their competition to extinction, possibly by imparting upon the mutated organisms the ability to alter the environment in which they and their competitors live. In these circumstances, the nondeterministic movement of subatomic particles can alter an organism, the species that organism belongs to, one or more neighboring species, and the environment itself. So, a seemingly innocuous, nondeterministic mutation like the one described here has the potential to change large swaths of the biosphere [2, 46]. In short, within biological systems, a nondeterministic event does not always get averaged out, and it can be potentiated.

It is hard to know just how many nondeterministic subatomic processes have shaped the evolution of a species [47–49]; however, the fact that these mutations do occur is sufficient to refute the idea that physical determinism governs evolutionary processes. Consequently, mutagenic DNA tautomerizations make predicting the exact path of evolution impossible, even for Laplace's demon, and this reality eliminates the logical necessity of humans reappearing if the "tape of life" was to be replayed. Furthermore, nondeterministic mutations illuminate the fact that the subatomic properties of atoms, which were established at the moment of the Big Bang, can impact the evolution of life in a surprising and profound manner.

IS NATURAL SELECTION CONVERGENT, CONTINGENT, OR BOTH?

While it is not a logical certainty that humans would appear again if the evolution of life were reenacted, it is still possible that the developmental and physiological limitations of biological systems, and the constraints of natural selection, would ensure that organisms like us would eventually emerge. Therefore, to know if humans were indeed destined to be, we must explore the evidence for and against this possibility.

Arguing in favor of the idea that biological systems would eventually produce something like a human was the University of Cambridge paleontologist, Simon Conway Morris. Conway Morris stated that "the constraints of evolution and the ubiquity of convergence make the emergence of something like ourselves a near-inevitability," and he believed that "the contingencies of biological history will make no long-term difference to the outcome" of the evolutionary process [50]. Conway Morris supports his statements with scores of examples of biological convergence, and he makes a compelling argument that biological systems are limited in how they can solve specific problems.

On the other side of this ideological spectrum was the Harvard University paleontologist, Stephen J. Gould. According to Gould, the evolution of life has been greatly influenced by historical contingencies, and the ability of an organism to survive, particularly in a changing and stressful environment, is largely a matter of luck. Gould believed that the type of life forms we see on Earth today would have been very different had the random vagaries and vicissitudes of life unfurled differently over the long course of time. Like Conway Morris, Gould supported his arguments with a plethora of biological examples. In Gould's case, the data he presented demonstrate what he believed were historical accidents that locked species into evolutionary pathways. He contends that a species' future is ultimately determined by whether or not these "frozen accidents" happened to be beneficial in a world that changed in unpredictable ways [51].

In his recent text, the Harvard evolutionary biologist, Jonathan Losos, takes a position that is between those offered by Conway Morris and Gould, and he argues that convergence and contingency are both important phenomena that routinely operate during evolution. In his own studies of anole lizards, Losos tested the hypothesis that, if the tape of life were rewound, the lizards would evolve differently. Ideally, one would test this idea by going back in time and watching the ancestors of the lizard evolve a second time. But, since that is not a possibility, Losos did the next best thing—he placed his lizards on different, but similar, Caribbean islands, and he watched them evolve in real time.

What Losos found was that the lizards did indeed tend to evolve in the same way on similar islands, and he writes that, "there are limited ways to make a living in the natural world, so natural selection drives the evolution of the same features time and time again" (p. xiii) [17]. However, Losos also notes that, despite the high frequency of convergence, there are also many "one-offs" within the natural world. He defines these "one-offs" as species (or groups of species) whose way of life evolved only one time over the entire four billion years of evolution. Examples of "one-offs" can be found across the globe, and they include giraffes, elephants, penguins, chameleons, koalas, platypus, kangaroos, and humans [17]. Furthermore, Richard Dawkins, George McGavin, and Robert Blankenship point out that, in addition to species-specific ways of life, there are also many traits and processes that appear to have evolved only once in the long history of biological systems. Among these are the water-splitting and oxygen-generating, photosynthetic reaction center [52], the freely rotating wheel found in Rhizobium, and the silk diving bell produced by the spider, *Argyroneta aquatic*. Other processes that are "one-offs" include the Bombardier beetle's talent for generating controlled chemical explosions, the archer fish's ability to take down prey using a stream of water, and the modern human's ability to transmit information using syntactic language [19].

The existence of "one-offs" raises the question, "if the products of evolution are inevitable, why did so many of them only arise once"? Why, for example, did the humanoid way of life not appear during a previous geological era, and why did humans themselves arise only once, and only in Africa? Why didn't the lemurs that drifted from Africa to Madagascar 40 million years ago also give rise to humans, and why did the African monkeys that floated to South America 36 million years ago also fail to produce a humanoid population?

To explain these peculiarities, Losos submits that natural selection may not be as predictable as some suggest. He argues that, in some situations, evolutionary pathways do not converge because there are several equally good ways to adapt to an environmental challenge, and in other cases, convergent evolution is subverted because two similar organisms have subtle, but important, genetic differences that prevent them from responding in a similar way to a given environmental challenge.

To support this contingency hypothesis, Losos points to studies performed with *Escherichia coli*. Bacteria such as *E. coli* are easy to grow, they have a very short reproductive cycle, and since they have a diameter of only 0.00004 inches, they can produce a very dense population in a relatively small space. As a result, *E. coli* is an excellent system for the study of long-term evolutionary processes.

Among the first to utilize *E. coli* to examine evolution over an extended period was the University of Michigan biologist, Richard Lenski. Lenski isolated identical bacteria from the surface of a petri dish, and then he transferred them to 12 flasks containing nutrient-rich liquid media. *E. coli* can grow well in such a solution, but they pollute the media with waste products and eventually deplete it of many key nutrients, so to keep each line of cells alive, some of the *E. coli* must be transferred daily to a new flask with fresh media. With notable patience, dedication, and persistence, Lenski and his team kept the 12 lines of *E. coli* growing for decades, and during that time, they watched the organisms adapt to their environment. They found that, after thousands of generations, the liquid media-cultured bacteria in each of the 12 lines were able to divide faster and grow larger than their ancestors, all of which grew on the surface of agar plates. However, exactly how much faster and larger they grew varied from one strain of *E. coli* to another, suggesting that each line acquired a different set of mutations over time. Nevertheless, while they did vary in some ways, all 12 strains of *E. coli* generally had very similar evolutionary trajectories (i.e., they all grew faster and larger), and through DNA analysis, the Lenski team was able to demonstrate that all 12 lines of microbes acquired similar gene alterations. Up to this point, Lenski's long-term evolution experiments suggested that Conway Morris was correct and that organisms tend to evolve in similar ways when confronted with similar environmental challenges. But, on a Saturday in January 2003, the Lenski team made a discovery that shook up the status quo, and it revealed that *E. coli*'s evolutionary adaptations did not always result in convergence.

On that fateful winter day, Tim Cooper, a graduate student in the lab, was busily transferring bacteria from the polluted, nutrient-depleted flasks to flasks with fresh media when he noticed that the liquid in the old media, which contained generation 33,127 of the microbes, was much more turbid than usual. After some investigation, Cooper and the other members of the lab discovered that the increase in turbidity was triggered by the ability of the *E. coli* in this strain to make use of a typically unutilized carbon-based molecule present in the solution. Specifically, they found that the densely growing *E. coli* had developed the capacity to use citrate as an energy source, and consequently, this group of bacteria could divide more often in the liquid media than could the microbes in the other 11 strains. This was a shocking finding because *E. coli* are perhaps the best-studied organisms on Earth, and prior to Cooper's finding, no one had ever observed a strain that had independently acquired the ability to use citrate when growing aerobically. So, after more than 33,000 generations, what caused this one strain of *E. coli* to start using citrate in the presence of oxygen?

Fortunately, Lenski and his group were able to answer this question because they had been freezing samples of *E. coli* throughout the many years of the experiment, and by 2003, it was possible to thaw vials of bacteria and sequence their entire genomes. After looking at the DNA changes in the altered strain of *E. coli*, Lenski's group determined that a gene called *citT*, which encodes a citrate transport protein, had mutated. For unknown reasons, *citT* is normally expressed only when the bacteria are growing anaerobically, and therefore *E. coli* usually cannot use citrate if they are in aerated liquid media, but, unlike its sister strains, the altered strain could.

It took many years of additional study, but eventually, the Lenski group determined how the *citT* mutations imparted this ability onto the altered strain. As it turned out, the bacteria in the strain that could use citrate under aerobic growth conditions had duplicated *citT*, and the mutated bacteria had placed the new copy of *citT* in front of an *E. coli* DNA regulatory domain that is activated by oxygen. The odds of this happening are vanishingly small, but even this highly unlikely event was not sufficient to impart a significant growth advantage to the mutated aerated *E. coli*. To achieve that end, the aerobically activated version of *citT* had to be duplicated many times over. Only then did the altered *E. coli* acquire the ability to overwhelm its aerobically grown competitors.

In the years since Tim Cooper first identified the unusual *citT* mutants, the Lenski laboratory has continued to grow their 12 strains of bacteria, and recently, they marked the appearance of the 60,000th generation in each of the 12 lines. Yet, although mutations have continued to accumulate within the trillions upon trillions of *E. coli* that have come into existence during the last 27,000 generations, no other microbes acquired the ability to use citrate when growing aerobically. What the Lenski group did notice; however, was that the *citT* mutant strain continued to diverge from its sister strains, and today, this line of bacteria has lost the ability to efficiently use glucose. Due to this divergent evolution, Lenski argues that the *citT* mutants have been in the process of becoming a new species, and he suggests that at this point, it may be reasonable to declare that they are different enough from their relatives to assert that they are indeed a new species [53].

Similar types of very rare mutations no doubt occurred in many other species, but attempting to document their existence would be difficult, as it would take about 1,000 years to generate 60,000 generations of fruit flies, 10,000 years to produce that many generations of mice, and about 1.2 million years to examine 60,000 generations of humans [17]. However, in the grand scheme of geological time, a million years is a relatively brief unit of time. In fact, since there have been about 5,000 years of written history and approximately 5,000 million years of Earth history, geologists often like to think of one year of written history as the functional equivalent of one million years of geological time [54]. When viewed in this way, it is easy to see that, over the eons, mutations that fundamentally alter the biology of a species likely appear routinely.

So, what can we make of Lenski's work? When taken as a whole, his long-term evolutionary studies demonstrate that species do indeed tend to converge if they are closely related and if they are evolving in similar environments over relatively short periods. However, the likelihood of convergence decreases if the species are not closely related or if they are in different environments. In addition, the tendency toward convergent evolution also tends to decline over long periods of time.

CONFOUNDING LAPLACE'S DEMON

The work of Lenski and others supports the idea that nondeterministic mutations could fundamentally change a species' evolutionary trajectory. In addition, since species interact with each other and their environments, it is not hard to imagine how nondeterministic mutations may have contributed to the modification of the biosphere. As an example, consider the evolution of cyanobacteria. These oxygen-generating, photosynthetic prokaryotes first appeared about 2.7 billion years ago [55], and the photosynthetic reaction centers they possessed helped transform light energy into chemical energy. Some of the cyanobacteria's reaction centers broke apart water molecules to harvest the electrons they needed to carry out photosynthesis, and as they did, they generated oxygen as a waste product. This waste product, however, was no ordinary garbage, and as we will see in the next chapter, it eventually accumulated in the water and the air. Ultimately, the liberation of oxygen by bacteria fundamentally altered the globe on which we live, and it made it possible for animals like us to exist.

Over the span of a billion years, cyanobacteria and their ancestors likely acquired countless nondeterministic mutations, and some of these may have helped these organisms survive. It is also possible that this type of mutation contributed directly to the formation of the cyanobacteria's globe-altering photosynthetic capabilities. If this is indeed the case, even Laplace's demon could not have predicted with certainty that some seemingly innocuous species of ocean-bound microbe would eventually evolve the ability to fundamentally change Earth and the life that dwells on it.

Given this reality, it seems to me that Gould was correct, and if you were to rewind the tape of organic evolution and let it run again, life on Earth would likely be very different indeed. For example, we could find that, if the tape were rewound, the ancestors of the cyanobacteria would be driven to extinction by some other prokaryote, and we could also find that no other organism acquired the ability to transform light energy into chemical energy while simultaneously splitting water and liberating oxygen. In this scenario, our planet would be very different, but we wouldn't know it, because we wouldn't be here.

In the next chapter, we will continue to examine how conditions that prevailed over geological time shaped life. In particular, we will focus on the events that gave rise to the cells that form the three superkingdoms of life, namely, the Bacteria, the Archaea, and the Eukarya. As we will see, there is much to consider, and although we won't be able to analyze every important event that transpired, we will be able to examine some of the circumstances that shaped our world today.

REFERENCES

1. Locey, Kenneth and J. Lennon, *Scaling laws predict global microbial diversity.* Proceedings of the National Academy of Sciences, 2016. **113**(21): p. 5970–5975.
2. Miller, K.R., *Finding Darwin's God: a scientist's search for common ground between God and evolution.* First edition. ed. 1999, New York, NY: Cliff Street Books. xiii, 338 p.
3. Branch, G. and E.C. Scott, *The latest face of creationism.* Scientific American, 2009. **300**(1): p. 92–99.

4. Richard Dawkins and J. Coyne, *One side can be wrong*. The Guardian, 2005.

5. Starr, et al., *Biology: the unity and diversity of life*. Sixth edition. ed. 1992, Belmont, Calif.: Wadsworth Pub Co. xxvii, 921, 12 p.

6. Marty, C., *Darwin, Cuvier and Lamark*. Scientific American, 2009. www.scientificamerican.com/article/darwin-cuvier-and-lamarck/#.

7. Gregory, F., *The Darwinian revolution*, in *Great courses variation: great courses (compact disc)*. 2008, Chantilly, VA: Teaching Co.

8. Darwin, C. and A. Wallace, *On the tendency of species to form varieties; and on the perpetuation of varieties and species by natural means of selection*. Zoological Journal of the Linnean Society, 1858. **3**(9): p. 45–62.

9. Darwin, C. *On the origin of species by means of natural selection or the preservation of favoured races in the struggle for life*. 1859, Beloit, KS: McGraft Publishing Company.

10. Masci, D., *For Darwin Day, 6 facts about the evolution debate*. Pew Research Center, 2017.

11. Darwin, E. *Zoonomia, or, the laws of organic life*. First published in 1794. Print publication year. 2009, Cambridge, MA: Cambridge University Press.

12. Charlesworth, B. and D. Charlesworth, *Darwin and genetics*. Genetics, 2009. **183**(3): p. 757–766.

13. Jenkin, F., *Review of the origin of species*. North British Review, 1867. **46**: p. 277–318.

14. Henig, R.M., *The monk in the garden: the lost and found genius of Gregor Mendel, the father of genetics*. 2000, Boston, MA: Houghton Mifflin.

15. Galton, D., *Did Darwin read Mendel?* QJM: An International Journal of Medicine, 2009. **102**(8): p. 587–589.

16. Lacey, R.D.D., *The year 1000: what life was like at the turn of the first millennium: an Englishman's world*. First edition. ed. 1999, Boston: Little, Brown. 230 p.

17. Losos, J.B., *Improbable destinies: fate, chance, and the future of evolution*. 2017, New York, NY: Riverhead Books. xv, 368 p.

18. Powell, A., *Harvard biologist's new book details a new era in the study of evolution*. Harvard Gazette, 2017.

19. Dawkins, R., *The ancestor's tale: a pilgrimage to the dawn of evolution*. 2004, Boston, MA: Houghton Mifflin.

20. Nield, D., *Scientists just figured out continental plates can move up to 20 times faster than we thought*. Science Alert, 2016.

21. Dubner, A., *Supercontinents—list of known historical Earth land masses*, 2020. www.ranker.com/list/the-earth_s-known-supercontinents/analise.dubner.

22. Wilf, H.S. and W.J. Ewens, *There's plenty of time for evolution*. Proceedings of the National Academy of Sciences of the United States of America, 2010. **107**(52): p. 22454–22456.

23. Weiner, J., *The beak of the finch: a story of evolution in our time*. 1st Vintage Books ed. 1995, New York, NY: Vintage Books. xvi, 332 p.

24. Uyedaa, Josef C., et al., *The million-year wait for macroevolutionary bursts*. Proceedings of the National Academy of Sciences, 2011. **108**(38): p. 15908–15913.

25. Riesch, R. and M. Plath, *Evolution at the limits*. Scientific American, 2017. **316**(4): p. 54–59.

26. Mayr, E., *What evolution is*. 2001, New York, NY: Basic Books. xv, 318 p.

27. Niles Eldredge and S.J. Gould, *Punctuated equilibria: An alternative to phyletic gradualism*. Models in Paleobiology, 1972. Freeman, Cooper and Company(San Francisco): p. 82–115.

28. Mayr, E., *Darwin's principle of divergence*. Journal of the History of Biology, 1992. **25**(3): p. 343–359.

29. Gewirtz, P., *On "I know it when I see it."* Yale Law Journal, 1996. **105**(4): p. 1023.

30. Wilkins, J., *How many species concepts are there?* The Guardian, 2010.

31. Hug, L., et al., *A new view of the tree of life.* Nat Microbiol, 2016. **1**: p. 16048.

32. Laplace, P.S., *A philosophical essay on probabilities.* Student's ed. 1952, New York, NY: Dover Publications. viii, 196 p.

33. Pauling, L. and W.E. Bright, *Introduction to quantum mechanics: with applications to chemistry.* 1985, New York, NY: Dover Publications. xi, 468 p.

34. Brumfiel, G., *Quantum uncertainty not all in the measurement.* Nature, 2012. doi. org/10.1038/nature.2012.11394.

35. Dennett, D.C., *Elbow room: the varieties of free will worth wanting.* 1984, Cambridge, Mass.: MIT Press. x, 200 p.

36. Hawking, S. and L. Mlodinow, *The grand design.* 2010, New York, NY: Bantam Books. 198 p.

37. Davies, P.C.W. and J.R. Brown, *The ghost in the atom: a discussion of the mysteries of quantum physics.* 1986, Cambridge: Cambridge University Press. 157 p.

38. Golo, V.L. and Y.S. Volkov, *Tunneling of protons and tautomeric transitions in base Pairs of DNA.* International Journal of Modern Physics C, 2002. **14**(1): p. 133–156.

39. Klug, W.S., et al., *Concepts of genetics.* Eleventh edition. ed. 2015, Boston: Pearson.

40. Slocombe, L., J.S. Al-Khalili, and M. Sacchi, *Quantum and classical effects in DNA point mutations: Watson–Crick tautomerism in AT and GC base pairs.* Physical Chemistry Chemical Physics, 2021. **23**(7): p. 4144–4150.

41. Popper, K., S.B. Raimund, and W. William, *The open universe: an argument for indeterminism.* The Postscript to The Logic of Scientific Discovery/as edited by W.W. Bartley, III; Variation: Popper, Karl Raimund; Sir, 1902–1994. Postscript to The Logic of Scientific Discovery. 1982, Totowa, NJ: Rowman and Littlefield. xxii, 185 p.

42. Weaver, R.F. and P.W. Hedrick, *Basic genetics.* Second edition. ed. 1995, Dubuque, IA: Wm. C. Brown Publishers. xiv, 498 p.

43. Will, K., et al., *A novel exon in the cystic fibrosis transmembrane conductance regulator gene activated by the nonsense mutation E92X in airway epithelial cells of patients with cystic fibrosis.* Journal of Clinical Investigation, 1994. **93**(4): p. 1852–1859.

44. Lorson, C.L., et al., *A single nucleotide in the SMN gene regulates splicing and is responsible for spinal muscular atrophy.* Proceedings of the National Academy of Sciences of the United States of America, 1999. **96**(11): p. 6307–6311.

45. Tuddenham, E.G., et al., *Haemophilia A: database of nucleotide substitutions, deletions, insertions and rearrangements of the factor VIII gene, second edition.* Nucleic Acids Research, 1994. **22**(22): p. 4851–4868.

46. McFadden, J., *Quantum evolution.* First American ed. 2001, New York, NY: W W Norton. 338 p.

47. Beck, F. and J.C. Eccles, *Quantum aspects of brain activity and the role of consciousness.* Proceedings of the National Academy of Sciences of the United States of America, 1992. **89**(23): p. 11357–11361.

48. Yamamoto, N. and G. Lopez-Bendito, *Shaping brain connections through spontaneous neural activity.* European Journal of Neuroscience, 2012. **35**(10): p. 1595–1604.

49. Brembs, B., *Towards a scientific concept of free will as a biological trait: spontaneous actions and decision-making in invertebrates.* Proceedings of the Royal Society, 2011. **278**(1707): p. 930–939.

50. Conway, M.S., *Life's solution: inevitable humans in a lonely universe.* 2003, Cambridge: Cambridge University Press.

51. Gould, S.J., *Wonderful life: the Burgess Shale and the nature of history.* First edition. ed. 1989, New York, NY: W.W. Norton. 347 p.

52. Blankenship, R.E., *Early evolution of photosynthesis.* Plant Physiology, 2010. **154**(2): p. 434–438.

53. Powell, A., *Evolution in real time.* Harvard Gazette, 2014.

54. Alvarez, W., *A most improbable journey: a big history of our planet and ourselves.* First edition. ed. 2017, New York, NY: W W Norton & Company. 246 p.

55. Editor, *Timeline of photosynthesis on Earth.* Scientific American, 2008. https://www.scientificamerican.com/article/timeline-of-photosynthesis-on-earth/.

8 The Rise of the Superkingdoms

Bacteria, Archaea, and Eukarya

For the first half of geological time our ancestors were bacteria. Most creatures still are bacteria, and each one of our trillions of cells is a colony of bacteria.

Richard Dawkins

The quickest way to gain entire, innovative new genes is to eat them.

Lynn Margulis

The Creation is never over. It had a beginning, but it has no ending. Creation is always busy with new scenes, new things, and new Worlds.

Immanuel Kant

Why did the first protocells to appear on Earth change, and why did they give rise to Bacteria, Archaea, and Eukarya? Why didn't the first protocells simply stay as they were? After all, they had the planet to themselves, and they were likely doing okay. Of course, based on the information covered in the last chapter, the answer is obvious; they changed because change is inevitable. Genetic material is innately mutable, and change is unavoidably introduced during the process of DNA and RNA replication. Consequently, once a cell divides, it must produce descendants that are genetically different, and once there are differences among cells, some will be better able to survive and reproduce than others. This sets the stage for competition, and those that excel at surviving and reproducing will eventually dominate—until a superior descendant replaces them. Over time, this reiterative process played out in a plethora of different environments produces remarkable results—just look at the life around you and you will see its power.

Current evidence suggests that, when protocells first appeared, they relied heavily on the catalytic properties of RNA, but it's likely that some RNA molecules, which just by chance started to associate with amino acids, functioned better than those that did not. Eventually, because of descent with modification and natural selection, within some protocells, the associations between RNA and amino acids likely became more complex, and RNA molecules began to affiliate with increasingly long strings of amino acids. This ultimately gave rise to the formation of RNA–protein complexes, and gradually, the role of RNA in many of these complexes began to diminish. Eventually, as this process continued to unfold, proteins became the primary catalysts of many biochemically important reactions.

DOI: 10.1201/9781003270294-11

167

With the advent of protein-based catalysis, the cell eventually evolved a mechanism to produce DNA from RNA, and with DNA, RNA, and proteins in place, the protocells of the ancient world evolved into the first prokaryotes (see Chapter 6 for more on this). Today, there are two major types of prokaryotes, the Bacteria and the Archaea. These two groups contain most of the species within the biosphere, and they constitute two of the three superkingdoms of life.

At first glance, the prokaryotes appear to be relatively "simple" genetically and morphologically, and they do not form large, complicated multicellular systems like those that make up you and me. However, I put the word simple in quotes in the previous sentence because many prokaryotes exhibit truly impressive and complex cellular chemistry, and their collective repertoire of biochemical tricks made them the most adaptable forms of life on the planet. Ultimately, prokaryotes colonized many of the most arduous and demanding environments imaginable, and during their approximately one and a half billion-year reign as the sole denizens of Earth, they spread around the globe.

In addition to being robust and widely distributed, the prokaryotes are also incredibly numerous. Even today, in a world containing multicellular organisms such as blue whales and giant redwood trees, the diminutive prokaryotes still constitute about 15% of the planet's biomass [1].

Using almost any standard, the prokaryotic lifestyle has proven to be highly successful. During their long existence, prokaryotes evolved mechanisms for transporting materials into and out of cells, they created specialized systems for capturing and utilizing energy, and they developed processes for recycling essential elements such as nitrogen, carbon, and phosphate [2]. During this period, the hardy and adaptive prokaryotes also evolved sensory systems to detect and respond to environmental stimuli, and they created highly efficient mechanisms for the storage, expression, and repair of genetic and epigenetic information. With their impressive cache of biochemical metabolic tricks, these deceptively modest creatures altered everything from the chemical composition of Earth's atmosphere to the mineral composition of the planet's crust [3].

Today, more than 3.5 billion years after their debut, the prokaryotes are still shaping our world and carrying out critical transformations upon which complex life depends. Of all the life to occupy this planet, none has enjoyed a longer existence, and given their versatility and extreme adaptability, they will almost certainly continue to endure as long as there is life on Earth.

THE LONG ROAD TO PROKARYOTIC VERSATILITY

Prokaryotic life forms are capable of truly remarkable feats, but many of the biochemical processes that make these achievements possible took hundreds of millions of years to evolve. This is an extraordinarily lengthy period, but it could have been much longer. For example, it is conceivable that evolution could have occurred half as fast as it did. Had this been the case, the earliest Eukarya would only now be appearing, and humans would be on schedule to emerge in about four billion years or so. Unfortunately, in another four billion years, Earth will not be able to sustain life, so had the process of organic evolution occurred at half the speed that it did, our kind would never have come into existence. Given this, it is worth pondering why evolution occurred at the rate it did.

Ultimately, the rate of organic evolution is due, in part, to the mutability of DNA and RNA. As I mentioned earlier, in some cases, this ability is an intrinsic feature of the chemistry of nucleic acids, and it is imparted by the elements that compose the molecules themselves. Therefore, if we follow the causative trail long enough, we will see that the rate of evolution is, in part, a product of the physics that arose during the Big Bang.

Another factor that affected the speed of the evolutionary process was the rate of prokaryotic replication. Members of the Bacterial and Archaean superkingdoms are small, and they reproduce rapidly, and this allows for the generation of large numbers of descendants in small amounts of time. Today, it is estimated that there are between 4×10^{30} and 6×10^{30} prokaryotes on our planet [2], and among the uncountable number of prokaryotes that lived in the past, there was a great deal of genetic variation for natural selection to act upon.

An additional circumstance that affected the rate of evolution of these single-celled organisms was the existence of a large amount of environmental variations. The existence of hostile environments and the creation of new ones greatly supercharged the process of evolution. For example, the geological processes that created land introduced wholly new territories, and the prokaryotic life that colonized land was forced to adapt to their new conditions or perish.

The immensely important process of terrestrial continent formation began during the Archean eon when the planet's internal heat flow was nearly three times what it is today [4]. During this period, the heat produced by the frictional processes within the planet's core and the decay of radioactive isotopes, as well as the heat trapped during the planet's formation, slowly seeped to the surface of our globe where it partially melted the basalt rock that encased the planet. Some of the melt within the mantle was eventually ejected back to the surface by the early Earth's highly active volcanoes, and as it oozed into bodies of water, a fraction of it piled up upon the underlying layers of basalt and peridotite, where it crystallized and formed granite. After an extended period, sheets of this comparatively buoyant rock became thick enough to emerge from the ocean, and once Earth's tectonic activity began, these granite islands collided and merged. Eventually, granite blocks called cratons formed, and these rock beds formed the nuclei of the planet's newly burgeoning continents.

By the time our planet was about 1.5 billion years old, continents were present, and these mounds of rock and other environmentally disruptive events combined to keep the early prokaryotic world in a highly dynamic and disrupted state [5]. When this environmental flux was further coupled with the intense intraspecies and interspecies competition, the selective pressure on prokaryotic populations was immense, and all this sped up the rate at which these creatures evolved.

EARTH'S VICISSITUDES—NOT TOO SEVERE, BUT NOT TOO GENTLE EITHER

Life on Earth has never been easy, but the cataclysmic horrors of nature, including those that generate terror in the minds of humans today, are ironically the same events that made our existence possible. Given this reality, it is worth noting that not every "potentially habitable" planet would expose life to the same challenges as those found on our planet. On a more placid and environmentally tranquil world, the rate of evolution would likely be slower, while on many other potentially habitable worlds, the trials associated with environmental variations may be so overwhelming that life could not long endure. Thus, even the physical demands that a potentially habitable world imposes on life matter greatly, and it is likely that in many cases, even these must fall into a narrow range if complex life is to emerge.

ADDITIONAL EVOLUTIONARY ACCELERANTS—THE EVOLUTION OF EVOLVABILITY AND THE EVOLUTION OF THE CHROMOSOME

For organisms exposed to the forces described earlier, change was not only unavoidable, but it was also the key to survival, and this reality ultimately led to the evolution of "evolvability" [6]. Evolvability is the capacity to generate adaptive traits that facilitate survival and reproduction, and this characteristic is prominently displayed by organisms that can effectively transmit genetic information horizontally (i.e., organisms that can acquire genetic information from other potentially unrelated cells and then productively assimilate it into their genomes).

As a result of productive horizontal genetic transmission, a prokaryote does not necessarily have to evolve a useful trait *de novo*; instead, it can acquire it from almost any organism in its environment. In some cases, this information is obtained tranquilly through processes such as conjugation, during which material is transferred between cells that are in direct physical contact. In other instances, genetic information is more forcefully acquired by ingesting the target. And in still other situations, genetic material is delivered to a host because of the invasion of nucleic acid-containing viruses [7].

For prokaryotes, another crucial step in enhancing the rate of evolution was the development of the chromosome. The appearance of this structure was a watershed event because once genes were tethered, they were forced to replicate together, and this attenuated the competition that occurred between them. As a result of this conjoining process, a community of genes could evolve, which when working together, maximized each other's replication [8].

In addition to facilitating the formation of cooperating groups of genes, the conjoining of genes on a chromosome increased the likelihood that each daughter cell would receive a complete complement of duplicated genetic information [9]. These advances in the faithful transmission of genetic information were important developments that ultimately impacted all life on the planet, and the prokaryotic organism that first evolved a chromosome was likely a "one-off, odd ball" of great importance [10].

As a result of the mutable nature of DNA, large populations, changing environments, the evolution of evolvability, and the evolution of structures such as chromosomes, prokaryotic populations generated many exceptionally unusual creatures during their first two billion years of existence. Among these was the lifeform that evolved the ability to store information in the form of a DNA-based genetic code. This cell became the common ancestor of all living creatures, and it was another member of the all-important, "one-off, odd ball" sect of unusual, but highly significant, prokaryotes. If, within the prokaryotic masses, there were other cells that developed a DNA-based genetic code that was different from the one employed by cells today, their lineage ended, and they left no evidence of their existence.

OXYGEN-GENERATING PHOTOSYNTHETIC BACTERIA—ONE OF THE MOST UNUSUAL CREATURES EVER TO INHABIT EARTH

One truly strange and usual group of prokaryotes that emerged from the early global prokaryotic population was the cyanobacteria. These creatures are exceptional because they gave rise to the first organisms to carry out oxygen-generating photosynthesis [11]. With this ability, the cyanobacteria oxygenated the biosphere, and they eventually gave rise to plant chloroplasts. Within a plant cell, the chloroplasts convert the energy of sunlight into chemical energy, and through this process, they allow eukaryotic organisms to efficiently tap into the most significant and reliable energy source available—the Sun. Because of these cyanobacteria, there are trees and plants, oxygen to breathe, and an atmosphere that limits the transmission of UV light. In short, because of the cyanobacteria, primates like you and me exist.

Given cyanobacteria's central role in the evolution of primate life, it is worth considering how they evolved and acquired the ability to carry out oxygenic photosynthesis. Unfortunately, fulsome answers to these questions remain elusive, but we do have some relevant data [11, 12]. For example, we know that the extremely complicated, multi-subunit reaction centers needed for photosynthesis evolved only *one time* in an ancient bacterium, and we know that these reaction center complexes were eventually acquired by the cyanobacteria as a result of horizontal gene transfer. In addition, data indicate that one group of photosynthetic cyanobacteria eventually evolved the ability to use water as an electron donor during photosynthesis. To do this, the cyanobacteria acquired the ability to use light to deconstruct water into electrons, hydrogen ions, and oxygen, and as a result, they became *the first and only organisms* to independently evolve the ability to carry out oxygenic photosynthesis.

It is worth noting, however, that the oxygen that these most unusual creatures produced was toxic to most of the other prokaryotes that were on the early Earth. Therefore, to carry out oxygenic photosynthesis, cyanobacteria first had to do what few others could: namely, they had to develop a tolerance for oxygen. So, how did the ancestors of the first oxygen-producing photosynthetic cyanobacteria develop this capacity?

One answer to this conundrum was proposed by the geochemist, Yuk Yung. Yung states that during the periods when the young Earth was in an ice age, UV radiation acted on ice and generated hydrogen peroxide (i.e., H_2O_2), and he suggests that this substance was converted to water and oxygen when the ice periodically melted. The oxygenated meltwater would have mixed with the anoxic seawater that surrounded the ice, and this would have triggered an ecological crisis for the prokaryotes in the area. As a consequence, the microbes in this region were under tremendous pressure to evolve protective enzymes that neutralized the destructive effects of oxygen [13], and if an oxygen tolerant microbe did arise, it would have potentially been capable of evolving oxygenic photosynthesis. Alternatively, an oxygen tolerant microbe may have donated genes that imparted oxygen tolerance via horizontal gene transfer to the ancestors of oxygenic phototrophs.

With the ability to protect itself from the devastating effects of oxygen, a microbe would also be in a position to potentially evolve the ability to use oxygen to carry out aerobic respiration. This process generates about 18 times more usable chemical energy than the anaerobic respiratory mechanisms that were used by the anoxic microbes of early Earth, so a microbe with the ability to tolerate oxygen may have eventually acquired many favorable traits.

Whether cyanobacteria did descend from organisms that evolved oxygen tolerance in the polar ice melt regions, or whether it acquired the genes that impart oxygen tolerance from creatures that did, remains a mystery. But, once oxygenic photosynthesis and aerobically metabolizing cyanobacteria did appear, they were in possession of an incredible one–two punch, which allowed them to dominate their environments. By dumping oxygen into their environment, the cyanobacteria initiated a global microbial mass extinction that eliminated much of their competition, and by capturing the energy of sunlight and efficiently transforming it into chemical energy through oxygenic photosynthesis and aerobic respiration, they were able to acquire the resources needed to multiply and thrive in the niches of the creatures they extinguished.

The good fortune of the oxygen tolerant cyanobacteria was further potentiated by the fact that oxygen enables cells to run hundreds of metabolic reactions that could not occur in its absence. The products of these reactions include novel metabolites that modulate the activities of membranes, as well as molecules that facilitate the development of intracellular and intercellular signaling pathways. In the process of enhancing their metabolic capabilities and their physiological signaling systems, the cyanobacteria would have added complexity to their genomes, and because of lateral gene transfer, this complexity would have been quickly incorporated into the genetic structure of other oxygen tolerant microbes.

Regardless of how they initially came to be, we know that the first aerobically respiring microbes lived approximately 2.9 billion years ago [14], and among these oxygen tolerant organisms, a subset evolved the ability to carry out oxygen-producing photosynthesis over a period of about two hundred million years [15, 16]. As a result, by 2.7 billion years ago, aerobically respiring and oxygen-generating photosynthetic cyanobacteria existed, and they rapidly spread around the globe [17].

The good news for the aerobically respiring and oxygen-generating photosynthetic cyanobacteria kept on coming because soon after they appeared, they benefited from many of the geographical and geochemical events occurring around them. Among these were the plate tectonic movements that resulted in the formation of the planet's first supercontinent, Kenorland. As this landmass was taking shape, enormous slabs of granite collided, and one tectonic plate slid under another to give rise to the supercontinent's great mountain chains. From the perspective of an organism like you and me, these mountains appear to last forever, but in the grand scheme of time, they are merely transient structures, and not long after their formation, the magnificent mountains of Kenorland began to erode. As they did, they released a flood of nutrients that flowed into the surrounding coastal regions, which benefited the cyanobacteria and further propelled their population explosion [18].

As the nutrient-fed colossal masses of cyanobacteria went about their business photosynthesizing, some of the oxygen they initially pumped into their aqueous surroundings oxidized dead organic matter, and much of the rest of it was sequestered by the rusting of iron ore deposits. Eventually, however, these oxygen sinks satiated, and about 2.5 billion years ago, the oxygen that the cyanobacteria released started to build up in the planet's seas. Over the next several hundred million years, some of this oxygen leaked out of the oceans and into the atmosphere producing the so-called "Great Oxidation Event," and during this period, Earth went from being an anoxic globe to one in which the oxygen concentration was about 1% of what it is today [19].

So, soon after the world-restructuring and life-destroying cyanobacteria appeared, life went on. However, the cyanobacteria and the Great Oxidation Event they triggered significantly pared down the number of obligate anaerobes and they set the stage for new life forms that were yet to evolve.

THE END OF THE GOOD TIMES

Good times never last forever, and eventually, the cyanobacteria and the organisms that survived the "oxygen catastrophe" witnessed an end to their heyday. The circumstances that initiated this ending became manifest about 2.4 billion years ago when the 300-million-year Huronian glaciation set upon the planet. Many events conspired to bring about this era, including the activity of the cyanobacteria themselves. As throngs of these photosynthetic organisms spewed oxygen into the environment, the gas emissions they generated reacted with methane, which was a major component of the atmosphere prior to the Great Oxidation Event. As a consequence, the methane in the atmosphere decreased, while the level of carbon dioxide and water increased (i.e., $CH_4 + 2O_2 \dashrightarrow CO_2 + 2H_2O$).

Earlier I mentioned that carbon dioxide traps heat, but methane is an even more powerful greenhouse gas, and so as the concentration of atmospheric oxygen increased and the concentration of methane decreased, the planet's surface temperature crashed. This plunge was augmented by the breakup of Kenorland, the supercontinent whose formation had earlier portended good news for the cyanobacteria. As this landmass ripped apart, weathering along its coasts increased (and therefore the amount of heat-trapping CO_2 in the atmosphere decreased), and large amounts of carbon-rich sediment were buried. Together, these events helped usher in the Huronian glaciation, which was one of the worst cold spells in Earth's history. By some accounts, during parts of Huronian, the average temperature hovered around -58°F [19], and the globe was completely, or almost completely, encased in ice, which in some regions, was miles thick. This event was likely the first time Earth existed in this extreme "snowball" or "slush ball" state, and these punishing environmental conditions took their toll on the life that existed.

Interglacial periods during this extended ice age provided some occasional intervals of reprieve, but during the Huronian era, life was locked down in ice for millions of years at a time. During the extended deep freeze, the rate of photosynthesis decreased, and the aerobically respiring and oxygen-generating photosynthetic prokaryotes struggled to persist. However, single-celled prokaryotes are remarkably resilient and robust, and some can survive in icy areas. Others likely found conditions that were more suitable to their existence in volcanic outflows or near deep-sea thermal vents. In any case, many prokaryotes did persevere through this extraordinarily severe environmental catastrophe.

The devastating Huronian glaciation was likely terminated by the buildup of carbon dioxide in the atmosphere. This occurred in part because, as the temperature plummeted, so did the rate of prokaryotic reproduction and weathering. Simultaneously, fewer carbon-based life forms were dying, so there was less carbon-rich sediment deposition and less limestone formation. In addition, despite being covered in sheets of ice that were miles thick, Earth's plate tectonic and volcanic activity continued unabated, and as a result, carbon dioxide unceasingly poured into the atmosphere. An additional catalyst to the warming of the planet also likely came from the activities of psychrophilic (i.e., cold-loving) methanogens (i.e., methane-producing Archaean), which managed to persist through this long glacial period. These organisms enhanced the process of global warming by releasing significant amounts of methane into the atmosphere [20]. So, after hundreds of millions of years, Earth slowly emerged from its long, grueling, and brutalizing deep freeze.

OXYGEN—A TERRAFORMING GAS

Importantly, in addition to helping to trigger the planet's longest and deepest ice age, the oxygen produced by the cyanobacteria also eliminated the hydrocarbon-based haze that lingered in the early Earth's atmosphere, and in so doing, it opened our world to a less obstructed influx of sunlight. This same gas also formed an atmospheric ozone (O_3) layer that filtered out much of the high-energy UV radiation from the Sun. Without atmospheric ozone, biologically important molecules such as DNA, RNA, proteins, and lipids would be severely damaged, and life on Earth would have been restricted to the protective confines of the subaqueous or subterranean world.

Oxygen also had a profound effect on the composition and redox state of the solutes in water. Dissolved iron, which is an essential component of many enzyme systems, was plentiful in the anoxic oceans of the early Earth, but once these waters became oxidized, the iron precipitated out of solution, and this produced an iron shortage, which affects organisms such as algae even today. Other elements, such as the highly radioactive metal uranium, became more soluble in the oxygenated waters, and so these elements now float in areas where they would have been absent in the ancient past [19]. Furthermore, the change in the oxidation state of metals such as iron and mercury triggered the formation of new minerals, which contributed to the proliferation of these biologically important compounds. Eventually, life's effect on Earth's geochemistry further increased the number of minerals from the dozen or so that were present in the pre-solar dust to the more than 4,500 that exist on the planet today [3].

As was mentioned earlier, the presence of oxygen also made it possible for life to extract energy more efficiently from the products of photosynthesis, and oxygen allowed for the production of hundreds of metabolic metabolites, some of which changed the fundamental operations of cells. These novel abilities allowed some prokaryote cells to colonize new territories and exploit new niches.

So, in retrospect, the oxygen waste product produced by cyanobacteria was the most consequential pollutant ever generated. As this gas continuously permeated the biosphere, it shaped the variety and diversity of life, and it profoundly altered the planet.

During the 2.9 billion or more years that cyanobacteria have existed, they have evolved a high degree of genetic diversity [21]. Currently, they can be found across the globe in freshwater, saltwater, hot springs, and hypersaline bays, as well as in terrestrial habitats ranging from desert rocks to ice-covered tundra. Today, we often encounter cyanobacteria in the form of slimy, smelly pond scum, but despite their sometimes less than appealing appearance, they are among the most noteworthy creatures ever to live. In addition to successfully contending with oxygen and carrying out oxygenic photosynthesis [22], they were also among the earliest cells to evolve aerobic respiration [16, 23]. Furthermore, these microbes are members of a relatively small subset of organisms that can extract relatively inert nitrogen gas (N_2) from the atmosphere and combine it with hydrogen to form compounds such as NH_3, NH_2^-, and NH_3^-, all of which have roles in various essential biosynthetic processes [24].

CYANOBACTERIA—ONE OF THE FIRST PROTOMULTICELLULAR ORGANISMS

Interestingly, to carry out the nitrogen fixation process, some filamentous cyanobacteria evolved specialized cells called heterocysts. Unlike the more common vegetative cells within a cyanobacterial filament, the heterocysts do not photosynthesize. Rather, they are a differentiated and oxygen-resistant form of the vegetative cyanobacteria. Within the heterocyst, there is only enough oxygen to power aerobic respiration, and consequently, the oxygen-sensitive, nitrogen-fixing enzymes can function, even when they are near vegetative cells that are liberating oxygen into their surroundings.

Within these nitrogen-fixing filamentous cyanobacteria, a continuous outer membrane holds the vegetative cells and heterocysts together, and the various linked cell types communicate and share their metabolic products. In particular, the heterocysts share the nitrogen-rich compounds they produce with the photosynthetic vegetative cells, and in return, the vegetative cells provide energy-rich molecules to the heterocysts. According to many biologists, multicellular organisms are formed by groups of differentiated and linked cells that share information and products, and consequently, it is arguably true that filamentous nitrogen-fixing cyanobacteria represent one of the first multicellular organisms to appear on the planet [25, 26].

CYANOBACTERIA—A STRONG CANDIDATE FOR MVP (MOST VALUABLE PROKARYOTE)

Given all the unique features and abilities of the cyanobacteria, as well as their importance in the restructuring of Earth's environment, one could make a strong case for the assertioin that the cyanobacteria are among the most important organisms to ever appear on our planet. They are the outcome of billions of years of evolution, and in many ways, they represent a pinnacle of prokaryotic life.

During the unimaginably long process that gave rise to the cyanobacteria, an uncountable number of organisms were subjected to the rigors of natural selection, and the appearance of the first oxygenic photosynthetic cyanobacteria was an indisputably historic event. However, it is important to keep in mind that the emergence of these most unusual creatures was never a certainty, and therefore life as we know it was never an inevitability.

In addition to the oxygen-generating photosynthetic cyanobacteria, there are other noteworthy candidates for the title of MVP "most valuable prokaryote," and the pantheon of Earth's most significant creatures would surely include other "one-offs" such as the first protocell; the first free-living, chromosome-containing, DNA-programmed, prokaryote; and the first microbe to develop the universal genetic code. However, if there were a hall of fame for creatures that occupied our planet, it would surely contain a prominent display of the seemingly lowly, but ultimately all-important, cyanobacteria.

THE DIVERGENCE OF THE PROKARYOTES AND THE FORMATION OF THE BACTERIA AND ARCHAEA SUPERKINGDOMS

The Bacteria and Archaea superkingdoms diverged from each other very early in the history of life [27]. However, until less than half a century ago, biologists did not even realize that there were two superkingdoms of prokaryotes. Part of the reason for this is that Bacteria and Archaea share a high degree of morphological similarity, and the remarkable degree of diversity that exists within the prokaryotic superkingdoms did not become evident until the recent unveiling of biochemical and genetic data. This information indicated that the Bacteria typically have a peptidoglycan-based cell wall surrounding their plasma membrane while most of the Archaea do not, and the data also illustrated that the biochemical composition of the plasma membranes of Bacteria and Archaea differs in important ways. Furthermore, the two groups of prokaryotes have very distinctive information processing systems, and their genes differ substantially [28].

In terms of their habitats, Bacteria and Archaea can be found in a very wide range of environments, including the human digestive tract, but within these two superkingdoms, the Archaea are, in general, more benevolent, and no pathogenic or parasitic Archaea have been identified. Metabolically, the Archaea cannot carry out oxygenic photosynthesis, but they do use a wider range of energy sources than Bacteria, and as was mentioned earlier, some can convert carbon compounds into methane. In the end, however, despite their numerous differences, the disparities between the Bacteria and Archaea pale in comparison to those that exist between the prokaryotes and the eukaryotes, and the emergence of eukaryotes marked a fundamental and remarkable change in the nature of life on our planet.

THE EUKARYA—CELLULAR LIFE 2.0

The eukaryotic superkingdom is often referred to as the Eukarya or Eukaryota. However, since this cell type "only" arose between 2.4 billion and 1.7 billion years ago [19, 29–31], the members of this clade are relative newcomers to the community of life.

When compared to the prokaryotes, the various types of eukaryotes are less diverse and fewer in number, but a typical eukaryotic cell's volume is about 100 times as large as their archetypal prokaryotic brethren. In addition, unlike prokaryotes, which are mostly obligatory unicellular creatures, eukaryotes exist as unicellular and multicellular organisms, and in their multicellular state, eukaryotic cells can organize into truly staggeringly large groups. The average human, for example, is a multicellular collective of many trillions of eukaryotic cells, and an adult blue whale is estimated to be composed of approximately 100 quadrillion (100×10^{15}) eukaryotic cells! Therefore, although there are relatively few eukaryotic species, their collective biomass is still greater than that of their prokaryotic relatives [2].

So, what makes eukaryotic cells so different from their prokaryotic ancestors? A quick comparison of the groups reveals that there are at least 30 distinct features that differentiate them. Some of the more prominent differences are that prokaryotes are typically more diminutive than their eukaryotic relatives, prokaryotes have fewer genes and fewer chromosomes, and prokaryotes lack membrane-bound substructures (i.e., organelles) while eukaryotes have many of them. Of course, there are exceptions. There are, for example, some prokaryotes that are bigger than some eukaryotes, and there are some prokaryotes that have more DNA than some eukaryotes. But again, these are exceptions.

As was mentioned, organelles distinguish eukaryotes from prokaryotes, and the organelle that is most often associated with the eukaryotic cell is the nucleus. This structure holds most of the DNA content of a eukaryote, and unlike prokaryotes, which usually have their DNA arranged in the form of a single, circular chromosome [32, 33], eukaryotes typically have many linear chromosomes composed of a complex mix of DNA and proteins.

Eukaryotic cells also have subcellular organelles called mitochondria, which are responsible for securing most of the chemical energy that a eukaryotic cell uses, and some plants and algae have organelles called chloroplasts that carry out oxygenic photosynthesis. Some of the other numerous features that differentiate the eukaryotes from the prokaryotes are summarized in Table 8.1. However, while the differences

TABLE 8.1
Differences between prokaryotes and eukaryotes.

Trait	Prokaryotes	Eukaryotes
Size	Small, typically 0.5–5 μm	Usually large, typically 10–100 μm
Nucleus	Absent	Present, has an inner and outer membrane
Cytomembrane system	Absent	Contains endoplasmic reticulum, Golgi body, lysosomes, microbodies, transport vesicles; all membrane bound
Other organelles	Absent	Several, for example, mitochondria and chloroplasts are present in plant cells

(*Continued*)

Trait	Prokaryotes	Eukaryotes
Cell wall	Peptidoglycan based in bacteria; many Archaea have proteins and glycoproteins arranged as an S-layer	Cellulose in plants, chitin in fungi, absent in animals
Metabolism	Various diverse processes	Usually based largely on mitochondrial aerobic respiration
DNA	Relatively small amounts of DNA (typically 130,000–14 million base pairs); DNA not bound to histones, no protein-rich chromosomes	Relatively large amounts of DNA (up to 152 billion base pairs); DNA bound to histones to form protein-rich chromosomes
Reproduction	Fission and budding	In many plants and animals, sexual reproduction occurs via haploid gamete formation and fertilization
Cell division	Via fission	Via mitotic and meiotic cell cycles
Genetic recombination	Can occur by unilateral gene transfer	Occurs during meiosis, and to a lesser extent, during mitosis
Multicellularity	Unicellular, arguably some multicellular photosynthetic bacteria (e.g., some species of nitrogen-fixing cyanobacteria [26])	Can exist as unicellular organisms or as complex, highly interactive, multicellular organisms
Cytoskeleton	Based on Ftsz, MreB, Par M, Crescentin, and a number of other proteins	Based largely on actin, tubulin, and in some organisms, intermediate filaments
Ribosomes	Small RNA–protein complexes	Large RNA–protein complexes
Spliceosomes	Absent, intron sequences not present	Present, RNA–protein complexes spice RNA introns

between these cell types are abundant and well documented, a question with an answer far less chronicled is "how did the eukaryotic cell develop its complexity?"

THE EVOLUTION OF THE EUKARYA

We know that the process of eukaryotic cellular evolution took place in many steps over an extended period, and by about 1.7 billion years ago, true eukaryotic cells existed [31]. However, exactly how the first eukaryotes evolved remains uncertain. Some have suggested that the process began with the genocidal activities of one bacterial species toward the others, and according to this hypothesis, at some point, a bacterium evolved a penicillin-like antibiotic that prevented competing species from developing cell walls with peptidoglycans. Presumably, the bacterium secreting these antibiotics was immune to the effects of these compounds, but these antibiotics would have interfered with the ability of nearby completing prokaryotes to assemble a cell wall, and consequently, these affected prokaryotes would have been in a fragile and vulnerable state. Eventually, some of the wall-less bacteria may have stumbled upon a way to survive the chemical attacks launched by their neighbors by evolving a cell wall and membrane that was immune to the antibiotic, and some of these creatures may have eventually given rise to the Gram-negative bacteria (i.e., those with membranes that contain a thin layer of peptidoglycan) [9]. This intense chemical warfare may have also given rise to Archaean without a cell wall [34], and within this group, those that shored up their cellular integrity likely did so by developing a robust, force-generating cytoskeleton. This development may have also given a wall-less prokaryote the ability to gather food in new ways since, unlike walled prokaryotes, which must transport all their food across a formidable barrier, a wall-less proto-Archaean would likely be flexible enough that it could potentially surround and engulf would-be food sources. Using this mechanism, a wall-less prokaryote could consume nutrient sources en masse and with great efficiency.

This hypothesis goes on to propose that wall-less Archaean occasionally attempted to capture cells they could not digest, and in rare instances, this may have led to the development of a symbiotic relationship between the predator and its intended prey. Recently obtained data suggest that just such a symbiotic relationship between a proto-eukaryotic, phagocytic Archaean and a purple bacterium did eventually give rise to the eukaryotic mitochondrion. A host cell with mitochondria would be able to acquire the energy it needed more efficiently, and the engulfed symbiont would gain the protection and resources that come with living in a successful predatory proto-eukaryote, so overall the host and symbiont both benefited from this arrangement. Over time, the predator, and its would-be meal, ultimately became intimately dependent on each other, and at some point, they could no longer live apart. At that moment a new type of cell was born.

NEW MECHANISMS FOR THE REPLICATION AND STORAGE OF GENETIC INFORMATION

The lack of a cell wall may have also forced ancient proto-eukaryotic cells to evolve new strategies to replicate their genetic information [9]. This is true in part because bacterial chromosomes are typically attached to the proteoglycan-containing cell wall at the site where chromosomal replication begins and ends, and once a chromosome is duplicated, the new chromosome's origin and terminus must be attached to the opposite side of the cell wall. This helps ensure that when the prokaryote divides by fission, each daughter cell receives one copy of the parent cell's genetic material. However, in the absence of a cell wall, the proto-eukaryotic cell would need to modify its cytoskeletal system for apportioning its chromosomal content to its offspring, and it is likely that this eventually led to the evolution of an actin-based spindle apparatus. In eukaryotes, the spindle apparatus works in conjunction with other cellular proteins to push chromosomes to opposite sides of a dividing cell during the mitotic and meiotic cell cycles, and by executing this process in a coordinated and highly regulated manner, the eukaryotic chromosomal segregation mechanism allows each daughter cell to receive the correct complement of genetic material.

The maintenance and replication of DNA require a considerable amount of energy, but with its spindle apparatus and highly efficient energy-harnessing mitochondria, the wall-less, proto-eukaryotic cell would have been able to increase the amount of DNA and chromosomes it possessed. In addition, the proto-eukaryote could have increased the number of places where DNA replication begins within any given chromosome, and this would increase the rate at which large amounts of genetic information could be replicated. Together, these changes would allow the proto-eukaryotic genome to become larger and more complex than many of its wall-bound, prokaryotic relatives.

THE DEVIL IS IN THE DETAILS

The aforementioned speculative hypothesis posits that the loss of a cell wall was a critical initial step in the evolution of eukaryotes because it allowed ancient single-cell organisms to develop complex symbiotic relationships and increase their genetic complexity. But, even if the outlines of this story are true, and the missing details eventually get filled in, there are still many other distinctive eukaryotic features whose genesis needs to be explained. Among the more prominent is the cyto-membrane system, which is composed of subcellular specialized organelles such as the nucleus, the endoplasmic reticulum, Golgi bodies, microbodies, lysosomes, and transport vesicles. How these and the other uniquely eukaryotic structures evolved remains largely a mystery and explaining when and why they emerged will be no small feat [35]. However, while the details of eukaryotic cell evolution are cloaked in a vail of unknowns, it is likely that many eukaryotic genes, proteins, and subcellular structures arose as a result of complex interactions and lateral gene transfers between prokaryotes, proto-eukaryotic cells, and their viruses [36, 37].

Another unanswered question in the transition from prokaryotes to eukaryotes relates to the nature of the cell that acquired the purple bacteria that led to the evolution of the mitochondria. Recently, a group of investigators provided data indicating that the proto-eukaryotic cell that engulfed a purple bacterium likely belonged to an Archaeal phylum similar to the Lokiarchaeota, or Loki, for short [38]. These unusual archaea have many so-called "eukaryotic signature genes" including some that encode

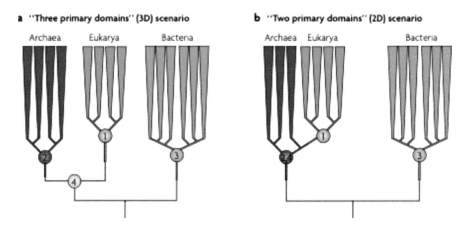

FIGURE 8.1 Two potential evolutionary paths to Eukarya [39].

cytoskeletal components and membrane-remodeling and vesicular-trafficking proteins. Consequently, it is likely that members of this group engulfed material, as well as entire cells, via endocytosis, and it is probable that an organism from this phylum, or one closely related to it, was the proto-eukaryote that developed a symbiotic relationship with a purple bacterium it attempted to consume. If these Archaea were the host cells that engulfed the proto-mitochondrion purple bacterium, then the eukaryotes likely did not arise as an independent domain of life (as depicted in the left-hand side of Figure 8.1). Rather, this hypothesis argues that eukaryotes arose from an Archean lineage that merged with a Bacterium (see the right-hand side of Figure 8.1).

A CLOSER LOOK AT THE MYSTERIOUS "LOKI"

As fascinating as the Loki are, the fact is that they have never been grown in a laboratory, nor has anyone ever seen a member of this group. We only know of their existence because their DNA was isolated from sediment located about 11,000 feet below the surface of the ocean in a region between Norway and Greenland [40]. The sample with the Loki DNA was obtained from an area near an ocean thermal vent, and it is possible that to grow in this hot, high-pressure area, the Loki evolved a robust cytoskeleton to reinforce their cell walls. It is also possible that there are other Loki-like archaea that are even more closely related to the eukaryotes that are yet to be discovered. If so, these prokaryotes may one day help us determine when and where the first eukaryotic common ancestor and the last eukaryotic common ancestor (LECA) lived.

Based on the hypothesis laid out earlier, what does seem likely is that once the LECA appeared, it only took about 300 million years for it to give rise to the five diverse supergroups of eukaryotes [31]. This suggests that the eukaryotic way of life allowed for the successful exploitation of many different ecosystems, and that, in turn, led to the rapid diversification of the eukaryotes.

Despite the excitement about the hypothesis that a Loki-like Archaean was the host cell that engulfed a pre-mitochondrial bacterium, the concept still does have its difficulties. For starters, bacteria and eukaryotes have a more similar membrane structure than do Archaean and Eukarya, and therefore if an Archaean was the proto-eukaryotic host of the mitochondrial endosymbiont, it is hard to explain why modern eukaryotes have a membrane that is similar to that of bacteria. Some have suggested that horizontal gene transfer from bacteria and their viruses may provide an answer to this dilemma [41, 42]. Others have noted that the Loki possess the ability to produce lipids similar to those found in the Bacterial and Archaean membranes [43], so in at least some Archaean phyla, such as that containing the Loki, the differences between the Archaean and Bacterial membranes may not be as great as was first assumed.

THE EUKARYOTIC REVOLUTION

Although many questions regarding the genesis of modern eukaryotes remain, what we can say with a high degree of confidence is that the modern eukaryotic cell is a chimera that was produced in part by the fusion of an Archaean (or Archaean-like ancestorial cell; see Figure 8.1) and a Bacterium. Recent evidence suggests that the

Archaean's contribution led to the development of the eukaryotic information processing systems including mechanisms that control gene expression, DNA replication, and DNA repair, while the Bacteria likely donated the raw material that led to the evolution of the modern eukaryote's operational systems, including many of its metabolic pathways. However, while the Eukarya are chimeras, over time they did evolve genetic information and processes that were unique to them, and in so doing, they became their own distinctive group. The specialized and unique materials that the chimeric offspring produced include the genes needed for organizing and packaging DNA into protein-rich structures (e.g., nucleosomes, solenoids, coiled loops, and chromosomes) and the genes necessary for partitioning the chromosomes and cellular cytoplasm by the mitotic or meiotic cell cycle. In addition, because eukaryotes evolved a well-developed set of organelles, they can efficiently separate many cellular reactions that prokaryotes cannot. For example, eukaryotes can temporally and spatially separate gene expression into transcription (a nuclear process that produces RNA), and translation (a cytoplasmic process that produces proteins). These distinctive abilities that early eukaryotes developed are responsible for much of the complexity that characterizes a modern eukaryotic cell.

WHAT ARE THE ODDS OF FORMING A EUKARYOTIC CELL?

The formation of the complex chimeric proto-eukaryotic cell required many hundreds of millions of years of evolution, and given the myriad environments present on Earth, and the extraordinarily large number of cells that were interacting during this period, it is safe to say that there were an incalculable number of events that influenced the formation of the first eukaryote. But how likely was it that a eukaryote would eventually emerge? That is difficult to calculate. However, consider the fact that the first prokaryotes formed within 0.75 billion years of Earth's accretion, but another 1.4–2.1 billion years was required before these prokaryotes gave rise to the first true eukaryotes. The extraordinarily long period that was required to generate eukaryotes suggests that the emergence of this type of life was a very low probability event. Furthermore, while eukaryotes were forming, it is likely that many other cell types emerged, but the study of modern eukaryotic cells indicates that all extant eukaryotes were derived from one primordial progenitor [44]. This suggests that the ancient cell that gave rise to the modern eukaryote was among the rarest and most important ever to live, and given its special status, if there were indeed a "cellular hall of fame," as I suggested there should be, this mother of all eukaryotes would surely be the recipient of the MVE (most valuable eukaryote) award. Furthermore, given my fondness for eukaryotic creatures, this very strange, highly improbable, odd amalgamation that we call a modern eukaryotic cell would also get my vote for MVC (the most valuable cell, of any kind, ever to live).

EUKARYOTIC LIFE IN THE MILKY WAY

The possibility of finding prokaryotic-like life forms on planets such as Mars excites many individuals. Yet, while such a discovery would be monumental, it certainly would not guarantee that complex life exists elsewhere in the Milky Way. On the other hand, if we ever find a planet with eukaryotic-like cells, that would be truly astonishing, because in such a world there would be at least the possibility of complex multicellular creatures.

Nevertheless, as we will see, once the first eukaryotes appeared, there were still more than a few transitions that needed to occur before complex multicellular life could come forth. And as is the case with most transitions within the universe, this sojourn was a protracted and arduous one. However, as we continue our voyage, we will soon begin to encounter creatures that resemble us, and as we analyze these organisms, we will gain additional insight into our own nature, as well as that of the life that currently surrounds us.

REFERENCES

1. Bar-On, Y.M., R. Phillips, and R. Milo, *The biomass distribution on Earth*. Proceedings of the National Academy of Sciences of the United States of America, 2018. **115**(25): p. 6506–6511.
2. William Whitman, D.C., and William Wiebe, *Prokaryotes: the unseen majority*. Proceedings of the National Academy of Sciences, 1998. **95**(12): p. 6578–6583.
3. Hazen, R.M., *The story of Earth: the first 4.5 billion years, from stardust to living planet*. 2012, New York, NY: Viking. 306 p.
4. Williams, G., *History of the Earth's obliquity*. Earth-Science Reviews, 1993. **34**(1): p. 1–45.
5. Bruno Dhuime, A.W.A.C.J.H., *Emergence of modern continental crust about 3 billion years ago*. Nature Geoscience, 2015. **8**: p. 552–555.
6. Dawkins, R., *The ancestor's tale: a pilgrimage to the dawn of evolution*. 2004, Boston, MA: Houghton Mifflin.
7. San Millan, A., et al., *Interactions between horizontally acquired genes create a fitness cost in Pseudomonas aeruginosa*. Nature Communications, 2015. **6**: p. 6845.
8. Lindholm, A.K., et al., *The ecology and evolutionary dynamics of meiotic drive*. Trends in Ecology & Evolution, 2016. **31**(4): p. 315–326.
9. Maynard Smith, J. and S. Eors, *The origins of life: from the birth of life to the origin of language*. 1999, Oxford: Oxford University Press.
10. Fields, H., *The origins of life*. Smithsonian Magazine, 2010. www.smithsonianmag.com/science-nature/the-origins-of-life-60437133/.
11. Blankenship, R.E., *Early evolution of photosynthesis*. Plant Physiology, 2010. **154**(2): p. 434–438.
12. Blankenship, R.E., *How cyanobacteria went green*. Science, 2017. **355**(6332): p. 1372–1373.
13. Liang, M.C., et al., *Production of hydrogen peroxide in the atmosphere of a snowball Earth and the origin of oxygenic photosynthesis*. Proceedings of the National Academy of Sciences of the United States of America, 2006. **103**(50): p. 18896–18899.
14. Kim, K.M., et al., *Protein domain structure uncovers the origin of aerobic metabolism and the rise of planetary oxygen*. Structure, 2012. **20**(1): p. 67–76.
15. Hohmann-Marriott, M.F. and R.E. Blankenship, *Evolution of photosynthesis*. Annual Review of Plant Biology, 2011. **62**: p. 515–548.
16. Soo, Rochelle M., et al., *On the origins of oxygenic photosynthesis and aerobic respiration in cyanobacteria*. Science, 2017. **355**: p. 1436–1440.
17. Kiang, N., *The color of plants on other worlds*. Scientific American, 2008. **298**: p. 48–55.
18. Campbell, I.H. and C.M. Allen, *Formation of supercontinents linked to increases in atmospheric oxygen*. Nature Geoscience, 2008. **1**(8): p. 554–558.

19. Lenton, T. and A.J. Watson, *Revolutions that made the Earth*. 2011: Oxford: Oxford University Press.

20. Tung, H.C., N.E. Bramall, and P.B. Price, *Microbial origin of excess methane in glacial ice and implications for life on Mars*. Proceedings of the National Academy of Sciences of the United States of America, 2005. **102**(51): p. 18292–18296.

21. Bryant, D.A. and N.U. Frigaard, *Prokaryotic photosynthesis and phototrophy illuminated*. Trends in Microbiology, 2006. **14**(11): p. 488–496.

22. Hamilton, T.L., D.A. Bryant, and J.L. Macalady, *The role of biology in planetary evolution: cyanobacterial primary production in low-oxygen Proterozoic oceans*. Environmental Microbiology, 2016. **18**(2): p. 325–340.

23. Hemp, J. and L.A. Pace, *Evolution of Aerobic respiration*. Astrobiology Science Conference, 2010: p. 5624.

24. Latysheva, N., et al., *The evolution of nitrogen fixation in cyanobacteria*. Bioinformatics, 2012. **28**(5): p. 603–606.

25. Flores, E. and A. Herrero, *Compartmentalized function through cell differentiation in filamentous cyanobacteria*. Nature Reviews Microbiology, 2010. **8**(1): p. 39–50.

26. Gould, S.E., *Bacteria with bodies—multicellular prokaryotes*. Scientific American Blog Network, 2011.

27. Gribaldo, S. and C. Brochier-Armanet, *The origin and evolution of Archaea: a state of the art*. Philosophical Transactions of the Royal Society, 2006. **361**(1470): p. 1007–1022.

28. Allers, T. and M. Mevarech, *Archaeal genetics—the third way*. Nature Reviews Genetics, 2005. **6**(1): p. 58–73.

29. Knoll, A.H., et al., *Eukaryotic organisms in Proterozoic oceans*. Philosophical Transactions of the Royal Society B: Biological Sciences, 2006. **361**(1470): p. 1023–1038.

30. Han, T.-M. and B. Runnegar, *Megascopic Eukaryotic Algae from the 2.1-billion-year-old negaunee iron-formation, Michigan*. Science, 1992. **257**(5067): p. 232–235.

31. Dacks, J.B., et al., *The changing view of eukaryogenesis—fossils, cells, lineages and how they all come together*. Journal of Cell Science, 2016. **129**(20): p. 3695–3703.

32. Fraser, Claire M., et al., *Genomic sequence of a Lyme disease spirochaete, Borrelia burgdorferi*. Nature, 1997. **390**: p. 580–586.

33. Egan, E.S., M.A. Fogel, and M.K. Waldor, *Divided genomes: negotiating the cell cycle in prokaryotes with multiple chromosomes*. Molecular Microbiology, 2005. **56**(5): p. 1129–1138.

34. Meyer, B.H. and S.-V. Albers, *Archaeal cell walls*. Wiley Online Library, 2014.

35. Lopez-Garcia, P. and D. Moreira, *Open questions on the origin of eukaryotes*. Trends in Ecology & Evolution, 2015. **30**(11): p. 697–708.

36. Yoshikawa, G., et al., *Medusavirus, a novel large DNA virus discovered from hot spring water*. Journal of Virology, 2019. **93**(8): p. e02130–18. DOI: 10.1128/JVI.02130-18.

37. Forterre, P., *The origin of viruses and their possible roles in major evolutionary transitions*. Virus Research, 2006. **117**(1): p. 5–16.

38. Spang, A., et al., *Complex Archaea that bridge the gap between prokaryotes and eukaryotes*. Nature, 2015. **521**(7551): p. 173–179.

39. Simonetta Gribaldo, et al., *The origin of eukaryotes and their relationship with the Archaea: are we at a phylogenomic impasse?* Nature Reviews Microbiology, 2010. **8**: p. 743–752.

40. Quammen, D., *The scientist who scrambled Darwin's tree of life*. New York Times, 2018.

41. Forterre, P., *A new fusion hypothesis for the origin of Eukarya: better than previous ones, but probably also wrong*. Research in Microbiology, 2011. **162**(1): p. 77–91.

42. Koonin, E.V., *Origin of eukaryotes from within Archaea, archaeal eukaryome and bursts of gene gain: eukaryogenesis just made easier?* Philosophical Transactions of the Royal Society B: Biological Sciences, 2015. **370**(1678): p. 20140333.

43. Villanueva, L., S. Schouten, and J.S. Damste, *Phylogenomic analysis of lipid biosynthetic genes of Archaea shed light on the 'lipid divide'.* Environmental Microbiology, 2017. **19**(1): p. 54–69.
44. Gribbin, J., *Are humans alone in the galaxy?* Scientific American, 2018. **319**(3). DOI: 10.1038/scientificamerican0918-94.

9 Partnering Up
The Journey From Single-Cell Eukaryotes to Multicellular Animals

The power of Selection, whether exercised by man or brought into play under nature through the struggle for existence and the consequent survival of the fittest, absolutely depends on the variability of organic beings. Without variability, nothing can be effected; slight individual differences, however, suffice for the work, and are probably the chief or sole means in the production of new species.

Charles Darwin

Life on earth is more like a verb. It repairs, maintains, re-creates, and outdoes itself.

Lynn Margulis

Here, on the edge of what we know, in contact with the ocean of the unknown, shines the mystery and beauty of the world. And it is breathtaking.

Carol Rovelli

The journey from single-cell eukaryotes to multicellular animals took about 1.5 billion years, and during this time, an uncountable number of natural experiments occurred. But even in a world where life can emerge and flourish, most of Nature's experiments turn out badly, and most mutants fail to thrive. Consequently, for life's variation and complexity to prosper, time and persistence are required. Fortunately, life is ancient, and Nature is nothing if not relentlessly persistent.

In the previous chapter, we left off with the appearance of the first eukaryotic cell. It took many hundreds of millions of years for Nature to generate this entity, but once it came into existence, it set the stage for the emergence of sexually reproducing multicellular organisms. From there, the evolution of a diverse group of developmentally regulated, complex multicellular creatures became possible, and Earth transformed from a planet occupied by unicellular and simple multicellular creatures to a world populated by a rich diversity of incredibly intricate and beautifully complex biological systems.

THE ORIGINAL SEXUAL REVOLUTION

During the process of developing into what we now recognize as a modern eukaryote, the early eukaryotic cell acquired the ability to reproduce sexually. This was a watershed event, and had early eukaryotic cells failed to develop this ability, Earth would be a very different place.

DOI: 10.1201/9781003270294-12

Exactly when the first eukaryotes started reproducing sexually is not clear. Some estimate this process originated as recently as one billion years ago; others say it was closer to 1.9 billion years ago [1]. But what's 900 million years one way or the other? What is clear is that one billion to two billion years ago, sexually reproducing organisms emerged, and today more than 99.9% of eukaryotes reproduce sexually [2]. How sexual reproduction first evolved is also not obvious. Some hypothesize that the antecedents of sexual reproduction can be traced to the proto-cells of the RNA world [3]. However, other investigators believe that the first true forerunners of sexually reproducing cells were the prokaryotes. We do know that ancient bacteria and archaea carried out DNA recombination and DNA exchange, which are hallmarks of eukaryotic sexual reproduction, and it is likely that these and other related processes were modified and pressed into service by the primordial sexually reproducing eukaryotes [4].

While the matter of when and how sexual reproduction came to be is worth pondering, arguably the most interesting question related to this topic is why sexual reproduction occurs at all given the many advantages of asexual reproduction. During asexual reproduction, every organism can reproduce, and each time this happens, all the parent's information is passed on to its offspring. Sexual reproduction, on the other hand, demands that each organism find and secure a mate, and for many organisms this is an energy-intensive process with a high failure rate. The process of sexual reproduction also requires the generation of energetically expensive gametes (i.e., pollen, sperm, and eggs), most of which go unused, and during sexual reproduction, just half of each parent's genes are transmitted to each descendant. When considered together, these facts suggest that sexual reproduction is not an ideal mechanism for the propagation of a species, and yet, as was mentioned earlier, virtually all eukaryotes can reproduce via sex and they have been doing so for the last one billion to two billion years. This simple reality strongly suggests that sexual reproduction imparts significant advantages to eukaryotes.

WHAT'S SEX GOOD FOR?

So, what are the benefits of reproducing sexually? Likely one of the most significant is that it generates individuals with novel combinations of genes. In the case of humans, the generation of new gene combinations begins at the moment of conception. In that instant, we each receive two sets of complementary genetic information. Specifically, we get 23 chromosomes from our fathers and 23 from our mothers. When we enter puberty, the sex cells in our bodies (i.e., the spermatogonium and oogonium) take the matching chromosomes from each of our parents and literally break them apart and recombine them. On average, each pair of structurally similar chromosomes engages in this recombination process one or two times [5], although the rate of chromosomal recombination is 1.65 times higher in females, and it is also affected by the environment [5, 6]. As a result of all this chromosome shuffling, genes that we received from our fathers are connected on a single chromosome to genes we inherited from our mothers, and in some cases, this may result in the formation of sex cells that contain chromosomes with truly unique combinations of gene pairs.

In addition to combining genes in new ways, our sex cells also randomly partition one copy of each pair of chromosomes to the cells that eventually form sperm or eggs. During this partitioning process, the 23 sets of chromosomes align randomly along the middle of the cell that will produce a sperm or an egg. In each case, a chromosome within a pair has an equal probability of being on the top or on the bottom of the midline in these cells, and eventually, because of the gamete formation process, the chromosomes on the top will end up in one gamete and those on the bottom will end up in a different gamete. As a result, the final genetic constitution of a gamete ultimately depends in part on how the 23 pairs of recombinant chromosomes align along the invisible midline within our gamete-forming cells. In organisms like us, there are 2^{23} possible ways in which our 23 pairs of chromosomes could be distributed to our newly forming sperm or eggs. Consequently, the process of randomly aligning the recombinant chromosomes along the midline of a gamete-forming cell can generate a truly staggering number of genetically different types of gametes. In fact, these processes can potentially produce 8,388,608 genetically different sperm, and they can theoretically produce the same number of genetically different eggs. Assuming that human sperm and egg randomly meet following copulation, a mating pair of humans could, in principle, produce in excess of 70 *trillion* genetically different children simply by randomly distributing their pairs of homologous chromosomes to sperm and eggs. Given that there have only been about 108 billion people who have lived on the planet [7], these calculations reveal that a single mating pair of humans can, in principle, generate far more genetically different individuals than the number of people who have lived. Clearly, the process of recombinant chromosome formation and the random partitioning of the chromosome pairs to sperm and eggs contribute significantly to the genetic diversity of individuals within a population.

Interestingly, the need to break up old gene combinations and generate new ones, i.e., the need for sexual reproduction, increases when the environment is unstable and demanding. This can be seen most easily in organisms that reproduce both asexually and sexually. In these creatures, the rate of sexual reproduction is markedly enhanced under stressful conditions [2, 8]. This stress-induced response among facultative, sexually reproducing organisms suggests that the evolution and maintenance of sexual reproduction within the eukaryotic clade are driven, in part, by the inherent difficulty of life on Earth.

IT PAYS TO BE DYNAMIC

Life on Earth is difficult in part because, as our planet travels the galaxy, the environment changes. Difficulties also arise because the very ground that we stand on moves with the shifting of tectonic plates, and as this is occurring gushing solar flares, erupting volcanoes, Earth-pummeling asteroids, supernova-generated high-energy gamma radiation, and a host of other potentially troublesome phenomena constantly alter the environment that living organisms occupy. Furthermore, because of life's ability to evolve in an unpredictable manner, and because living organisms incessantly alter the biosphere, the challenges associated with living on this planet increase as the number and types of creatures multiply. In such a world, with all its uncertainty and competition, it pays to generate offspring with many different variations, and sex allows eukaryotic organisms to accomplish this.

The emergence of sex ensured that eukaryotic life forms like us are unique combinations of gene variants, and therefore, we are all essentially experimental biological systems whose goal is to survive and reproduce in a world that knows no long-term constancy. This ability to generate vast numbers of gene combinations took 200 million to 1 billion years to evolve, and it emerged only once in the history of life [9], so it appears that sexual reproduction is yet another example of a significant "one-off" event. Considering that emergence of sexual reproduction was not necessarily inevitable, it is worth pondering what our world would be like if sex had not arisen. Although it is always hard to know the answer to a counterfactual question like this, what we can say with certainty is that our planet would be very different biologically, geologically, and environmentally. We can also state, that in the absence of sexual reproduction, animals, as we know them, would not exist.

Among animals, the desire to engage in sex is an integral part of existence because, within the animal kingdom, the yearning to participate in sexual reproduction was hardwired into the nervous system. As a result, the longing for sexual contact has molded and shaped the very essence of the nature of an animal. In humans, our sexual nature becomes evident early in life, and in our peak reproductive years, most of us think about sex at least once a day. Occasionally, the drive to engage in sex is even stronger, and some individuals report thinking about sex hundreds of times a day [10]. For many of us, the pursuit of sexual satisfaction consumes a great deal of time and energy, and within our species, the desire for sexual contact and its attending emotions has inspired great poetry, art, music, and literature. This same desire has also led to protracted bloody wars among sprawling empires, and it has motivated feelings ranging from passionate love to murderous rage. And imagine, all of this began one billion to two billion years ago with some very unusual, one-off, single-celled eukaryotes that were simply trying to leave behind a few descendants and eke out an existence within the ancient ocean in which they lived.

JOINING TOGETHER—THE EMERGENCE OF MULTICELLULAR LIFE

To move from sexually reproducing, single-cell eukaryotes (i.e., protozoans) to animals (i.e., metazoans), it was also necessary for single-cell eukaryotes to come together to form multicellular conglomerates. The antecedents of multicellular eukaryotic life appeared in the nitrogen-fixing cyanobacteria, and the ability to form multicellular aggregates evolved independently at least 22 times within the eukaryotic superkingdom [11]. These facts suggest that, in a world of single-celled eukaryotes, multicellularity was destined to occur. Nevertheless, it took hundreds of millions of years to evolve this state of being, and multicellular eukaryotes did not appear until about 1.2 billion years ago.

What triggered the formation of the first multicellular eukaryotes? Some suggest that the appearance of unicellular eukaryotic predators that could phagocytize (i.e., ingest) other unicellular eukaryotes drove the selection of relatively large, and therefore hard to capture and ingest, multicellular creatures. However, multicellular organisms also exhibit a number of physiological traits that cannot be duplicated by unicellular creatures [12], and these likely impart significant survival advantages.

Two examples of processes that multicellular organisms execute differently from their unicellular counterparts are those related to cellular motility (i.e., cell movement) and cellular division. Both functions require the spatially specific distribution of long, thin, protein-based tubules, which appropriately enough are called "microtubules." Within some unicellular eukaryotic organisms, the organization of these microtubules depends on the cell's location and whether it is about to partition its chromosomes and undergo cellular division. If the cell is not dividing, and if it needs to move to a new location, a specialized, protein-based complex called the microtubule organizing center (MTOC) assembles the microtubules into a whip-like structure called a flagellum, which can propel and maneuver the cell. On the other hand, if the cell needs to partition its chromosomes in preparation for cellular division, these same MTOCs arrange the microtubules into a "spindle apparatus" that can participate in the partitioning of the recombinant chromosome pairs by pushing each member of a pair to one of the poles of the dividing cell. Interestingly, within a single cell eukaryote, the MTOCs cannot generate both a flagellum and a spindle apparatus at the same time, and this fact has significant biological consequences. To survive, unicellular eukaryotic organisms must be able to move and divide, but during cell division, a unicellular eukaryote must disassemble its flagellum, and therefore, during this period, its ability to move is compromised. However, within a multicellular eukaryotic system, there are specific cells that divide while others attend to the processes associated with movement. This allows multicellular creatures to move and divide simultaneously, which is an extremely useful ability.

Initially, the many advantages associated with multicellularity may have led to the formation of colonies of genetically different, single-celled eukaryotes. Within these groups, individual cells take turns executing specific cellular processes such as those required for movement or cell division. As a rule, this arrangement works well—as long as all of the cells in the collective participate equally in the various tasks [13]. However, the success of this system diminishes when members do not share the group's responsibilities equally. If, for example, some of the colony members forgo the work associated with moving and instead focus their energy on reproducing, the dividing cells eventually take over the colony, and then the entire collective fails because of its inability to effectively move from one area to another. Unfortunately for a colony, each cell within the group is a unique genetic entity, and therefore each initially benefits by reproducing at the expense of its fellow colony members. Without the ability to avoid the long-term consequences of this selfish behavior, the cells of a colony tend to die off as the number of selfish freeloaders within the colony increases.

The problem of the freeloader can be countered, however, by forming groups of genetically identical cells. In such conglomerates, all the cells have the same genes, and therefore all equally benefit from the formation of a reliably functioning complex in which each member shares in the work that needs to be done to survive. As a result, multicellular conglomerates of genetically identical cells tend to outcompete the less cooperative, multicellular colonies of genetically unique cells.

We humans, and many other eukaryotic creatures, are produced by the clonal expansion of a fertilized egg, and consequently, each of us is a highly regulated, multicellular complex of nearly genetically identical cells. By finding a way to cooperate, our cells essentially solved the "tragedy of the commons" problem. The fact that groups of cells overcame this difficulty more than a billion years ago is something to remember the next time you read about a philosopher or politician who states that he or she was among the first to devise a plan to successfully solve a thorny philosophically engaging, "tragedy of the commons" dilemma.

THE ROGUE CELL—AN OMEN OF DEATH IN A CLONAL COMPLEX

In some cases, a cell within a clonal complex fails to cooperate with its neighbors. When this happens, the biologically imposed solution to selfishness fails, and the rogue threatens the existence of the entire multicellular complex. In humans, the most studied examples of uncooperative, scallywag cells are those that give rise to cancer. Ultimately, the cancerous cell fails to properly carry out its specific function, and it often breaks free of the interactions that bind it to its fellow clones. Unfortunately, once the oncogenic cell acquires this "independent actor" status, it replicates and moves about in an uncontrolled manner, and in so doing, it compromises the function of the well-organized cellular collective that is the individual. Often, the damage done by a cancerous defector and its descendants is so great that it kills the individual, and ironically, because of their activity, the cancerous cells also eliminate themselves. In the end, the fate of the cancer cell serves as a cautionary tale about the dangers of self-centeredness, and the destiny of a rogue cancer cell can be held in vivid contrast to that of the well-coordinated group of individual clones that exist in many multicellular organisms.

ACQUIRING A FUNCTIONAL IDENTITY—DEVELOPMENTAL BIOLOGY EMERGES WITHIN A EUKARYOTIC MULTICELLULAR COMPLEX

While it is easy to understand how genetically different cells acquire distinct functions, it is less obvious as to how cellular clones do so. The answer, in part, lies in the ability of an individual cell to differentially express some fraction of its genetic endowment, because by selectively expressing the genes that it shares with its sister clones, each cell can become a specialist.

To initiate differential gene expression, the identical cells within a multicellular complex must have regulatory proteins and signaling systems that can impose unique patterns of regulation upon their genomes. As it turns out, bacteria and archaea evolved regulatory proteins and signaling systems long before the first eukaryotes appeared. With these systems, they were able to detect changes in the environment, alter their gene expression, and respond appropriately to stimuli. To generate their specialized functions while working cooperatively and efficiently, the clones of eukaryotic cells in a multicellular complex had to modify and expand the gene networks that they inherited from their prokaryotic ancestors. With their enhanced apparatus, the cells of early clonal complexes determined their spatial positioning within a group, and then they activated specific genes in response to this information. Over time, this process of selective gene expression became more sophisticated, and groups of cells that were able to differentiate themselves in a manner that benefited the multicellular collective were advantaged, i.e., they were more likely to survive and reproduce.

Eventually, the strategic regulation of gene expression contributed to the evolution of many sophisticated developmental processes such as gastrulation. During gastrulation, some cells move to the center of the multicellular complex, and once there, they begin to form part of a trilayer structure that leads to the production of tissue and organ systems like those found in humans [13].

It is worth noting that, as the developmental systems that coordinate gene expression became more complex, the probability that these systems would malfunction increased. Central organizing events such as gastrulation are particularly critical developmental stages that we must all successfully transit before we are born, and unfortunately, in humans, as many as half of all pregnancies end, often before a woman knows she is pregnant because the newly forming individuals fail to successfully negotiate this or some other essential developmental process [14]. Given this reality, many developmental biologists believe that the successful completion of early developmental stages constitutes one of our greatest achievements, and to quote Lewis Wolpert, "It is not birth, marriage, or death, but gastrulation, which is truly the most important time in your life."

Successfully executing developmentally regulated gene expression programs is of the utmost importance to an individual; however, the evolution of sophisticated developmental processes is also one of the most essential sets of events to occur in the history of life. Ultimately, due to the existence of positional and coordinately controlled gene expression systems, simple complexes of interacting cells eventually gave rise to incredibly intricate and beautifully complex multicellular creatures like you and me.

THE DIVERSIFICATION OF EUKARYOTIC MULTICELLULAR ORGANISMS

Once the first simple clonal multicellular organisms appeared, many others soon followed. This was inevitable because new genes are constantly being generated in a population as a result of environmental mutagens, internal physiological processes, and spontaneous errors that occur during DNA replication. Once these new genes arise, they often affect the unfolding of developmental programs. Furthermore, in multifaceted ecological systems, sexually reproducing organisms help to optimize evolvability (see Chapters 7 and 8 and [15, 16]), and this further accelerated the genesis of novel developmental programs and species.

One of the first groups of complex multicellular species to appear on the planet was the photosynthetic red algae of 1.2 billion years ago. The arrival of these and other photosynthetic creatures was particularly important because it contributed to the oxidation of the environment [17–19], and that, in turn, helped to produce the stratospheric ozone located 6–30 miles above the planet (for more on this see the previous chapters). Ozone is a highly reactive layer of gas that filters out 97–99% of the incoming, DNA-damaging, UV light (i.e., light with wavelengths of 200–315 nm), and by 600 million years ago, the presence of this gas in the atmosphere allowed many forms of life to emerge from their subterranean and aqueous environments and safely colonize the planet's surface.

Despite the improving environmental conditions, animals were relatively slow to appear, and among the first to do so was a creature similar to the modern-day ocean sponge. Although this is still an issue of contention, some in the scientific community argue that this organism first evolved about 640 million years ago, and if this is correct, then the humble sponge is our oldest animal forbearer [18, 20, 21]. Earlier I suggested that, if we were to assemble a photo album of our predecessors, we should assign a place of prominence to several of the primogenital members, including the first protocell, the first prokaryote, and the first eukaryote. To this list, I would like to add the ancient ocean sponge (see Figure 9.1 for examples of modern-day sponges).

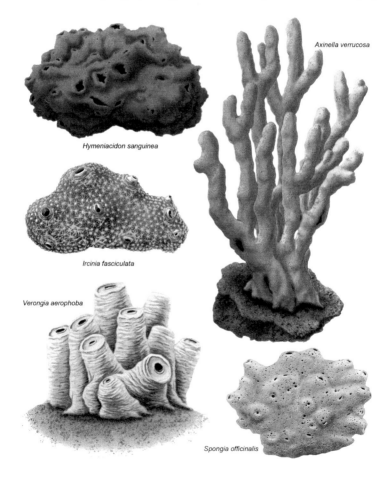

FIGURE 9.1   An artistic rending of modern porifera (i.e., sponges).

Together, these four entities spanned the first 85% of our lineage's existence, and in comparison, all the remaining animal ancestors in our direct line of descent are relative newcomers to our family tree.

THE GREAT BLOSSOMING OF THE ANIMALS

Sponges themselves arose from single-celled eukaryotes that are related to modern-day choanoflagellates ([13, 18] and see Figure 9.2), and it took 2.8 billion years of "descent with modification" and "natural selection" for nature to generate these wonderfully simple animals. But while they were a long time coming, once they finally arrived, they quickly gave rise to a great deal of diversity and morphological complexity.

One particularly noteworthy epoch of animal diversification is the "Cambrian explosion," which occurred 541–516 million years ago. During this period, diversification of the animal kingdom appears to have been exceptionally rapid, and by the end of this era, most of the phyla (i.e., major groups) of animals were in existence. The rapid diversification of the animal kingdom gives rise to the question of what drove this process, and currently, it appears that the answer incorporates a number of geological, ecological, and genetic phenomena that occurred around the time of the Cambrian explosion.

One of the most significant geological events of this period was the rifting of the supercontinent of Rodina. As this large landmass ripped apart, a new coastline was generated and freshly exposed continental rocks rich in phosphate and other nutrients began to weather more quickly. The rifting of Rodina also triggered the formation of active volcanoes along thousands of miles of the seafloor, and the liquid rock that poured from these structures shifted seawater onto the continental landmass by forming a buoyant crust that elevated some sections of the ocean floor. The inland

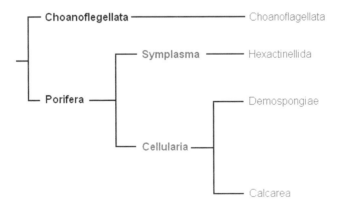

FIGURE 9.2 The porifera (i.e., sponge) family lineage. The extant porifera and their relationship to each other are illustrated. The choanoflagellate are extant free-living unicellular and colonial eukaryotes that share a common ancestor with the porifera. (This image is from Wlodzimierz, Public domain, via Wikimedia Commons.)

seas generated because of this volcanic activity enhanced the rate of water evaporation, and this in turn increased the volume of precipitation and the rate at which Rodina weathered.

As phosphate and other growth-limiting nutrients from the weathering and disintegrating rocks leached into the waters in and around Rodina, the population of oxygenic photosynthetic algae flourished and released oxygen into the environment. Interestingly, the oxygen itself potentiated the release of phosphate, iron, and other biologically important elements from marine and coastal rocks [22–24], and as all these growth-limiting nutrients poured into the waters, the population of oxygenic photosynthetic algae further skyrocketed. So, weathering triggered the growth of the oxygenic algae population, and that increased the release of oxygen, which in turn led to more weathering, and more algae growth. With this positive feedback cycle in place, the number of photosynthetic organisms in the seas quickly expanded, and the environment on the planet rapidly changed in a manner that favored the metabolic diversification of the aerobes [23].

Under these conditions, the planet's aerobic organisms were able to colonize a wide range of pre-Cambrian niches. However, this relatively prosperous period for aerobes was fated to end, because while the phototrophs were releasing oxygen, they were also consuming CO_2, and many of these phototrophs took the carbon they captured while living to their underground graves. So, ultimately, all this carbon hoarding by dead phototrophs triggered a drastic drop in global temperature [25], and eventually, Earth underwent the "snowball" or "slush ball" glaciations of 720–630 million years ago.

Once the planet was entombed in ice, processes similar to those that occurred during the earlier Huronian glaciation played out again. Specifically, the number of photosynthetic organisms and the rate of weathering decreased, but the planet's volcanoes kept pumping out CO_2. Eventually, this led to a spike in CO_2 and global temperatures, and as the surface temperature climbed, a significant amount of methane was likely released from the Arctic permafrost and the bottom of the ocean floor. This gas was produced over extended periods by the action of geological forces on dead and buried organisms, and prior to the periods of post-glacial warming, much of this gas was trapped in ice-like crystal lattices. However, as Earth's surface temperature rose, some of the crystal structures that entrapped the methane melted, resulting in the release of this potent greenhouse gas into the biosphere. Consequently, after the snowball or near snowball events, the planet's surface temperature quickly rocketed to hothouse levels.

Following this particularly trying period in Earth's environmental history, the climate eventually became less volatile. This was likely due in part to the partitioning of Rodina and the reestablishment of the photosynthetic algae populations. The plate tectonic-driven dissolution of the Rodian supercontinent repositioned the high albedo land and ice masses at the poles, and low albedo ocean waters flowed into the equatorial regions of the globe. These geophysical changes helped to trap heat in the equatorial oceans, and they attenuated rapid decreases in Earth's surface temperature. Meanwhile, the reestablishment of photosynthetic life decreased the amount of carbon in the atmosphere, and this also buffeted the planet against rapid temperature increases. So, the planet eventually regained its equilibrium; however, the Neoproterozoic era (which occurred 1 billion to 542 million years ago) was not an easy time to be alive, and it is necessary to go back as far as the Huronian ice age of 2.4 billion years ago ([25] and see also Chapter 8) to find temperature changes that rivaled those that occurred during this period. Ultimately, most Neoproterozoic life forms were not well equipped to handle the dramatic swings in the climate that the geological and environmental forces conspired to bring about, and consequently, many of the organisms of this era went extinct.

For those that did endure, however, better days were ahead. The trials and tribulations of the Neoproterozoic era eliminated entire populations, and as a result, the survivors had a plethora of niches that were ripe for exploitation. In fact, the ancestor of all modern-day animals was one of the creatures that managed to persist through the difficult days of the Neoproterozoic, and once the first animal appeared, it was in the enviable position of existing in a world rich in resources and low in competition.

As the beneficiaries of this good fortune, animals were able to disperse around the globe, and many adapted to their new environments by relying heavily on genetic systems that fostered evolvability. With these systems, the animals quickly generated novel genes [26], and they altered the expression of many of their ancient genes [27]. As a result, during the Cambrian explosion, the animals diversified and adopted a myriad of different forms.

Our direct ancestors were among the fortunate organisms to emerge during this period. As it turned out, their genetic endowments, along with some good luck, allowed them to take advantage of the geological and environmental occurrences of the pre-Cambrian period. This was not the first stroke of luck our forebearers experienced, and as we go forward, we will see that it was also not the last time our ancestors ended up on the favorable side of luck's ledger.

REFERENCES

1. Eme, L., et al., *On the age of eukaryotes: evaluating evidence from fossils and molecular clocks.* Cold Spring Harbor Perspectives in Biology, 2014. **6**(8).
2. Otto, S., *Sexual reproduction and the evolution of sex.* Nature Education, 2008. **1**(1): p. 182.
3. Bernstein, H., et al., *Origin of sex.* Journal of Theoretical Biology, 1984. **110**(3): p. 323–351.
4. Lode, T., *Sex and the origin of genetic exchanges.* Trends in Evolutionary Biology, 2012. **4**(1).
5. Reich, D., *Who we are and how we got here: ancient DNA and the new science of the human past.* First edition. ed. 2018, New York, NY: Pantheon Books. xxv, 335 p.
6. Kong, A., et al., *A high-resolution recombination map of the human genome.* Nat Genet, 2002. **31**(3): p. 241–247.
7. Toshiko, K. and H. Carly, *How many people have ever lived on earth?* Population Reference Bureau, 2018.
8. Bernstein, H. and C. Bernstein, *Evolutionary origin of recombination during Meiosis.* BioScience, 2010. **60**(7): p. 498–505.
9. Goodenough, U. and J. Heitman, *Origins of eukaryotic sexual reproduction.* Cold Spring Harbor Perspectives in Biology, 2014. **6**(3).
10. Fisher, T.D., Z.T. Moore, and M.J. Pittenger, *Sex on the brain? An examination of frequency of sexual cognitions as a function of gender, erotophilia, and social desirability.* Journal of Sex Research, 2012. **49**(1): p. 69–77.

11. Knoll, A.H., *The multiple origins of complex multicellularity.* Annual Review of Earth and Planetary Sciences, 2011. **39**(1): p. 217–239.

12. Stanley, S., *An ecological theory for the sudden origin of multicellular life in the late precambrian.* Proceedings of the National Academy of Sciences, 1973. **70**(5): p. 1486–1489.

13. King, N., *The unicellular ancestry of animal development.* Developmental Cell, 2004. **7**: p. 313–325.

14. Editors, *How many people are affected by or at risk for pregnancy loss or miscarriage?* NIH, 2013. https://web.archive.org/web/20150402093633/http://www.nichd.nih.gov/health/topics/pregnancyloss/conditioninfo/Pages/risk.aspx.

15. Eors, S. and J. Maynard Smith, *The major evolutionary transitions.* Nature, 1995. **374**: p. 227–232.

16. Klug, W.S., et al., *Concepts of genetics.* Eleventh edition. ed. 2015, Boston, MA: Pearson.

17. Lücking, R., et al., *Fungi evolved right on track.* Mycologia, 2017. **101**(6): p. 810–822.

18. Sebe-Pedros, A., B.M. Degnan, and I. Ruiz-Trillo, *The origin of Metazoa: a unicellular perspective.* Nature Reviews Genetics, 2017. **18**: p. 498–512.

19. Heckman, D.S., et al., *Molecular evidence for the early colonization of land by fungi and plants.* Science, 2001. **293**(5532): p. 1129–1133.

20. Feuda, R., et al., *Improved modeling of compositional heterogeneity supports sponges as sister to all other animals.* Current Biology, 2017. **27**(24): p. 3864–3870, e4.

21. Wood, R.A., *The rise of the first animals.* Scientific American, 2019. **320**(6): p. 24–31.

22. Anbar, A.D. and A.H. Knoll, *Proterozoic ocean chemistry and evolution: a bioinorganic bridge?* Science, 2002. **297**(1137–1142).

23. Koch, L.G. and S.L. Britton, *Aerobic metabolism underlies complexity and capacity.* Journal of Physiology, 2008. **586**(1): p. 83–95.

24. Catling, D.C., et al., *Why O_2 is required by complex life on habitable planets and the concept of planetary "oxygenation time."* Astrobiology, 2005. **5**(3): p. 415–438.

25. Hazen, R.M., *The story of Earth: the first 4.5 billion years, from stardust to living planet.* 2012, New York, NY: Viking. 306 p.

26. Paps, J. and P.W. Holland, *Reconstruction of the ancestral metazoan genome reveals an increase in genomic novelty.* Nature Communications, 2018. **9**(1): p. 1730.

27. Valentine, J. and D.H. Erwin, *Fossils, molecules and embryos: new perspectives on the Cambrian explosion.* Development, 1999. **126**: p. 851–859.

10 The Blossoming of Terrestrial Life

All earth was but one thought—and that was death.

Lord Byron

In the land of skunks, he who has half a nose is king.

Chris Farley

The race is not to the swift, nor the battle to the strong, neither yet bread to the wise, nor yet riches to men of understanding, nor yet favour to men of skill; but time and chance happeneth to them all.

Ecclesiastes 9:11

Throughout most of history, life thrived primarily in aqueous environments, and if life exists elsewhere, it too likely has its origins in water. However, creatures that live in the seas are severely constrained, and if there are aqueous planets that support life, it is almost certainly life without advanced technologies or highly complex, human-like societies. For the written word, computer technology, modern medicine, advanced transportation systems, and all the other complexities of modern societies to develop, dry land is required, and creatures capable of living on the terrestrial sections of the globe must emerge. But, for most creatures, transitioning to land is extremely difficult. Obviously, on Earth, some life forms did acquire the ability to live on dry land, and consequently, we, and the advanced cultures we have created, exist. In this chapter, we will explore some of the history of this all-important sea to land transition, and we will examine how life on Earth fundamentally changed after various organisms established a foothold on dry land.

THE PHANEROZOIC EON

Life first moved ashore during the Archean eon, but terrestrial life, and in particular animal terrestrial life, did not flourish and rapidly diversify until the Phanerozoic eon. The Phanerozoic eon began at the start of the Cambrian explosion 541 million years ago and it continues to this day.

Geologists have divided up the Phanerozoic into three major divisions or eras: the Paleozoic (541–252 million years ago), the Mesozoic (252–66 million years ago), and the Cenozoic (66 million years ago until the present). Collectively, these three periods represent less than 15% of the time life has been in existence; however, despite this fact, most of our information about living organisms comes from an analysis of the Phanerozoic eon. In part, this is because the rocks from this period are more numerous, more easily accessible, and better preserved than those from earlier times.

In addition, before the Phanerozoic, life consisted largely of tiny, single-celled, water-bound, morphologically simple creatures that did not fossilize easily, but during the Phanerozoic, creatures evolved hard structures that were simpler to recognize and more likely to fossilize. Accordingly, an immense amount of information from the last 541 million years has been uncovered, and in this chapter, we will focus on a few of the major developments that occurred during this most recent of eons.

THE INVASION OF THE COASTS BY BACTERIA AND EUKARYOTIC ALGAE

As was mentioned earlier, the transition from the seas to land was one of life's most noteworthy developments, and one of the first creatures to encroach upon the shores of the continents were the mat-forming prokaryotes of 3.2 billion years ago [1].

Multicellular eukaryotes' land invasion occurred long after this, and among the first to attempt the feat were the fungi and algae. There is debate as to when exactly they did so, but some indicate that the green algae and fungi colonized the waterlogged outcrops of the continental landmasses approximately 1 billion years ago [2, 3].

Over a period of about 500–700 million years, some of the descendants of the first fresh water algae evolved into land plants [4], and soon after these creatures appeared they began altering the terrestrial landscape. The plants achieved this in part by laying down roots, which shored up the soils and altered the flow of ancient streams and rivers [5]. The root systems of the plants also released acids that potentiated the rate of rock weathering, and this led to the production of soils that deep-rooted trees would later colonize.

Three hundred sixty million years ago, the terrestrial landmass of the planet had been thoroughly transformed by earlier land invaders, and by this time, Pannotia, which was the supercontinent in existence at this point in Earth's history, was home to the first forestlands. But, in addition to bringing new life to many desolate and barren surfaces, the appearance of the phototrophs also prefigured an onslaught of death and destruction. This is because the thriving phototrophs transformed rocks into nutrient-rich soils, and these soils, and the nutrients they contained, spilled into the oceans. This triggered algae blooms, and when the algae died, their decomposition depleted large segments of ocean of oxygen. This, in turn, led to the suffocation and mass extinction of oxygen-dependent marine life [6]. Furthermore, as they had done earlier in Earth's history, the phototrophs removed carbon from the atmosphere, and that induced yet another devastating glaciation event [7]. Collectively, these conditions led to the Devonian mass extinction, which is the second of the five great mass extinctions that occurred after the emergence of animals. In each of these disasters, more than half of the extant species on the planet were annihilated, and during the Devonian mass extinction, more than 70% of the planet's species vanished.

THE ANIMALS FINALLY COME ASHORE

When the algae, fungi, and plants first secured a foothold on the continent some 1,000 million to 700 million years ago, the animals were still residing in the seas.

However, about 480 million years ago, the first members of this group ventured onto land. In the vanguard were creatures similar to modern insects, spiders, and centipedes [8, 9], and it is likely that other invertebrates similar to extant worms, tardigrades, and nematodes were also among the earliest continental settlers [10] (see Figure 10.1). Having found new environments to exploit, these creatures diversified in numerous ways, and by 400 million years ago, some insects had taken to the skies, making them the first animals to fly [11].

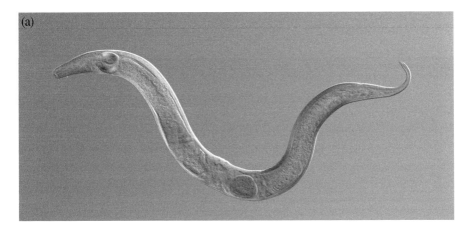

FIGURE 10.1   Modern-day relatives of some of Earth's first terrestrial animal inhabitants. Top: a nematode, bottom: a centipede.

FIGURE 10.1　On the top, is a tardigrade, and at the bottom, there is an image of a spider.

FIGURE 10.2 An artist's rendition of the lobe fish, Tiktaalik. (From: Obsidian Soul, CC BY 4.0 <https://creativecommons.org/licenses/by/4.0>, via Wikimedia Commons.)

While the intrepid invertebrates were moving onto dry land, our animal ancestors were still offshore. But, about 25 million years after the insects first took flight, some of our forbearers emerged from the sea and took their first tentative steps on solid ground. The animal pioneer that led the way was likely a lobe fish similar to Tiktaalik (see Figure 10.2). This organism had a vertebral column, and it also had primitive lungs and fin-like feet—features that allowed it to spend short periods on semidry ground. However, Tiktaalik, and other creatures like it, were far from being land lovers, and they awkwardly ambled about on semidry land for short periods, probably in an effort to find food and avoid predators.

By 365 million years ago, which was about 10 million years after the lungfish first sauntered along a continental landmass, Amphibia similar to Acanthostega entered the fossil record (see Figure 10.3). These water-loving ancestors of ours were better at exploiting land-based resources than were the lobe fish, but like all Amphibia, Acanthostega and its ilk were still essentially aquatic creatures that stayed close to the wetlands.

FIGURE 10.3   An artist's depiction of the amphibian, Acanthostega. [From: Nobu Tamura <http://spinops.blogspot.com>, CC BY-SA 3.0 <http://creativecommons.org/licenses/by-sa/3.0/>, via Wikimedia Commons.]

FIGURE 10.4   An artist's rendering of Hylonomus, one of the planet's first reptiles. (From: Nobu Tamura <http://spinops.blogspot.com>, CC BY-SA 3.0 <http://creativecommons.org/licenses/by-sa/3.0/>, via Wikimedia Commons.)

The first of our predecessors to spend significant time in the dry terrestrial uplands were the scaly and skittery reptiles. The first members of this clade evolved from Amphibia about 312 million years ago, and some of the early representatives of the group likely resembled Hylonomus which was an eight-inch-long lizard-like creature (see Figure 10.4). These hearty animals had a number of adaptations that made it possible for them to live on dry land including hard-shelled, and desiccation-resistant eggs that contained a cache of nutrients and internal gas exchanging membranes.

Soon after they appeared, the reptile clade broke into two groups: the sauropsids and synapsids. The sauropsids eventually gave rise to the dinosaurs and the modern-day reptiles while the synapsids generated the reptilian-like forerunners of modern mammals. Given what we know about dinosaurs, it may be surprising to learn that, in the initial match-up of these two lineages, the synapsids came out on top. When these synapsids first came to dominance, they did so on what was the Pangaean supercontinent, and within this large contiguous landmass, fearsome synapsid apex predators such as Dimetrodon and Ophiacodon hunted synapsid herbivores of all shapes and sizes (see Figure 10.5). Running around in the background, biding their time, and struggling to stay alive, were the sauropsid ancestors of the terrifying Mesozoic era dinosaurs. As it turned out, the sauropsids' rise to prominence would not occur until after yet another great mass extinction event.

FIGURE 10.5   Synapsid apex predators, Dimetrodon (top) and Ophiacodon (bottom). (From: ДиБгд, Public domain, via Wikimedia Commons.)

LIFE'S NEAR-DEATH EXPERIENCE

Life for the dominant reptilian synapsids was apparently just fine until about 252 million years ago, when the dreadful happenings of the Permian extinction began. This event, which is also known as "The Great Dying," was the most harrowing mass eradication of multicellular life in our planet's history, and during a geologically brief period of less than 60,000 years, this extinction event nearly sterilized the globe. By the time it was over, 70% of the terrestrial species and 96% of the marine species had vanished. Even the insects, which fared reasonably well during the previous mass extinctions of the Phanerozoic, lost eight of their 27 orders [12].

When faced with evidence regarding the demise of the reptilian synapsids and all the death and destruction that attended, the first question that comes to mind is, "what caused this dreadful and horrific outcome"? For a long time, the answer to this simple question was not clear. However, most paleontologists now believe that the proximate cause was yet another disruption of the planet's carbon cycle. In this case, the crisis was precipitated not by phototrophs, but rather by the release of volcanic material from the subterranean "Siberian Traps" [7]. As the name suggests, the Siberian Traps are below what is now Siberia, and when the molten rock erupted from this region, the torrents of lava produced covered wide swaths of Earth. In some places, the total discharge was 2.5 miles deep, and by the time the eruptions were over, the Siberian Traps had generated enough molten rock to blanket the entire continental United States in a layer a half a mile deep.

The sheer volume of lava that gushed from the Traps was sufficient to wreak havoc on many of the planet's ecosystems, but the most lethal killer released during this event was not the molten rock per se, but rather invisible gases. Of these, the most significant was likely CO_2, which immediately started to warm the planet and alter the chemistry of the oceans and atmosphere. This environmental tragedy was potentiated when the lava from the Traps flowed into the surrounding Tunguska basin deposits of coal, oil, gas, and salts. The resultant mixing triggered sustained explosions and polluted the atmosphere with methane and ozone-damaging chemicals such as methyl chloride, methyl bromide, and halogenated butane.

When the volcanic eruptions were over, the materials released from the Siberian Traps included a litany of toxins and approximately 10,000–48,000 gigatons of carbon. In addition to poisoning the oceans and atmosphere, these substances likely increased the temperature of the tropical waters from 77°F to 104°F [13], and some of the immense amounts of atmospheric carbon that were liberated by the volcanic activity also diffused into the oceans where it lowered the pH and decreased the amount of dissolved carbonate.

Ultimately, many marine organisms died due to exposure to the toxins, while others perished because of their inability to adapt to the temperature or pH changes, or because they could no longer access the carbonate they needed to build their exoskeletons and shells. The setbacks for marine life were further compounded by the downstream events of global warming. These included the temperature-induced depletion of dissolved oxygen in the water column, which in some coastal areas, was magnified because heat-induced weathering stimulated the growth of oxygen-depleting algae blooms.

Although most creatures on the planet were devastated by the horrors of the "Great Dying," some sulfide producing anaerobic bacteria did well under the new circumstances. Being one of the few beneficiaries of this period, this group was lucky enough to experience an increase in their collective metabolic activity. As a result, large amounts of the hydrogen sulfide they produced were released into the oceans. This gas pushed many marine species that were already struggling to survive into extinction. So, because of the Permian extinction, death came to life in the sea in an unprecedented manner, and whether a creature met its end because of temperature or pH changes, nutrient depletion, starvation, suffocation, or exposure to deadly gases, the result was the same.

On Pangaea, the situation during the end-Permian was equally grim. As a result of the Siberian Trap activity, millions of square miles of land were covered in lava, forests were burnt, and the atmosphere was polluted with a concoction of toxic emissions. Among these pollutants were greenhouse gases, which contributed to the establishment of heat waves that routinely approached temperatures of 140°F. In addition to baking the continents, the extreme heat enhanced the rate of evaporation, triggering torrential rainfalls in some regions and droughts in others. In areas where the rainfall was high, the plant-based root systems that shored up the soils and kept the rivers and streams flowing in defined routes were largely decimated, and because of this, the rainwater-fed waterways moved in undisciplined, sprawling patterns. This frequently caused flooding in the surrounding landscapes, which initiated new ecological disasters.

Meanwhile, within the largely barren interior of the supercontinent, many of the shallow inland seas dried up. This devastated the ecosystems that depended on these water sources, and in place of a lively community, the droughts left behind large piles of ocean salts.

Making matters worse, the hot ocean waters and extreme heat created by the volcanic emissions produced continent-spanning "hypercanes." These seemingly otherworldly storms had wind speeds in the neighborhood of 500 mph, and they generated storm surges that are hard to envision, even with the most vivid imagination. The volcanic gases in the atmosphere also potentiated the devastation by acidified water vapor and toxic acid rain.

The acid rain produced stripped the ground soil of its ability to support many forms of plant life, and it also lowered the pH of lakes, ponds, rivers, streams, and ground water, further stressing aqueous-based lifeforms. While all of this was occurring, the ozone-damaging gases released by the Siberian Traps were allowing large amounts of UV light to reach the planet's surface, where it mutated DNA and negatively impacted other biologically important cellular compounds. So, during the end-Permian mass extinction, the whole planet was shaken up, and there was no place on Earth immune from nature's wrath.

Collectively, the effects of the environmental and climatic changes that the end-Permian catastrophe produced were crushing, and when these cataclysmic events finally abated, nearly 90% of the planet's species were gone. Furthermore, the planet was so badly damaged that it took up to 30 million years for some parts of the biosphere to recover [14]. But as the planet sweltered, and as toxic gases and the smell of decaying corpses oozed into the atmosphere, new niches and opportunities were opening for those that somehow managed to survive this unbelievable hell on Earth.

And as was the case in the past, the lucky survivors of these catastrophic events limped into the future where they eventually created the dynamic ecosystems that renewed the face of Earth.

THE RISE OF THE DINOSAURS

As a group, the reptilian synapsids did not emerge from the Permian extinction event in good shape. Having been one of the planet's dominant groups, the synapsids were laid low by the Great Dying, and by the time the Paleozoic faded into the Mesozoic, most of the members of this clade were extinct. However, within 25 million years of the Permian extinction, at the start of the Mesozoic, the sauropsid line of reptiles did give rise to a new type of lizard-like creature, and in 1842, the English paleontologist Sir Richard Owen labeled these animals "dinosaurs," or "fearfully-great lizards" [15]. From this group, a new dynasty of dominant terrestrial vertebrates would eventually arise. However, the word "eventually" in the last sentence is an important modifier, because despite the popular image of the dinosaurs, these creatures were not all that fearsome when they first appeared, and they did not immediately assume the status of the dominant terrestrial animal. In fact, when the initial, relatively small, egg-laying, and nest-building dinosaurs arose, they were quickly subjugated by the ferocious, crocodile-related, sauropsid reptiles that also appeared after the Great Dying [16, 17]. So, for a period, the ancestors of the dinosaurs were once again relegated to the sidelines of life. But Earth is a lively place, and sooner or later geophysical realities always knock dominant species out of their position.

For the crocodile-like sauropsids, the good times came to an end about 201.6 million years ago when the supercontinent of Pangaea was splitting into two large continents called Laurasia and Gondwanaland. As this was occurring, Earth opened yet again, and that set into motion an event reminiscent of the Siberian Traps explosions. In four great pulses that occurred over a period of 600,000 years, a volcanic area called the Central Atlantic Magmatic Province (CAMP) spewed lava over the landscape, and by the time these volcanoes went quiescent, they had deposited molten rock over an expanse of more than three million square miles. In addition, the CAMP volcanic events flooded the atmosphere with globe-warming, lethal gases which once again poisoned the land, sea, and air. So, just 50 million years after the Permian mass extinction, life on Earth was profoundly altered yet again by volcanic events, and by the time the planet recovered from this "end-Triassic mass extinction," the dominant crocodile-like sauropsids were gone, and the dinosaurs assumed their position as one of Earth's dominant groups.

OUR DISTANT ANCESTORS—MASTERS OF HOW TO SURVIVE IN THE MOST DIFFICULT OF ENVIRONMENTS

While the apex synapsids, the crocodile-like sauropsids, and the dinosaur sauropsids were taking turns as the globe's dominant terrestrial life forms, the mammalian-like synapsids that gave rise to us were keeping a low profile. During the end-Permian extinction, these relatives of ours (which Richard Dawkins estimates were our 160 million great-grandparents) were warm-blooded, mammal-like reptilian creatures

FIGURE 10.6 An artistic depiction of *Lycaenops ornatus*, an early mammal-like reptile. (From: Dmitry Bogdanov, CC BY 3.0 <https://creativecommons.org/licenses/by/3.0>, via Wikimedia Commons.)

roaming about somewhere on the supercontinent of Pangaea [18, 19]. However, unlike their sauropsid cousins, the mammal-like synapsids had four legs positioned vertically below their torso, and consequently, their gait was more like that of a modern mammal. These synapsids also had jaws that resembled those of extant mammals, and it is likely that they had scale-free, naked skin covered in some places by hair [20] (see Figure 10.6).

As I mentioned earlier, most of the synapsids did not survive the Permian extinction, so how exactly our synapsid relatives managed to do so remains a mystery. Some have suggested that their ability to maintain a constant body temperature was helpful because the ambient temperature during this event fluctuated. However, it is likely that many factors played a role in the survival of these synapsids, and the most significant was almost certainly "luck." This is true because evolution does not shape creatures for future events, and the mammalian synapsids in our lineage that survived the Great Dying were not designed to do so. Instead, they endured because of happenstance—essentially, they happened to be in the right place at the right time, and they happened to have the traits needed to survive. Furthermore, it is worth noting that, although our ancestors had traits that allowed for exaptation (i.e., they had traits that could be repurposed and used in new ways), they were still nevertheless lucky in a statistical sense, because when only a small number of individuals have a favorable set of adaptive traits, there is no guarantee they will survive. In fact, there is a very real possibility that, over the course of a generation or two, a small group of individuals with favorable adaptations will simply die off by chance [21]. So, our synapsid ancestors survived because they enjoyed extraordinary levels of good fortune, and although life during the greatest of the great mass extinctions was no doubt exceedingly unpleasant and difficult, they endured.

THE RISE OF THE MAMMALS

By 210 million years ago, the struggling Mesozoic synapsids had given rise to the first mammals. These creatures were likely similar in size to a shrew or a house mouse, and they were bit players on the world stage. As a group, they spent most of their days hiding and most of their nights scurrying along the terrain hunting insects. Having arrived in time for the end-Triassic mass extinction of 201.6 million years ago, these ancient relatives of ours witnessed the fall of the crocodile-like sauropsids, and then they endured a 135-million-year period of subjugation by the dinosaurs.

But, despite their plight, the early mammals did manage to diversify and colonize specific niches, and as they adapted to their challenging world, they evolved superb hearing, specialized jaws and teeth, and a high level of intelligence. By 160 million years ago, the rate at which the mammals were evolving was nearly double what it was at the beginning of the Mesozoic era, and by this point, some mammalian groups had acquired the ability to swim, burrow, climb trees, and glide through the air [22, 23]. Fast forward another 15 million years, and by then some of our warm-blooded, highly active mammalian ancestors had evolved molars that allowed them to grind food.

With their new specialized teeth, our relatives were able to take full advantage of the angiosperms (which are flowering and fruit-producing plants) when they appeared about 120 million years ago, and ultimately molars gave our forebearer a distinct survival advantage. Using their unusual dentition, our ancestors outlasted many other mammals such as the dinosaur-eating, 3.5-foot-long, *Repenomamus* (see Figure 10.7).

Despite its small size, *Repenomamus* was a ferocious, intimidating, and combative little beast, and without knowing the outcome of the evolutionary process, one might have guessed that it would have navigated an existence in this difficult period, even without molars. However, the evolution of a seemingly trivial trait can have outsized effects on the big picture of life, and the presence or absence of a few mutations can significantly impact even the mightiest and most tenacious creatures.

Interestingly, modern humans still have molars to grind food, and we now know that specialized teeth did indeed have an outsized role in our eventual appearance on the planet. So, the next time you look in the mirror at your pearly whites, remember they are more important than they perhaps appear.

Despite our ancient relatives' good luck in the dental department, great teeth were not all that mattered in the struggle to survive and reproduce, and that was

FIGURE 10.7 Repenomamus *giganticus*, a Mesozoic mammal. (From: Nobu Tamura <http://spinops.blogspot.com>, CC BY 3.0 <https://creativecommons.org/licenses/by/3.0>, via Wikimedia Commons.)

particularly true during the amply challenging Mesozoic. Perhaps the nature of some of the Mesozoic's trials can best be understood by looking at the ecosystems that existed 66 million years ago, right before the non-avian dinosaurs joined the ranks of the extinct. During this period, one of the most fearsome land creatures ever to live, *Tyrannosaurus rex* (see Figure 10.8), roamed freely over what is now Asia and the western part of North America. At a length of over 40 feet, and standing 12 feet tall at the hips, *T. rex* weighed in at about 14,000 pounds. In addition to its immense size, this mighty killing machine also possessed binocular vision, an excellent sense of smell, good hearing, a reasonably large brain, and an estimated bite force of 12,800 pounds. Compared to *T. rex*, today's largest land predator, the 10-foot long, 1,000-pound polar bear, is downright puny and unimpressive. For the mammals of the Mesozoic, there was no possibility of taking on *T. rex* or other similarly large and ferocious hypercarnivores, and there was also no hope of displacing the giant herbivores of the day, such as the 115-foot long, 220,000-pound *Puertasaurus reuili* (see Figure 10.9). Due to their diminutive size, our forebearers were largely incapable of accessing the riches of the reptile-dominated world, and their best strategy for survival was to stay small and inconspicuous. Given this situation, it is not surprising that, at the end of the Mesozoic era, the largest mammal was similar in size to a modern-day, 3-foot long, 30-pound wolverine. Such a creature was no threat to the giants of its day, and the nutritional value gained by ingesting one of these little beasts was

FIGURE 10.8   *T. rex*, one of the world's most prominent terrestrial carnivores.

FIGURE 10.9   Puertasaurus—an unusually grand herbivore within the dinosaur lineage.

hardly worth the attention of a 14,000-pound predator. But as the Mesozoic came to a close, the days of the mammals playing second fiddle to the dinosaurs were about to end. Life on Earth was soon to become truly horrible once again, and when the worst of this period had passed, all the non-avian dinosaurs would be gone. Thanks to yet another unimaginable great disaster, the mammals would finally have their chance to rule.

THE END-CRETACEOUS MASS EXTINCTION

The reign of *T. rex* and its kin ended abruptly 66 million years ago, when, on what was likely a warm summer day, a rock larger than Mount Everest shot out of the sky at a speed of more than 31,000 mph. As it descended, the asteroid created a pressure wave that temporarily heated the area below to temperatures several times hotter than the surface of the Sun, and when it finally slammed into the shallow waters of what is now the Yucatan Peninsula near the town of Chicxulub, Mexico, it may have released as much as 5.8×10^{25} joules of energy [24]. This amount of energy is many billions of times that liberated by the atomic bomb that destroyed the city of Hiroshima, Japan, and the consequences of the nearly instantaneous discharge of this force are almost inconceivable.

Immediately upon impact, the asteroid carved a hole in Earth that was more than 110 miles in diameter and 12 miles deep, and it pushed the terrain around the impact site into mountain ranges as high as the modern-day Himalayas. The collision also ejected about 100 billion tons of dust and debris into the atmosphere, and it acidified the oceans [25]. At the same time, the force released by the asteroid strike transformed much of the nitrogen and oxygen around Chicxulub into nitrogen oxides, and it converted large amounts of sulfate and carbonate-rich rocks into atmospheric sulfuric compounds and CO_2. Furthermore, the asteroid deposited enough energy in the area to cause the Yucatan Peninsula to radiate substantial amounts of heat for more than two million years [26].

For living organisms, the effects of this collision were immense, devastating, and immediate. Before the silently approaching, the mountain-sized fragment of cosmic debris even hit the planet, the heat, and the air blast it generated wiped out all life within thousands of square miles of the impact site. Even 1,800 miles away, in what is now Boston, Massachusetts, 90% of the trees were ripped up by the air blast. When the asteroid did hit, the sound of the collision in present-day Boston was as loud as an approaching subway train in the city's Park Street Station [13]. Immediately after the crash, earthquakes, some of which were hundreds of times more powerful than any humanity has ever experienced, violently shook the entire planet, and very soon thereafter, tsunamis rushed the shores, bringing waves of 330 feet to the areas where Texas and Florida are now located [26].

Within an hour of the asteroid impact, much of the displaced earth and asteroid debris began to fall back to the planet, and as the mountains of refuse descended, they heated the atmosphere to the point at which the surface below literally broiled for about 20 minutes. When this incoming blizzard of ejected rock and debris finally did hit the planet's surface, the violent pummeling set fires ablaze around the world. But all of this was just the beginning, and for life on Earth, the existential horror that directly followed the asteroid strike would only grow more intense.

During the first few days after the collision, Earth continued to reverberate from the trauma of the asteroid impact, and tsunamis repeatedly slammed into the coasts, inundating, and destroying, large swaths of coastal regions and inland areas. Meanwhile, the nitrogen oxides, sulfur compounds, and CO_2 that were released into the atmosphere generated acid rain, and these compounds, along with the dust and debris, poisoned the air and blocked approximately 20% of the incoming sunlight. This triggered an "impact winter," which cooled the oceans and devastated marine organisms [27]. Modeling studies suggest that the impact winter also decreased Earth's average surface air temperature by 47°F for a period of at least three years [28, 29], and during the coldest year, the annual average global temperature over land may have dropped from 59°F to 26°F [30]. In addition, the planet's ozone layer was severely compromised, and consequently, large amounts of biologically destructive UV light inundated the planet [31]. Collectively, the acid rain, acidified bodies of water, attenuated incident sunlight, impact winter, and intense UV light extracted a heavy toll on the planet's vegetative life, and many plant species that survived the initial trauma of the asteroid strike eventually perished. The death of the phototrophs was yet another profound calamity for the animals and the other creatures that depended on these organisms, and their demise led to the collapse of many ecosystems.

Eventually, the dust and debris in the atmosphere settled, but the greenhouse gases the asteroid generated remained aloft, and they began to increase the temperature on the planet's surface. So, after enduring an impact winter, life on Earth was subjected to a 100,000-year warming event, during which the average global temperature rose 9°F [32]. This wild swing in climatic conditions was extremely difficult for most creatures to endure, and not surprisingly, it was responsible for adding numerous species to the asteroid-generated casualty list.

To make matters worse, there is now evidence that, at the time of the asteroid strike, life was also contending with volcanic activity emanating from the Deccan Traps. This area is located near what is now Mumbai, India, and the volcanic activity was fueled by a magma chamber 620 miles wide and 30–60 miles deep [33]. Some believe that, at the time of the Chicxulub asteroid strike, these Traps were already active, and the intensity of their activity increased as a consequence of the asteroid impact [34–36]. Others doubt this hypothesis, and they believe the asteroid alone was largely responsible for the devastation of this period. However, regardless of the relationship between the Chicxulub asteroid and the Deccan Traps, we do know that, around the time the impactor hit the planet, these Traps were belching an enormous amount of toxic gases and lava, and it is estimated the basalt floods they produced covered at least 200,000 square miles of landscape in up to 6,000 feet of molten rock. Ultimately, the Deccan Traps explosions likely made the unimaginably bad situation generated by the asteroid strike even more grim [37].

By the time it was over, the end-Cretaceous extinction had claimed approximately 75% of the extant plant and animal species, a death toll second only to that of the end-Permian extinction of 252 million years ago. And as was mentioned earlier, among its many victims were all the non-avian members of the dinosaur clan. So, after more than 135 million years, the dinosaurs' reign as the dominant terrestrial animal was brought to a violent end largely by a giant space rock that abruptly appeared out of nowhere. Amazingly, after millions of years of "sitting pretty" and carrying on for what was—in human timelines—a functional eternity, the dinosaurs were suddenly no more.

Clearly, the end-Cretaceous mass extinction was disastrous for the dinosaurs, but it was also extremely hard on our relatives. Soon after the asteroid strike, nearly all the animals larger than a medium-sized dog were gone, and it has been suggested that as many as 93% of mammalian species went extinct during this event [38]. One

FIGURE 10.10 *Rhynchocyon petersi* (aka the black and rufous elephant shrew) likely resembled our end-Cretaceous ancestor of about 66 million years ago [42]. (From: Joey Makalintal from Pennsylvania, USA, CC BY 2.0 <https://creativecommons.org/licenses/by/2.0>, via Wikimedia Commons.)

research group estimates that only 15–20 mammalian species survived the end-Cretaceous extinction, and if that is true, all of the mammals found on Earth today evolved from these founder populations [39].

Obviously, our ancestors were members of one of the few mammalian species to survive this terrible time, and our direct ancestor during the end-Cretaceous likely resembled a modern-day rodent. This little creature was probably a scavenger and an insect-eating, non-placental, tree-climbing animal that weighed between 0.2 and 8.6 ounces (see Figure 10.10) [40]. To endure the horrors of the end-Cretaceous extinction, this ancient forbearer of ours likely ate whatever it could find, and it probably spent large amounts of time hibernating in underground burrows. As a result of its subterrestrial lifestyle, some members of this species were spared the gruesome fate of being broiled alive during the immediate aftermath of the asteroid impact, and if they used insects and dead organisms as their primary food sources, they would have fared much better in the post-asteroid world than their larger plant and meat-eating contemporaries. In addition, by hibernating, these ancient ancestors of ours decreased their metabolic rates and they probably increased their chances of surviving the drastic temperature swings that the asteroid impact and volcanic eruptions initiated [41].

TIME AND PLACE COUNTERFACTUALS

Exactly how many of our ancestors made it through the end-Cretaceous extinction is not known, but it isn't hard to imagine how small changes in survivability could have generated very different outcomes for our species. For instance, the extent of the volcanic activity after the asteroid strike would have been even greater if the giant hunk of space debris had hit in a location that triggered an even more vigorous set of volcanic explosions. Such an uptick in the volcano-induced carnage may have been sufficient to push the small number of mammalian species that did survive the end-Cretaceous into oblivion.

While scenarios like the aforementioned one may have ensured the elimination of the mammals, others may have increased the likelihood of the dinosaurs surviving. If, for example, the Chicxulub asteroid had taken a slightly different path, it might have completely missed Earth. In that case, the Deccan Trap activity may have simmered at a relatively low level, and the dinosaurs may have survived and extended their reign to the present day. In this scenario, the mammals likely would not have expanded and diversified, and if they managed to survive until today, they would probably still be small, unobtrusive creatures of minor significance.

There are numerous other scenarios that may have spared the dinosaurs. For instance, if the Chicxulub asteroid arrived a few seconds earlier or later, it may have hit the planet in an area that contained fewer hydrocarbon and sulfur-containing rocks. Eighty-seven percent of the planet's surface has less of this type of rock than the Chicxulub area does, and if the asteroid hit elsewhere, it would have likely generated less black carbon (i.e., soot), less sulfur gas, and less damage to the environment. Ultimately, this may have decreased the length and depth of the impact winter, and the long-term global warming that ensued after the asteroid strike may not have been as severe. In this situation, some of the non-avian dinosaurs may have been spared [43]. Furthermore, had the asteroid hit during a period when the Deccan Traps were less active, the extra push to extinction that the Traps volcanic activity provided would have been eliminated or attenuated, and this too may have increased the probability of survival for some of the non-avian dinosaurs. In the end, these last several scenarios suggest that the timing of the asteroid strike may have been extremely important, and had it been even slightly different, the dinosaurs might still be the dominant land animals, and the mammals might still be scurrying about under their feet.

Of course, the counterfactuals described earlier are simply descriptions of what may have happened if events had played out differently. But in the world of actual events, the Chicxulub asteroid strike and the Deccan Traps activity occurred, and they both took place during the same period. Furthermore, in the real world, the non-avian dinosaurs did go extinct, and our mammalian ancestors did survive.

When reflecting upon what did occur, one might note that the seemingly endless subjugation of the mammals by the dinosaurs may have been exactly the "good luck" our ancestors needed to make it through the end-Cretaceous extinction. From our modern perspective, we can see that, had the dinosaurs not successfully circumscribed our forbearers' evolutionary potential, our end-Cretaceous ancestors might have been more substantial than the small, minor lifeforms they actually were, and in that case, the giant rock from the sky would have almost certainly eliminated them as well. Some will no doubt argue that, in the absence of our actual ancestors, some other small burrowing mammal would have survived, and it would have eventually given rise to a highly intelligent lifeform like us. However, the organisms these other mammals might have given rise to wouldn't necessarily be human, for when it comes to the process of evolution, the starting point is very significant, and a different beginning often results in a different end.

So, the dinosaurs' existence mattered for humanity, and since their discovery in the 1840s, many of us have enjoyed a deep fascination with these frightening, lizard-like creatures of the Mesozoic. The fascination children and adults like me have for these beasts may, in fact, be justifiable, for it is hard to deny the relevance of the dinosaurs to our existence.

THE CENOZOIC—THE ERA OF THE MAMMAL FINALLY ARRIVES

The great Chicxulub asteroid strike that marked the end of the non-avian dinosaurs also delineated the end of the Mesozoic and the start of the Cenozoic era. Unlike all the previous periods in the history of Earth, this new epoch, which continues until this day, has ultimately belonged to mammals.

As the Cenozoic unfolded, the continents as we now know them began to take shape, and with the dinosaurs gone, new niches were wide open for the taking. Given all these new opportunities, the rate at which the mammals diversified increased greatly [44], and within 270,000 years of the appearance of the Chicxulub asteroid, the size of the mammals ballooned, with some becoming as large as modern pigs. By 60 million years ago, most of the 18 orders of the modern mammals had evolved, and by 56 million years ago, primates roamed the planet. Interestingly, these primates likely first appeared in Eurasia [45], but as a result of climate change, some of

FIGURE 10.11 *Catopitheus browni*—an artist's rendition of this early primate. (From: Xiphactinus88, CC BY-SA 4.0 <https://creativecommons.org/licenses/by-sa/4.0>, via Wikimedia Commons.)

the early members of this clade made their way to Africa, where about 34 million years ago, they gave rise to a monkey-like species called Catopitheus [46] (see Figure 10.11). Soon after this, the African continent became home to a number of additional anthropoid species, and at that point, the evolution of the modern human was just a few tens of millions of years away.

In the next chapter, we will begin to explore the rise of humans in general, and our species, i.e., *H. sapiens*, in particular. While doing so, we will pay particular attention to some of the events that shaped our nature. As we do, we will begin to see how we finally came to take our place as one of Earth's dominant species, and we will learn more about what made us the unusual creatures that we are.

REFERENCES

1. Martin Homann, et al., *Microbial life and biogeochemical cycling on land 3,220 million years ago.* Nature Geoscience, 2018. **11**: p. 665–671.
2. Heckman, D.S., et al., *Molecular evidence for the early colonization of land by fungi and plants.* Science, 2001. **293**(5532): p. 1129–1133.

3. Loron, C.C., et al., *Early fungi from the Proterozoic era in Arctic Canada*. Nature, 2019. **570**(7760): p. 232–235.
4. Furst-Jansen, J.M., S. de Vries, and J. de Vries, *Evo-physio: on stress responses and the earliest land plants*. Journal of Experimental Botany, 2020. **71**(11): p. 3254–3269.
5. Gibling, M.R. and N.S. Davies, *Palaeozoic landscapes shaped by plant evolution*. Nature Geoscience, 2012. **5**(2): p. 99–105.
6. Baraniuk, C., *The Devonian extinction saw the oceans choke to death*. BBC Earth, 2015.
7. Brannen, P., *When life on earth was nearly extinguished*. New York Times, 2017.
8. Jeram, Andrew J., P.A. Selden, and D. Edwards, *Land animals in the Silurian: Arachids and myriapods from Shropshire, England*. Science, 1990. **250**: p. 658–661.
9. Engel, M. and D. Grimaldi, *New light shed on the oldest insect*. Nature, 2004. **427**: p. 627–630.
10. Davide Pisani, et al., *The colonization of land by animals: molecular phylogeny and divergence times among arthropods*. BMC Biology, 2004. **2**(1).
11. California-Academy-of-Sciences, *Landmark study on the evolution of insects*. ScienceDaily, 2014.
12. Labandeira, C. and J. Sepkoski, *Insect diversity in the fossil record*. Science, 1993. **261**(5119): p. 310–315.
13. Brannen, P., *The ends of the world: volcanic apocalypses, lethal oceans, and our quest to understand Earth's past mass extinctions*. First edition. ed. 2017, New York, NY: Ecco, an imprint of HarperCollins Publishers. x, 322 p.
14. Sahney, S. and M.J. Benton, *Recovery from the most profound mass extinction of all time*. Proceedings of the Royal Society, 2008. **275**(1636): p. 759–765.
15. Editors, *Where does the word "dinosaur" come from?* 2005. Chicago, IL: Wyzant Resources. https://www.wyzant.com/resources/lessons/english/etymology/words-mod-dinosaur-info/.
16. Nesbitt, S.J., et al., *The oldest dinosaur? A Middle Triassic dinosauriform from Tanzania*. Biology Letters, 2013. **9**(1): p. 20120949.
17. Brusatte, S., *The unlikely triumph of dinosaurs*. Scientific American, 2018. **318**(5): p. 28–35.
18. Dawkins, R., *The ancestor's tale: a pilgrimage to the dawn of evolution*. 2004, Boston, MA: Houghton Mifflin.
19. Rey, K., *More than 252 million years ago, mammal ancestors became warm-blooded to survive mass extinction*. The Conversation, 2017.
20. Bajdek, P., et al., *Microbiota and food residues including possible evidence of pre-mammalian hair in Upper Permian coprolites from Russia*. Lethaia, 2016. **49**(4): p. 455–477.
21. Berwick, R.C. and N. Chomsky, *Why only us: language and evolution*. 2016, Cambridge, MA: MIT Press.
22. Brusatte, S. and Z.X. Luo, *Ascent of the mammals*. Scientific American, 2016. **314**(6): p. 28–35.
23. Meng, Q.J., et al., *New gliding mammaliaforms from the Jurassic*. Nature, 2017. **548**(7667): p. 291–296.
24. Durand-Manterola, H.J. and G. Cordero-Tercero, *Assessments of the energy, mass and size of the Chicxulub impactor*. arXiv:1403.6391 [astro-ph.EP], 2014.
25. Henehan, M.J., et al., *Rapid ocean acidification and protracted Earth system recovery followed the end-Cretaceous Chicxulub impact*. Proceedings of the National Academy of Sciences of the United States of America, 2019. **116**(45): 22500–22504.
26. Bryant, E., *Tsunami: the underrated hazard*. 2001, Cambridge: Cambridge University Press.

27. Vellekoop, J., et al., *Rapid short-term cooling following the Chicxulub impact at the Cretaceous Paleogene boundary*. Proceedings of the National Academy of Sciences of the United States of America, 2014. **111**(21): p. 7537–7541.

28. Artemieva, N. and J. Morgan, *Quantifying the release of climate-active gases by large meteorite impacts with a case study of Chicxulub*. Geophysical Research Letters, 2017. **44**(20): p. 10,180–10,188.

29. Union, A.G., *Dinosaur-killing asteroid impact cooled Earth's climate more than previously thought*. ScienceDaily, 2017.

30. Brugger, J., G. Feulner, and S. Petri, *Baby, it's cold outside: climate model simulations of the effects of the asteroid impact at the end of the Cretaceous*. Geophysical Research Letters, 2017. **44**(1): p. 419–427.

31. Guterl, F., *Waiting to explode*. Scientific American, 2012. **306**(6): p. 64–69.

32. MacLeod, K.G., et al., *Postimpact earliest Paleogene warming shown by fish debris oxygen isotopes (El Kef, Tunisia)*. Science, 2018. **360**(6396): p. 1467–1469.

33. Richards, M., *Twice doomed?* Harvard Gazette, 2015.

34. Brusatte, S., *What killed the dinosaurs*. Scientific American, 2015. **313**(6): p. 54–59.

35. Brusatte, S.L., et al., *The extinction of the dinosaurs*. Biological Reviews of the Cambridge Philosophical Society, 2015. **90**(2): p. 628–642.

36. Renne, Paul R., et al., *State shift in Deccan volcanism at the Cretaceous-Paleogene boundary, possibly induced by impact*. Science, 2015. **350**(6256): p. 76–78.

37. Bosker, B., *The nastiest feud in science*. The Atlantic, 2018.

38. Longrich, N.R., J. Scriberas, and M.A. Wills, *Severe extinction and rapid recovery of mammals across the Cretaceous-Palaeogene boundary, and the effects of rarity on patterns of extinction and recovery*. Journal of Evolutionary Biology, 2016. **29**(8): p. 1495–1512.

39. Quach, K., *You lucky creatures! Mammals only JUUUST survived asteroid that killed dinosaurs*. The Register, 2016.

40. O'Leary, M.A., et al., *The placental mammal ancestor and the post-K-Pg radiation of placentals*. Science, 2013. **339**(6120): p. 662–667.

41. Perkins, S., *Why some species thrived when dinos died*. Science, 2013. https://www.science.org/content/article/why-some-species-thrived-when-dinos-died.

42. Hecht, J., *Meet our last common mammalian ancestor*. New Scientist, 2013. https://www.newscientist.com/article/dn23148-meet-our-last-common-mammalian-ancestor/.

43. Kaiho, K. and N. Oshima, *Site of asteroid impact changed the history of life on Earth: the low probability of mass extinction*. Scientific Reports, 2017. **7**(1): p. 14855.

44. Gore, R., *The rise of mammals*. National Geographic Magazine, 2019. https://www.nationalgeographic.com/science/article/rise-mammals.

45. Williams, B.A., R.F. Kay, and E.C. Kirk, *New perspectives on anthropoid origins*. Proceedings of the National Academy of Sciences of the United States of America, 2010. **107**(11): p. 4797–4804.

46. Simons, E., *Skulls and anterior teeth of catopithecus (Primates:Anthropoidea) from the eocene and anthropoids origins*. Science, 1995. **268**: p. 1885–1888.

11 The Emergence of the Genus *Homo*

Earth's First Humans Come Into Being

The very large brain that humans have, plus the things that go along with it—language, art, science—seemed to have evolved only once. The eye, by contrast, independently evolved 40 times. So, if you were to "replay" evolution, the eye would almost certainly appear again, whereas the big brain probably wouldn't.

Richard Dawkins

I have to argue about flying saucers on the beach with people, you know. And I was interested in this: they keep arguing that it is possible. And that's true. It is possible. They do not appreciate that the problem is not to demonstrate whether it's possible or not but whether it's going on or not.

Richard Feynman

We explore the frontiers of our universe: to discover our origins and, ultimately, to find ourselves.

Daniel Coe

To have tree-dwelling primates, you obviously must have trees, and about 55 million years ago, the planet passed through a period during which the surface temperature increased by about 14°F in just 10,000 years. This global warming led to the appearance of new forestlands, some of which extended up to the high latitudes of the northern hemisphere. It was in these forests that our most distant tree-dwelling primate ancestors first appeared [1], and much of our body plan evolved as adaptations to our ancestor's arboreal lifestyle. Some of our most prized traits such as our strong limbs, binocular color vision, and our highly flexible joints arose in our ancient tree-dwelling forbearers.

By the time Catopitheus and other early monkey-like primates arrived on the scene approximately 54 million years ago, Earth's tectonic plates were moving in a manner that would alter the climate and the planet. Specifically, the Indian and Eurasian plates were merging, and over a period of millions of years, the Himalayan Plateau and the mountains that surround it formed. At around the same time, the Andes mountains were developing [2]. These new peaks, as well as the planet's other mountainous regions, altered the circulation pattern of the atmosphere, and this triggered an increase in the amount of rainfall on the new rock that the recently formed mountains exposed. Ultimately, these events resulted in an increase in weathering,

DOI: 10.1201/9781003270294-14

a decrease in atmospheric CO_2, and a drop in the planet's surface temperature (see Chapter 5) [3].

As the temperature declined, the tropical forests that our ancestors lived in receded to lower latitudes, and it was during this period of diminishing forestlands and global cooling that the first apes came into existence about 23 million years ago. Initially, these creatures were limited to tropical areas of Africa, but around 19 million years ago, the African and Arabian tectonic plates collided, and Africa and Eurasia became connected by land bridges. Over the next five million years, these bridges would occasionally become flooded, but when they were accessible, the apes were able to move north from Africa to Eurasia.

Despite the existence of the transient land bridges, the apes didn't avail themselves of the opportunity to move into Eurasia until around 16.5 million years ago. The 2.5-million-year delay in their migration northward likely occurred in part because the African apes lacked the necessary dentition to exploit the food sources that were available in the newly opened territories. However, once these primates acquired the anatomical and physiological modifications they needed to eat the nuts and other types of foods that were plentiful on the Eurasian continent, they began to take advantage of the opportunities the Eurasian woodland climes afforded. But, as always, the good times eventually came to an end, and about nine million years ago, a change in the climate limited the tropical forests that the apes inhabited to parts of Africa and Southeast Asia [3].

During the 23 million years they have been on the planet, the apes of Africa and Eurasia have generated various genera, including several that still exist today (see Figure 11.1). Among the first to split off from the common ancestral line were the creatures that gave rise to the orangutans. This separation is believed to have occurred 13–15 million years ago, and today, the orangutan can be found in the wild only in the rainforests of Borneo and Sumatra in Southeast Asia. The next apes to diverge from the main ancestral line were those that evolved into the modern gorilla. This split is estimated to have occurred about eight million years ago [4], and currently, members of this genus, which can be found in Sub-Saharan Africa, constitute the planet's largest primates. The most recent ape groups to form are those containing chimpanzees and humans. The speciation events that gave rise to these clades likely began 7–13 million years ago, but repeated interbreeding between these creatures delayed their final separation, and complete divergence did not occur until about seven million years ago [4, 5]. Interestingly, of these two groups, it was the ancestral population that gave rise to the chimps that dominated the prime tropical forest territories. The primate branch that produced humans was largely confined to the less desirable interface between the forests and grasslands.

One especially interesting group of apes that emerged about four million years ago in the borderlands of Africa's forests and grasslands was the *Australopithecus*. There were many species within this group; however, paleontological data suggest that some of these apes may be of particular interest to us because they may have given rise to modern humans. One of the most likely candidates for the designation of "recent pre-human ancestor" is *Australopithecus afarensis*, although it is also possible that a closely related species, such as *Australopithecus garhi, gave* rise to modern humans instead [6, 7]

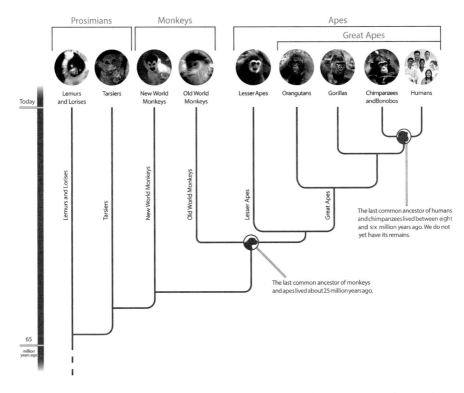

FIGURE 11.1 Phylogenetic relationships among humans, non-human apes, monkeys, and prosimians. (This figure is from the Human Origins Program, NMNH, Smithsonian Institution.)

THE GENESIS OF THE FIRST "HUMANS"

Given how long ago they lived, what can we say about *A. afarensis* and *A. garhi*? As it turns out, we can't say all that much about *A. garhi* since we have not found many fossilized remains of this species. However, the fossilized remains of "Lucy" and more than 300 other members of *A. afarensis* have been discovered, and consequently, this clade is among the best characterized ancient hominins [8].

Based on our trove of paleontological data, we know that male *A. afarensis* stood about 4 feet 11 inches tall and weighed approximately 92 pounds. Females were considerably smaller, and on average, they stood 3 feet 5 inches tall, and they tipped the scales at about 64 pounds (see Figure 11.2). In comparison to modern humans, the *A. afarensis* were intellectually limited and they had a brain that was approximately one-third the mass of our own. However, like us, *A. afarensis* were bipedal, and their ability to walk effectively on two legs distinguished them from many of their relatives. The fact that they, and perhaps some of the creatures from which they descended (such as Sahelanthropus and Orrorin, which lived about six million to seven million years ago, and *Danuvius guggenmos*, which lived 11.6 million years ago [9]), were bipedal indicates that this style of locomotion long preceded the evolution of advanced cognition.

FIGURE 11.2   A depiction of *A. afarensis*. This animal, or something similar to it, is likely a distant ancestor of ours. (From: Wolfgang Sauber, CC BY-SA 4.0 <https://creativecommons. org/licenses/by-sa/4.0>, via Wikimedia Commons.)

Why *A. afarensis* and various other hominins adopted a bipedal gate is a matter of debate. Some have suggested that, within the *A. afarensis* population, this trait arose as a suite of skeletal, muscular, and neurological adaptations to arboreal life, and it allowed arboreal apes to move efficiently on very flexible branches [10]. Others submit that *A. afarensis* spent a considerable amount of time in the grasslands, and their bipedalism permitted them to move more rapidly and efficiently over this terrain. It has also been suggested that bipedalism was adaptive because it freed the hands for carrying tools and offspring, or because it allowed for greater visual acuity in the grassland by making it easier to see both predators and prey, or because it minimized the surface area on the body that was exposed to the Sun, thereby decreasing the likelihood of heat exhaustion and dehydration [11, 12]. Of course, there may have been several benefits associated with bipedalism, and they may have all acted in concert within early hominid populations.

Over the long term, bipedalism also offered other important benefits. For example, it set the stage for the evolution of hands like ours, which are unique in part because they are capable of ulnar opposition (this occurs when the thumb on the hand can contact the tip of the small finger on the same hand). Bipedalism may have also facilitated the evolution of our wrists, which are particularly well suited to the transmission of force across the hand [13]. Together, these adaptations have given us strong and precise grips, and they have allowed us to exercise the highest level of fine motor coordination of any living creature [14]. Throughout our history, our unusual wrists and hands have been essential to our tool-making ability, and they have allowed us to construct instruments that far exceed what other creatures could produce, regardless of their brainpower and conceptual abilities.

In summary, the development of bipedalism changed our fate. But it is worth noting that, when this bipedalism first appeared, it did not impart the gait that we now employ. In creatures such as *A. afarensis,* this difference was due, in part, to the fact that, relative to us, these animals had shorter legs and longer toes. Short legs ensured that *A. afarensis* moved relatively slowly on land in comparison to modern humans, and their long toes, while useful for climbing, made the mechanics of walking and running difficult [15]. However, since the *A. afarensis* were able to travel through trees and on grasslands, they could gather food from both locations. This made the diet of these apes, which was rich in soft foods such as plants and fruits, relatively diverse. However, whether *A. afarensis* spent most of their time in the forests or in the grasslands is not clear. At a minimum, they would have likely spent their nights sleeping in the safety of the tree canopy, and they would have probably also used the trees as a refuge when they were under attack by one of the many hominid-eating predators that patrolled the African plains at this time.

Overall, *A. afarensis* were well adapted to their environment, and consequently, this species managed to survive for 900,000 years (i.e., from about 3.9 million to 3.0 million years ago [16]), which is about three times longer than our species has currently been in existence. Moreover, as a group, the australopithecines did quite well, collectively occupying the planet for about 2.5 million years (i.e., from about 4.5 million to about 2 million years ago). But, despite their successes, the *Australopithecus* taxon eventually succumbed to climate changes that began about 2.58 million years ago. This transformation was catastrophic for the *Australopithecus* because it devastated the forestlands upon which they depended.

As the trees disappeared, the *Australopithecus* were forced to spend more time in grassy areas. As a result, there was less adaptive value associated with traits that facilitated tree climbing and more selective pressure on attributes that allowed for survival and reproduction in non-arboreal regions. Some have speculated that this combination of selective pressures led some populations of *Australopithecus* in West Africa to seek food, water, shelter, and protection from predators in complex hilly terrains, and that ultimately led to their adoption of a fully terrestrial lifestyle [17]. However, even in hilly terrain, adapting to a world with fewer trees would have been no small feat, and in the absence of sophisticated tools or controlled fire, these apes would have been particularly susceptible to predation by cats, hyenas, and other carnivores. Furthermore, the loss of the forestlands would have deprived the *Australopithecus* of a significant food source, which was not easily replaced [18]. In response to these realities, Steven Stanley suggested that some australopithecines adapted to their new conditions through the evolution of enhanced intellectual functions [15].

According to Stanley, for most of their existence, the *Australopithecus'* intellectual abilities were limited by their need to rapidly develop the ability to cling to their mothers at a very young age. If a newborn ape had this capability, its mother could quickly usher it into the protective realms of the treetops when a predator appeared. On the other hand, if an ape matured more slowly due to the time-consuming maturation of its brain, it would not have been able to cling to its mother soon after birth, and that would place it in great peril. As a result, within the australopithecines, quick physical maturation was more valuable than a large, sophisticated, and slowly developing brain. However, Stanley argues that this situation changed when the environment became cold and dry, and the forestlands receded. Under these conditions, the apes would have spent more time in largely non-arboreal, complex hilly terrain, and these areas provided different types of safe havens than those available in forests. Consequently, having the ability to cling to one's mother while being whisked to the safety of the treetops was less important in the climate-altered world of the last remaining australopithecines. In their new environment, there was less selective pressure placed on rapid physical development and more opportunity for the evolution of larger, more slowly developing brains. Consequently, within the non-arboreal hilly areas, an infant ape that had a slowly maturing, large brain might survive, and if it managed to become semi-independent (i.e., if it made it to early childhood), the benefits it realized because of its superior intellect may have been substantial.

Under tremendous selective pressures, any adaptation that offered substantial survival advantages would have likely spread quickly through the small, isolated populations. Within a group, natural selection presumably would have favored those that were intelligent, equipped with the wrists and hands capable of making tools, and in possession of a skeleton that allowed for rapid and efficient bipedal locomotion. Arguably, at some point around 2.5 million years ago, these and other changes (more on this later) became manifest within a sequestered group, leading to the establishment of what could potentially be called the first humans (i.e., *Homo habilis* or a species like it).

While the speculative hypothesis outlined earlier is plausible, one of its tenets is that, in the absence of forestlands, the australopithecines were able to outmaneuver the predators of their day. This is hard to envision, given that even modern humans would be in significant jeopardy if they were placed, sans fire or sophisticated tools, among the many carnivorous social predators that roamed Africa 2.5 million years ago. Stanley countered this argument by pointing out that other species, including some antelopes, successfully transitioned from a life in the forests to one in the grasslands. However, antelope never climbed trees, and they were likely more amenable to this transition than were the australopithecines of the time. Nevertheless, the idea that some *Australopithecus* may have taken advantage of complex hilly terrain to increase their probability of survival is intriguing, and if it is correct, it may help explain how the initial transition from *australopithecines* to humans unfolded.

HISTORY REPEATS ITSELF

By now, the pattern of evolution described above is familiar to us, and once again, we encounter a situation in which the demise of one successful group coincides with the beginning of another. This time around, however, the group that would ascend and eventually move to the top of life's dominance hierarchy is our kind, i.e., members of the genus *Homo*. Worth noting is the fact that this transition from australopithecines to modern humans was not a simple one, and it certainly did not occur all at once. This is reflected by the fact that modern humans differ from apes in many ways. For example, relative to our ape ancestors, we have extended periods of development, narrow pelvic regions, large craniums and brains, longer linear bodies with elongated hind limbs and shorten forelimbs, small faces and teeth, altered foot and hand structures, relatively low levels of sexual dimorphism, enhanced long-distance mobility, increased carnivory, long life spans, markedly more complex social interactions, and significantly more advanced cultures that are rooted in the ability to communicate using syntactical language. These and many other alterations arose over an extended period in response to unpredictable environmental change, habitat instability, and resource volatility [19–21].

Of all the traits that distinguish us from our ape-like ancestors, the one that we are most proud of is our outsized brain. However, our large brains, and the craniums that house them, make the head of a modern human infant extraordinarily large relative to the rest of its body, and this, along with the narrowing of the pelvic girdle (to facilitate bipedalism), has made childbearing for humans difficult and perilous. Ultimately, during human evolution, these obstetric constraints necessitated what is effectively the premature delivery of the child, and as a result, modern humans are the most altricial mammals on the planet. At birth, human babies are years away from being able to function even semi-independently.

Together, delivery complications and the long period of neonatal helplessness impose significant challenges for the survival of the baby and its mother, and therefore, these situations pose a threat to the species as a whole. However, the benefit of this developmental pattern is that our brains can grow and develop for a longer period than they otherwise would, and this has allowed us to acquire extraordinary cognitive abilities. With these abilities, we facilitate highly choreographed social interactions, and we develop new and unique methods to procure, process, and utilize a wide variety of food sources [21]. In the end, at least for our species, the benefits of our large brains outweighed the liabilities, and consequently, we, and our big brains, persist.

To gain insight into when our sizable brains came into existence, it is helpful to review the endocranial volume of various human species as a function of time. As was mentioned earlier, *H. habilis,* or a similar species, were the first humans on the planet, and they had an endocranial volume of about 600 cm³. However, by 1.8 million years ago, *H. habilis* had given rise to *Homo erectus,* and the endocranial volume of this younger species was approximately 800 to 1,100 cm³. So, within several hundred thousand years of its appearance, the endocranial volume of the *Homo* genus expanded significantly, and by the time that *H. erectus* evolved, its brain volume was, on average, more than 80% larger than *A. afarensis* and about 40% larger than that of *H. habilis* [22].

This initial increase in brain volume is impressive, but brain growth in the *Homo* genus didn't stop there. By 1.5 million years ago, *H. habilis* was extinct, but *H. erectus* still existed, and it ultimately gave rise to many species including *Homo heidelbergensis, Homo antecessor, Homo neanderthalensis, Homo denisova, Homo floresiensis, and H. sapiens* (see Figure 11.3). *H. erectus* also likely generated other human species such as *Homo rohodesiensis* and *Homo luzonensis* [23]. Within this group of *H. erectus*-derived organisms, *H. neanderthalensis* stands out because it had an

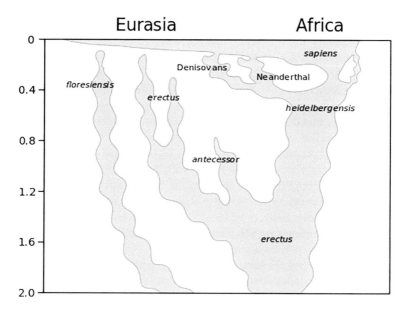

FIGURE 11.3   The ancestral relationships of several species of humans and the continents to which they migrated [28]. (From: User:Conquistador, User:Dbachmann, CC BY-SA 4.0 <https://creativecommons.org/licenses/by-sa/4.0>, via Wikimedia Commons.)

endocranial capacity of 1,300–1,600 cm^3, which is even larger than the 1,350 cm^3 of endocranial volume that the average modern *H. sapiens* possesses. So, over the last 2.5 million years, the endocranial volume of some members of our genus more than doubled. This is particularly noteworthy considering that, during a similar span of time, the endocranial capacity of Australopithecus only increased from 450 cm^3 to about 475 cm^3 [24].

Of course, intellectual ability depends on more than just endocranial volume and brain size. We know, for example, that neuronal node formation, neural wiring, and the environment also influence cognition [25–27]. However, paleoanthropologists have demonstrated that, as the endocranial capacities of various human species increased relative to their overall body size, so did the complexity of their behavioral repertoire. This suggests that at least part of the reason for the enhanced intellectual function of our species is that the engine of cognition (i.e., the brain itself) dramatically increased in size. But, while these data show that the size of the human brain increased relatively quickly, they do not address what drove this phenomenon. To investigate this issue, we need to consider the environmental conditions that prevailed during the evolution of our ancestors.

CLIMATE FLUCTUATION AND THE EVOLUTION OF THE HUMAN BRAIN

The vast majority of human evolution occurred during the Pleistocene (2.59 million to 11,700 years ago). This period covered most of the latest glaciation events, and it ended during the last glacial period of the current ice age. Most people think of ice ages as having occurred long ago; however, geologists consider an ice age to be a time during which large sheets of ice cover land, and for the last 2.59 million years, there has been at least one massive sheet of ice covering the Antarctic. So, by definition, we emerged during, and we currently live in, an ice age. Prior to the present period, it would be necessary to go back 360 million to 260 million years to the Karoo ice age to find enormous ice sheets covering land, and, in general, ice ages are relatively few and far between.

Often, the onset of an ice age is triggered by a significant decrease in atmospheric CO_2 or by the movement of tectonic plates. But, in at least one case, an ice age was initiated when the dust generated by a collision between a 93-mile wide asteroid and another large, unidentified object in the Kepler belt lowered the influx of sunlight on Earth for a period of about two million years [29]. However, ice ages are not necessarily constantly cold, and within an ice age, periods of intense glaciation are often interrupted by interglacial events, during which time the planet warms, and the ice sheets recede. In many cases, these relatively short-term intra-ice age climate fluctuations are influenced by alterations in Earth's elliptical orbit around the Sun, by variations in the planet's obliquity and precession (see Chapter 5), or by Earth-bound events, such as those that modify mountain chains or the flow of the oceans. So, during an ice age, the climate can fluctuate significantly over geologically short time periods (i.e., intervals of 10,000–100,000 years), and this can have a destabilizing effect on life.

As humans spread from one area to another during the Pleistocene ice age, some groups became isolated and they perished [3]. Others survived by adapting to the challenges that they faced.

Since the populations that did survive were often no longer interacting, they evolved largely independently of each other, and this sometimes resulted in speciation. In addition, the continuously changing climate likely selected individuals who generated creative and innovative solutions to the problems that arose in their own specific, environmentally unstable region [3, 22, 30]. So, the climatic instability of the Pleistocene may have favored people who were able to think their way out of the latest climate-induced dilemma, and this, in turn, may have privileged individuals with bigger and better brains. These superior brains presumably arose as a result of various genetic processes such as spontaneous mutation, environmentally driven mutation, the founder's effect, genetic drift, sexual selection, niche construction, gene-culture interaction, natural selection, epigenetic modification, and to some extent, gene flow between largely secluded populations [19, 21, 31].

To examine the validity of the hypothesis that climate induced salient changes in human populations, investigators correlated environmental instability with critical events in human evolution. Interestingly, they found a higher rate of co-occurrence than one would expect as a result of chance (see Figure 11.4 and [32, 33]), and therefore the data they analyzed suggest that human evolution was driven, at least in part, by climatic and environmental instability.

THE UNIQUE EVOLUTIONARY DESTINY OF THE GENUS *HOMO*

The aforementioned story outlines some of the conditions that may have led to the evolution of our large brains. However, the question of why other genera didn't also develop large brains and enhanced intellects during periods of rapid climatic change remains. This question is particularly perplexing, given that we know that many useful traits, such as the ability to perceive light, arose independently in various genera many times over [19].

When pondering this issue, it is important to consider the evolutionary trajectory a population can take. Initially, there are many options, but once a route is embarked upon, the window of possibility narrows. Eventually, taxa become developmentally and physiologically constrained, and at some point, there are only a limited number of ways in which a population can productively alter its genome. This is especially true when a situation necessitates that a large number of adaptations occur over a short period of time. During these circumstances, a species may not be able to quickly alter its developmental program in response to an unforeseen event.

In the case of humans, the adaptations required to increase brain size in response to rapidly changing climatic conditions were manageable. As a result, our genus evolved along a path that led to enhanced intellectual function and significant technological advances. In contrast, other genera may not have been able to modify their body structures and physiology in the same way, and therefore, despite experiencing similar environmental conditions, their evolutionary trajectory differed from that of our forebearers.

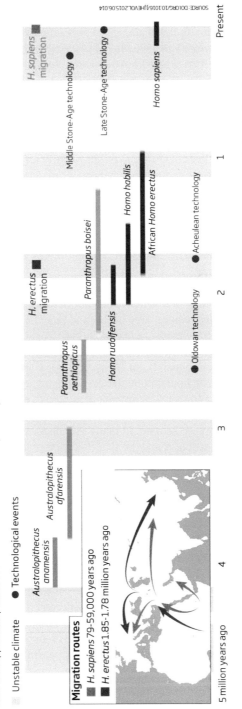

FIGURE 11.4 The climate-driven evolution of humans. Key events in human evolution correlate with the periods of climatic instability at a higher-than-expected rate [33].

That there is often more than one way in which a population can evolve is not a surprise, and there are frequently many evolutionary tracks that can generate a successful and stable solution to the challenges a population encounters. This reality is reflected by the great diversity that we see in ecosystems, and we know that there are indeed many ways for life to exist within our world. So, we should expect to see variation in adaptive strategies, even among organisms that share an environment. But still, during the course of evolution, there have been many genera that have populated the planet, and at first glance, it is surprising that only one successfully transited the evolutionary pathway that led to civilization-producing intelligence.

One possible explanation for that reality is that the conditions required for the evolution of advanced intellects may simply be statistically unlikely to occur. Our large brains evolved in a climate-and-geology-dependent manner, and the circumstances that generated our brains were ephemeral. We are here, in part, because of the way that processes we rarely think about (such as plate tectonic movements generated by Earth's internal heat and planetary oscillations caused by the gravitational interactions between Earth and other celestial objects) unfolded. If, for whatever reason, these processes had transpired differently, our big brains and the civilizations they produced may not be here.

It is also possible that enhanced intellectual function is rare because, despite our beliefs to the contrary, large brains may often not be worthwhile [34]. We know, for example, that large brains require large heads, and large heads complicate the birthing process, putting both the mother and child at risk. We also know that large brains are associated with a protracted period of helplessness, which endangers the young and their caretakers. Consequently, in many situations, the ability to survive and reproduce may actually be hindered by the development of a large brain. Moreover, outsized brains require a tremendous amount of energy to operate. In the case of our species, our brains constitute about 2% of our body weight, but they utilize approximately 20% of the energy we acquire. In many instances, when it comes to the quest for survival and reproduction, it may be that this energy could be better utilized in other ways.

In addition to these limitations, complex structures like our brains also frequently malfunction. There is a myriad of ways this can happen, and when it does, negative consequences often ensue. In our day-to-day lives, we can likely identify individuals whose brains are clinically malfunctioning, and we know that in some circumstances, these malfunctioning brains can threaten the possessor's survival. In addition, neurologically based problems are not limited to those with clinically diagnosed pathologies. Even those of us with supposedly "normal brains" often engage in illogical and flawed thought processes that can put us, and others, in danger [35]. Moreover, "typical brains" produce somewhere between 25,000 and 50,000 thoughts a day, and it has been estimated that, on average, 60–70% of these are negative. This means that most of us have 15,000–35,000 negative thoughts per day, and while this isn't necessarily bad [36, 37], it can be. This fact is painfully obvious in the United States, where clinically normal people acting upon their negative thoughts commit the majority of mass murders [38]. Of course, the ability to act on negative thoughts in a decisively destructive manner is amplified and democratized because, in some modern societies, individuals have access to high-powered weapons that are designed to quickly kill large numbers of individuals [39]. Throughout the vast majority of our evolution, access to this type of killing power was not available, especially to the average citizen. Still, during the course of our evolution, normal people acting on their negative thoughts have repeatedly demonstrated that they can kill many of their fellow humans [40]. Fortunately, in most cases, negative thoughts don't precipitate violent acts, but even when these thoughts don't result in physical violence, they can still cause significant harm. So, in short, large brains don't always lead to good outcomes for the possessor of that large brain, nor for those around him or her.

Why our thinking is often illogical, flawed, and negative is largely a mystery, but perhaps it is because our brains evolved in a less than ideal manner. If this happened, it would not be surprising, since nature often produces incredibly complex entities with less-than-optimal functioning. One example of such a structure is the human eye, which has a blind spot that limits our perceptual abilities. No competent engineer would purposely develop a visual detection system with such a limitation, but unlike an engineer, evolutionary processes do not possess foresight or preordained endpoints. Instead, they unfold in a manner that is guided by natural selection, and they build on what already exists. As a result, the products of evolution occasionally function in a nearly optimal or optimal manner (the octopus' eye, for example, does not have a blind spot), and sometimes, they do not.

Finally, it may be that Earth has only one species capable of producing a complex culture because, if more than one was on track to develop this trait, warfare between them would inevitably ensue, and one or the other would be eliminated. Thus, the dark side of beings with advanced culture may ensure that only one highly intelligent species can exist on our planet at any given moment (for more on this see [40] and the next chapter).

So, highly complex brains may be rare because they may be unlikely to form, they may be less helpful than we would like to believe, and they may be murderously competitive. If this is true, then the number of evolutionary success stories facilitated by large brains is going to be limited. In keeping with this possibility, we know that the survival of the human genus was anything but a certainty, and when the first humans appeared, the small groups they formed struggled to persist. Even a million years after it emerged, the human genus was still struggling, and approximately 1.2 million years ago, the total population of all the various species of humans alive was estimated to be under 56,000 [41, 42]. The difficulties encountered by individuals with large brains are also reflected by the fact that, today, only one species within the genus Homo remains. So, remarkably, despite all its intellectual promise, the genus Homo has now dwindled down to just us.

THE LAST OF THE *HOMO* GENUS

Our species first appeared about 300,000 years ago [43, 44], and for a very long time, we, like many of our predecessors and congeners, failed to thrive. For most of our existence, our population was likely under 1.3 million [45], and some suggest that for most of our past, it was under 100,000 [24]. During our history, it is also likely that we experienced population bottlenecks, and one such event may have occurred about 74,000 years ago, when the number of *H. sapiens* was estimated to be as low as 2,000–40,000 [41, 46, 47]. Some in the scientific community doubt that a bottleneck actually did occur at this time, and they don't believe that *H. sapiens* were recently on the edge of extinction [24, 48]. However, even if they are right, it is still true that, up until recently, the number of modern humans on the planet was very small, and if we were not on the edge of extinction, we weren't that far from it. Unfortunately, our continued existence is threatened even today, and whether we will be here in several millennia is far from certain (see Chapter 13).

In the light of this information, it is worth comparing our situation to the evolutionary history of the Limulus horseshoe crabs, which have carried on in a nearly stable state for 145 million years [49]. Alternatively, consider the crocodilians, which have existed largely in their current form for the last 85 million years [50]. These creatures, and many other evolutionarily stable species, have not produced advanced civilizations, but for tens of millions of years, they have excelled at the game of survival and reproduction. From an evolutionary perspective, these organisms, and not necessarily the ones with large brains, are the true champions of complex eukaryotic life.

THE FATE OF INTELLIGENT CREATURES ON OUR PLANET AND IN OUR GALAXY

Over the last several thousand years, *H. sapiens* have expanded in number, and we are now exhibiting some of the characteristics of an evolutionarily successful species. Our rapid population increase began about 10,000 to 12,000 years ago with the invention of agriculture. When agrarian societies began, there were fewer than 5 million people on Earth, but by the year 1800, our number increased to about 1 billion. This was a remarkable surge, but it was soon dwarfed by the growth that occurred following the industrial revolution. In just 200 years (which is the cosmic equivalent of a blink of the eye), the human population skyrocketed to more than 7.5 billion (see Figure 11.5). So, we have demonstrated that we can increase in number at an impressive rate; however, the question that remains is "can our species survive for an extended period?" If we can, we will one day rival, or even outshine, the horseshoe crabs and the crocodiles. But we have a long, long way to go before we, the only civilization-producing species to arise on the planet, can hope to obtain that distinction.

Before we move on, it is also worth pondering whether life on a planet like Earth would likely generate beings with advanced civilizations. Assuming there are other planets just like Earth, and given what we know about the genus Homo, there are good reasons to doubt that it would (see the aforementioned text and [51, 52]). However, some astrophysicists appear to be more optimistic [53], and a group of them have convinced the United States government and various philanthropic organizations to invest in a project designed to find intelligent, civilization-producing alien beings. By 1985, a team of these would-be alien hunters raised enough money to start their quest, and they opened The SETI (Search for Extra Terrestrial Intelligence)

The evolution of the world population

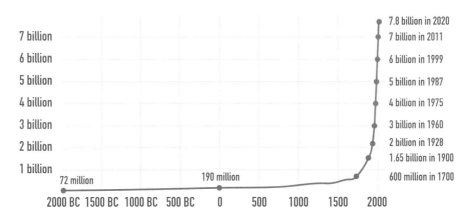

FIGURE 11.5 Human population growth through the last 4,000 years.

Institute in Mountain View, California. But, after several decades, the folks at SETI have not found any evidence of the beings they set out to discover, and today, they are struggling financially. In response, the leaders at SETI recently tried to generate excitement and support for their work by starting a program called "Earth Speaks." Through this initiative, they have asked the public to construct messages that can be sent to aliens [54]. However, given humanity's record of finding these beings, I don't think anyone needs to rush to finish their letter.

FINDING OURSELVES

While others search for intelligent aliens, we will continue to focus on the one species we know to have produced a complex civilization. To this end, we will examine, in more detail, the genesis, distribution, and characteristics of *H. sapiens*, and we will explore how our recent evolution shaped us. In the process, we will also consider the fate of some of the other members of the genus *Homo*, and we will reflect upon how these other humans influenced our internal and external worlds.

REFERENCES

1. Williams, B.A., R.F. Kay, and E.C. Kirk, *New perspectives on anthropoid origins*. Proceedings of the National Academy of Sciences of the United States of America, 2010. **107**(11): p. 4797–4804.
2. Evenstar, L.A., et al., *Slow Cenozoic uplift of the western Andean Cordillera indicated by cosmogenic 3He in alluvial boulders from the Pacific planation surface*. Geophysical Research Letters, 2015. **42**(20): p. 8448–8455.

3. Finlayson, C. *The humans who went extinct: why Neanderthals died out and we survived*. Variation: OUP E-Books. [Internet Resource; Computer File] 2009; 1 online resource (xi, 273 pages): illustrations, maps]. Available from: http://public.eblib.com/choice/publicfullrecord.aspx?p=472254 Materials specified: Ebook Library: http://public.eblib.com/choice/publicfullrecord.aspx?p=472254.

4. Glazko, G.V. and M. Nei, *Estimation of divergence times for major lineages of primate species*. Molecular Biology and Evolution, 2003. **20**(3): p. 424–434.

5. Patterson, N., et al., *Genetic evidence for complex speciation of humans and chimpanzees*. Nature, 2006. **441**(7097): p. 1103–1108.

6. Alemseged, A.D.A.Z., *Temporal evidence shows Australopithecus sediba is unlikely to be the ancestor of Homo*. Science Advances, 2019. **5**(5): p. eaav9038.

7. Asfaw, B., *Australopithecus garhi: a new species of early hominid from Ethiopia*. Science, 1999. **284**(5414): p. 629–635.

8. Smithsonian Institute, *Australopithecus afarensis*. Smithsonian Institution's Human Origins Program, 2019.

9. Madelaine Böhme, et al., *A new Miocene ape and locomotion in the ancestor of great apes and humans*. Nature, 2019. **575**(7783): p. 489–493.

10. Thorpe, S.K., R.L. Holder, and R.H. Crompton, *Origin of human bipedalism as an adaptation for locomotion on flexible branches*. Science, 2007. **316**(5829): p. 1328–1331.

11. David-Barrett, T. and R.I. Dunbar, *Bipedality and hair loss in human evolution revisited: The impact of altitude and activity scheduling*. Journal of Human Evolution, 2016. **94**: p. 72–82.

12. Videan, E.N. and W.C. McGrew, *Bipedality in chimpanzee (Pan troglodytes) and bonobo (Pan paniscus): testing hypotheses on the evolution of bipedalism*. American Journal of Physical Anthropology, 2002. **118**(2): p. 184–190.

13. Tocheri, Matthew W., et al., *The primitive wrist of Homo floresiensis and its implications for hominin evolution*. Science, 2007. **317**(5845): p. 1743–1745.

14. Gazzaniga, M.S., *Human: the science behind what makes us unique*. 2008, New York, NY: HarperCollins.

15. Stanley, S., *An ecological theory for the origin of homo*. Paleobiology, 1992. **18**(3): p. 237–257.

16. Wong, K., *The face of the earliest human ancestor, revealed*. Scientific American, 2019. **321**(6): p. 58–59.

17. Winder, I.C., et al., *Evolution and dispersal of the genus Homo: a landscape approach*. Journal of Human Evolution, 2015. **87**: p. 48–65.

18. Laden, G. and R. Wrangham, *The rise of the hominids as an adaptive shift in fallback foods: plant underground storage organs (USOs) and Australopith origins*. Journal of Human Evolution, 2005. **49**(4): p. 482–498.

19. Christakis, N.A., *Blueprint: the evolutionary origins of a good society*. First edition. ed. 2019, New York, NY: Little, Brown Spark. xxi, 520 p.

20. Berwick, R.C. and N. Chomsky, *Why only us: language and evolution*. 2016, Cambridge, MA: MIT Press.

21. Anton, S.C., R. Potts, and L.C. Aiello, *Human evolution. Evolution of early Homo: an integrated biological perspective*. Science, 2014. **345**(6192): p. 1236828.

22. Shultz, S. and M. Maslin, *Early human speciation, brain expansion and dispersal influenced by African climate pulses*. PLoS ONE, 2013. **8**(10): p. e76750.

23. Detroit, F., et al., *A new species of Homo from the late Pleistocene of the Philippines*. Nature, 2019. **568**(7751): p. 181–186.

24. Hawks, J., et al., *Population bottlenecks and Pleistocene human evolution*. Molecular Biology and Evolution, 2000. **17**(1): p. 2–22.

25. Sherwood, C., *Are we wired differently*. Scientific American, 2018. **319**(3): p. 60–63.

26. Bertolero, Max and D.S. Bassett, *How the mind emerges from the brain's complex networks.* Scientific American, 2019. **321**(1): p. 26–33.
27. Bertolero, M.A., et al., *A mechanistic model of connector hubs, modularity and cognition.* Nature Human Behaviour, 2018. **2**(10): p. 765–777.
28. Stringer, C., *What makes a modern human.* Nature, 2012. **485**: p. 33–35.
29. Birger Schmitz, et al., *An extraterrestrial trigger for the mid-Ordovician ice age: Dust from the breakup of the L-chondrite parent body.* Science Advances, 2019. **5**: p. eaax4184.
30. Timmermann, A. and T. Friedrich, *Late Pleistocene climate drivers of early human migration.* Nature, 2016. **538**(7623): p. 92–95.
31. Laland, K.N., J. Odling-Smee, and S. Myles, *How culture shaped the human genome: bringing genetics and the human sciences together.* Nature Reviews Genetics, 2010. **11**(2): p. 137–148.
32. Potts, R. and J.T. Faith, *Alternating high and low climate variability: the context of natural selection and speciation in Plio-Pleistocene hominin evolution.* Journal of Human Evolution, 2015. **87**: p. 5–20.
33. Slazak, M., *Key moments in human evolution were shaped by changing climate.* New Scientist, 2015, https://www.newscientist.com/article/mg22730394-100-key-moments-in-human-evolution-were-shaped-by-changing-climate/
34. Mayr, E., *Does it pay to acquire high intelligence?* Perspectives in Biology and Medicine, 1994. **37**(3): p. 337–338.
35. Kahneman, D., *Thinking, fast and slow.* First edition. ed. 2011, New York, NY: Farrar, Straus and Giroux. 499 p.
36. Raghunathan, R., *How negative is your "mental chatter"?* Psychology Today, 2013. https://www.psychologytoday.com/us/blog/sapient-nature/201310/how-negative-is-your-mental-chatter.
37. Marano, H.E., *Depression doing the thinking.* Psychology Today, 2001. https://www.psychologytoday.com/us/articles/200107/depression-doing-the-thinking.
38. Carey, B., *Are mass murderers insane? Usually not, researchers say.* New York Times, 2017.
39. Ingraham, C., *We're now averaging more than one mass shooting per day in 2015.* Washington Post, 2015.
40. LeBlanc, S.A. and K.E. Register, *Constant battles: the myth of the peaceful, noble savage.* 2003, New York, NY: St. Martin's Press.
41. Huff, C.D., et al., *Mobile elements reveal small population size in the ancient ancestors of Homo sapiens.* Proceedings of the National Academy of Sciences of the United States of America, 2010. **107**(5): p. 2147–2152.
42. Storrs, C., *Humans might have faced extinction.* Scientific American, 2010. **302**(4): p. 18.
43. Richter, D., et al., *The age of the hominin fossils from Jebel Irhoud, Morocco, and the origins of the Middle Stone Age.* Nature, 2017. **546**(7657): p. 293–296.
44. Schlebusch, Carina M., et al., *Southern African ancient genomes estimate modern human divergence to 350,000 to 260,000 years ago.* Science, 2017. **358**: p. 652–655.
45. Angela, P., *The extraordinary story of human origins.* 1993, Buffalo, NY: Prometheus Books. 328 p.
46. Ambrose, S.H., *Did the super-eruption of Toba cause a human population bottleneck? Reply to Gathorne-Hardy and Harcourt-Smith.* Journal of Human Evolution, 2003. **45** p. 231–237.
47. Whitehouse, D., *When humans faced extinction.* BBC News, 2003.
48. Hawks, J., *The so-called Toba bottleneck didn't happen.* John Hawks weblog, 2018.

49. Kin, Adrian and B. Błazejowski, *The horseshoe crab of the genus Limulus: Living fossil or stabilomorph?* PLOS One, 2014. **9**(10): p. e108036.
50. Black, R., *The top 10 greatest survivors of evolution.* Smithsonian.com, 2012.
51. Ward, P.D. and D. Brownlee, *Rare Earth: why complex life is uncommon in the universe.* 2000, Göttingen: Copernicus Publications.
52. Mayr, E., *Can SETI succeed? Not likely.* Bioastronomy News, 1995. **7**(3).
53. Sagan, C., *The abundance of life-bearing planets.* Bioastronomy News, 1995. **7**(4).
54. Speigel, L., *SETI 'earth speaks': want to say hello to an et?* Huffington Post, 2017.

12 The Genesis of Behaviorally Modern *Homo sapiens*

A Cognitively Advanced Human That Can Reflect Upon Its Existence

All men are by nature equal, made all of the same earth.

Plato

Man with all his noble qualities, with sympathy which feels for the most debased, with benevolence which extends not only to other men but to the humblest living creature, with his god-like intellect which has penetrated into the movements and constitution of the solar system—with all of these exalted powers—man still bears in his bodily frame the indelible stamp of his lowly origin.

Charles Darwin

A long and complex train of thought can no more be carried out without the aid of words, whether spoken or silent, than a long calculation without the use of figures or algebra.

Charles Darwin

THE VALUE OF A GOOD TOOLSET

When humans first evolved approximately 2.4 million years ago, they made use of tools similar to those first invented by the australopithecines 200,000 years earlier [1]. These implements formed what is referred to as the Oldowan toolkit, which consisted of relatively simple hammerstones, stone cores, and stone flakes (see Figure 12.1).

Simple though they may be, these modest devices molded our lineage and our fate. Unlike many of the fierce carnivores of the African plains, our ancestors lacked powerful muscles, bone-crunching jaws, and large, sharp teeth, and so to succeed, they needed a different kind of advantage: they needed tools. The Oldowan toolkit delivered this advantage, and with these tools, our ancestors could break bones, work wood, and manipulate hides. Ultimately, these implements dramatically increased our predecessor's ability to survive and reproduce.

DOI: 10.1201/9781003270294-15

FIGURE 12.1 Tradition Oldowan choppers. (From: José-Manuel Benito Álvarez (España)—> Locutus Borg, CC BY-SA 2.5 <https://creativecommons.org/licenses/by-sa/2.5>, via Wikimedia Commons.)

About 600,000 years later, soon after *Homo erectus* evolved, a new and improved line of tools emerged. Collectively, these devices made up the Acheulean toolkit, and they include large cutting, digging, and hunting instruments such as the biface, almond-shaped hand axe (see Figure 12.2). With tools like a hand axe, *Homo erectus* could tenderize meat, cut food into small pieces, and remove cartilage and other material that is difficult to chew and digest. This food processing decreased the amount of time and energy required to eat meat and it enhanced our forbearers' net calorie and nutrient intake. With extra energy and nutrients, *Homo erectus* could power its large brain and increase its foraging range. In addition, these tools lowered the selective pressure for strong jaw muscles, large teeth, and extensive energy-intensive digestive systems, and that made it possible for the human head and neck to evolve in a way that facilitated thermoregulation, speech, and the development of still larger brains [2].

Acheulean technology proved to be remarkably stable, and the tools it produced continued to be used by various species of humans until as recently as 130,000 years ago. In total, the Acheulean toolset was employed daily for more than 1.5 million

FIGURE 12.2 An Acheulean hand axe. (From: Osama Shukir Muhammed Amin FRCP (Glasg), CC BY-SA 4.0 <https://creativecommons.org/licenses/by-sa/4.0>, via Wikimedia Commons.)

years, and in the history of humankind, there has never been a similarly long-enduring technology.

FORGED BY FIRE

Another key advancement in the evolution and the expansion of humankind was the controlled use of fire. *Homo erectus* first acquired this ability about one million years ago [3], and at that time, humans likely used fire as protection from predators and as a source of heat.

By 0.5 million years ago, humans also routinely used fire to cook food [2], and the impact this development had on our evolution is hard to overestimate. Cooking decreased the likelihood of being sickened by food-borne pathogens, it greatly enhanced the caloric and nutritional value of foodstuffs, and it made many otherwise inedible plant foods suitable for consumption. These outcomes permitted our fore-bearers energy-intensive brains to evolve into larger and more complex structures.

By helping to break down food, cooking also decreased the need for a long diges-tive process. This allowed for the alteration of our jaw, teeth, digestive tract, and body trunk, and it also led to the modification of our physiology. Moreover, our new food preparation processes significantly decreased the amount of time our ancestors spent chewing. Our ape cousins dedicate up to seven hours of each day to this task [4], but with their Acheulean toolkit and culinary skills, our ancestors were able to spend less time chewing and more time acquiring additional resources.

Finally, our forebear's fire-based methods for preparing food led to the develop-ment of complicated and nuanced eating rituals. These practices influenced inter-personal bonding, group adhesion, gender dynamics, and other dimensions of our culture, and that, in turn, shaped our biological evolution [5, 6].

While fire was highly influential in shaping our past, its role in molding who and what we are continues to this day. As I write this paragraph, a car is idling in my driveway; pilot lights are flaming in my furnace, gas fireplace, and stove; and power plants are generating the electricity upon which my modern home relies. Furthermore, as I contemplate the structure of this sentence, I am also eagerly antic-ipating the cooked food and meal-oriented social interactions I will enjoy later this evening. All this, and much more, we owe to our ancestors' discovery of how to control fire.

THE FIRST MIGRATION OF HUMANS OUT OF AFRICA

With its large brain, modest stone tools, and the ability to harness fire, *Homo erectus* spread out across many regions of the globe. By 1.8 million years ago, these humans had left Africa [7], and within 200,000 years, they had migrated as far as the north-ern latitudes of northeast Asia [8]. Archaeological evidence also indicates that *Homo erectus* made its way into Europe approximately 1.2 million years ago [9].

As this species expanded its geographical range, groups splintered and drifted apart, and they began to encounter challenges unique to their location. In Europe, this process resulted in *Homo erectus* giving rise to *Homo antecessor* approximately 800,000 years ago, and in Africa, *Homo erectus* evolved into *Homo heidelbergensis,* which first appeared about 600,000 years ago. Some of the *Homo heidelbergensis* stayed in Africa, and they eventually gave rise to *Homo sapiens*. Others migrated to Eurasia where they evolved into *Homo neanderthalensis, Homo denisova*, and sev-eral other human species. Interestingly, there is evidence of mating between many of these clades, so some argue that these groups do not constitute separate species as defined by the BSC (see Chapter 7 for more on this). Yet, if two groups rarely mate, and if they display unique characteristics, many biologists still consider the popula-tions to be examples of different species, even if they do occasionally produce viable, reproductively competent offspring. In this debate, it's worth once again noting the confusion that often arises because of the many ways in which biologists define a species. However, in this text, I am going to consider the groups of humans men-tioned earlier to be members of different species. Having said this, it is important

FIGURE 12.3   Middle Stone Age tools. For additional examples of this technology, see [11]. (From: Vincent Mourre/Inrap, CC BY-SA 3.0 <https://creativecommons.org/licenses/by-sa/3.0>, via Wikimedia Commons.)

to keep in mind that these various human clades are so closely related that they did sometimes mate and produce offspring.

While different taxa of humans were arising in Eurasia, new groups were also appearing in Africa. One interesting faction arose about 315,000 years ago in a region known as Jebel Irhoud, which is part of the North African country of Morocco [10]. Members of this population were among the first to transition from Acheulean tools to Middle Stone Age technology (see Figure 12.3). As a result, this group of humans could produce small, pointed stone flakes, as well as stone awls that could be used to fashion hides and wooden objects. With this new toolkit, these people became less vulnerable prey, more efficient game hunters, and more significant threats to those around them.

In addition to having an unusual toolkit, the humans of Jebel Irhoud were also different because, unlike earlier members of our genus, their endocranial volume was similar to that of modern humans, and their small faces were also comparable in structure to those we possess. But, unlike us, the Jebel Irhoud population had large teeth, and they lacked a prominent chin and forehead. They also had an elongated brain case, which suggests that they had a smaller cerebellum and smaller parietal lobes than those found in modern-day humans. As a result of these differences in brain structure, the Jebel Irhoud population probably processed sensory information differently than we modern humans do, and it is likely that their communication and social skills were not as well developed as ours [12].

Despite the numerous disparities between the people of Jebel Irhoud and modern-day humans, some have argued that the Jebel Irhoud population is the oldest group of *Homo sapiens* discovered to date. Others disagree, and they believe that these individuals were not members of the *Homo sapiens* clade [13, 14]. However, regardless of how we choose to label the people of Jebel Irhoud, it is important to note that they clearly had a mixture of modern and archaic traits. Interestingly, these people were not alone in this distinction, and the evidence indicates that other human admixes were present near this time in various parts of Africa.

The fact that there were genetically different bands of *Homo sapiens* dispersed throughout Africa is not particularly surprising. The African continent is about 5,000 miles long, and at some points, it is more than 4,300 miles wide. With an area of more than 11.6 million square miles, it is the second-largest continent, and it is unusually diverse in its ecology and climate. Africa also has remarkable physical barriers, such as dense tropical forests, raging rivers, towering mountains, and hot, dry deserts. In such a place, groups that moved apart tended to remain separate, and once they were isolated, they were apt to diverge genetically, phenotypically, and culturally.

THE AFRICAN INCUBATOR

Among the physical barriers in Africa, the Sahara Desert is particularly striking, in part because it is the largest hot desert in the world. In total, the Sahara composes almost one-third of the African continent, and it has an area similar to that of China or the continental United States. Prehistoric human populations on opposite sides of this region independently adapted to environments ranging from marine coasts and rainforests to arid forests and savanna grasslands.

Had populations of *Homo sapiens* remained on opposites sides of the Sahara, they likely would have eventually speciated. However, throughout much of human history, Africa's environments have been unstable [15], and so for tens of thousands of years a region could be hot and dry, and then it could become humid and grassy for tens of thousands of years. In the case of the Sahara, a cycle of desertification and greening repeated over and over, and as a result, the desert lands were arising, fading away, and then reappearing throughout the Pleistocene. Consequently, human populations on different sides of the hot, dry Sahara could often be reunited when this area became verdant, which frequently occurred before the populations had speciated. Occasionally, during these periods of reunification, matings between previously separated groups produced "transgressive hybrids," which are individuals who exceeded either of their parents' ability to adapt to their surroundings. Once these unique individuals joined a group, their genes likely spread quickly throughout the population.

After being brought together by the greening of the Sahara, a population could be fractured once again when the area reverted to desert. As a result, the cycle of independent evolution and eventual reunification repeated, and with each round, new transgressive hybrids emerged, making the group as a whole better able to adapt. Ultimately, Africa's dynamism made it an efficient evolutionary incubator, and within this continent, individuals with new genes, gene combinations, and traits appeared.

According to those who support this view, about 50,000–80,000 years ago, Africa produced human populations that possessed most of the traits that modern humans exhibit today [16]. If this hypothesis is correct, then our species was shaped primarily within Africa over a period of hundreds of thousands of years [11, 17–19].

It is important to note that, within what I am referring to as the "African incubator," most groups that contributed to our genetic identity consisted of individuals whom most biologists would label *Homo sapiens*. However, it is likely that some of our genetic information also came from the genomes of other human species, such as *Homo naledi* and *Homo heidelbergensis*. At the time *Homo sapiens* were evolving, these species were also present on the African continent, and the pan-African amalgamation hypothesis outlined earlier leaves open the possibility that these other species also mated with *Homo sapiens* [20, 21]. Consequently, the modern human genome may be a composite that includes genetic information from several now-extinct human species of Africa.

THE AFRICAN *Homo sapiens* INVADE EURASIA

Eventually, some African *Homo sapiens* migrated to Eurasia, although exactly when this first occurred is not clear. There are fossil data that suggest *Homo sapiens* were in Greece more than 210,000 years ago, and by 180,000 years ago, they could be found in Israel [15, 22]. Evidently, when they did arrive, the *Homo sapiens* mated with the Neanderthals that were already in the region [23]. However, the evidence also suggests that these early *Homo sapiens* invaders did not fare well in their new home, and soon after arriving in Eurasia, they vanished from the region.

About 50,000–80,000 years ago, the *Homo sapiens'* situation improved, and a permanent population of *Homo sapiens* was finally established in the territory north of Africa [24]. The individuals who succeeded in constructing these long-enduring settlements were likely behaviorally modern *Homo sapiens*, and while the exact routes on which they traveled out of Africa and throughout Eurasia are debated, some have suggested that they traversed one or more of the geographical passageways displayed in Figure 12.4.

Like their ancestors, some of the modern *Homo sapiens* who entered Eurasia during this period mated with the Neanderthals that were already present, and today, the genome of people whose forebears lived outside of sub-Saharan is typically 1.5 to 2.1% Neanderthal. Individuals whose ancestors never left Africa appear to have a more limited amount of Neanderthal DNA, and it is likely that the African *Homo sapiens* acquired this Neanderthal DNA when a group of Eurasian *Homo sapiens* returned to Africa [25].

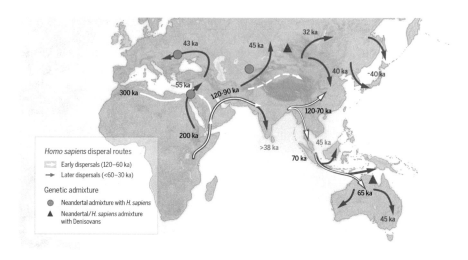

FIGURE 12.4 Geographical routes tracing the dispersion of early modern humans. Sites of genetic admixturing between *Homo sapiens* and the Neandertals and Denisovans are also depicted. (From: Katerina Douka & Michelle O'Reilly, Michael D. Petraglia, CC BY-SA 4.0 <https://creativecommons.org/licenses/by-sa/4.0>, via Wikimedia Commons.)

In addition to mating with the Neanderthals, some groups of early modern *Homo sapiens* apparently intermixed with other ancient Eurasian populations such as the Denisovans. As a result, many people whose lineage can be traced to parts of Asia inherited 3 to 5% of their genetic information from this species of now-extinct humans [26, 27].

The continued presence of Neanderthal and Denisovan DNA in the genomes of modern humans suggests that this genetic information helped those who inherited it survive and reproduce in the territory outside of Africa. We know, for example, that some of the Neanderthal DNA that is in the human gene pool imparts immunological resistance to Eurasian pathogens, and other segments of this DNA affect our skin [17, 28]. The utility of having genes that impart resistance to life-threatening pathogens is obvious, but the benefits of retaining Neanderthal genes that affect the skin are not as easy to discern. However, some Neanderthal skin genes likely resulted in less skin pigmentation, and one potential benefit of lighter skin is that it can facilitate the production of vitamin D in areas with low sunlight intensity. The dark skin pigmentation of those who first left Africa absorbed much of the incoming, DNA-damaging UV light, which protected our ancestors from the destruction caused by the intense African sun, but in Eurasia, where the sunlight intensity is lower, a similar level of UV light absorption by the skin hampered vitamin D production, and in a world that lacked vitamin D enriched foods, that could have led to a potentially lethal vitamin D deficiency. As for the Denisovan genes, they likely influenced many processes, but we know that at least some were useful in low oxygen environments, such as the high-altitude Tibetan Plateau.

As was mentioned earlier, in addition to the Neanderthal and the Denisovan DNA sequences, it is likely that modern *Homo sapiens* retained DNA sequences that they acquired from other Eurasian species of humans [29]. Consequently, because of these introgressive events, the now-extinct humans of Eurasia are, in a very real sense, still with us today. As the genetic analysis of our genomes continues, we will undoubtedly learn more about how the DNA of our now extinct sister species shaped our biology and our nature.

THE AFRICAN INCUBATOR VERSUS THE EURASIAN INCUBATOR

While our species was evolving in the "African incubator," the Eurasian humans were doing the same in what could be called the "Eurasian incubator." As was true in Africa, within Eurasia, various human species mated, and there is evidence that *Homo neanderthalensis* (which were concentrated in western and central Eurasia) and *Homo denisova* (which were primarily found in the central and eastern Eurasia) produced fertile, hybrid offspring [30]. However, despite the interactions between human Eurasian species, the genetic diversity of the early Eurasian populations did not reach that found among the *Homo sapiens* in Africa [30]. Furthermore, unlike in Africa, where *Homo sapiens* eventually dominated the landscape, no one indigenous group gained control of the entirety of Eurasia.

When trying to make sense of the fate of the modern *Homo sapiens* and early Eurasian populations, it is worth pondering how each group's home continent may have played a role in their destiny. If, as I suggest, Africa was a better incubator of modern humans than was Eurasia, then during their long occupation of Africa, the modern *Homo sapiens* may have acquired adaptations that gave them a survival advantage over their Eurasian counterparts.

THE HOME-FIELD ADVANTAGE

So, what is special about Africa, and how might it have given our species an edge in the competition to survive and reproduce? Well, to start with, the human genus first emerged in Africa, and our early evolution occurred on that continent. As a result, humans were particularly well suited to life in Africa, and up until about 100,000 years ago, the vast majority of humans were concentrated in the middle latitudes of that continent (see Figure 12.5).

Within the *Homo* clade, *Homo sapiens* are a particularly young species, and as is the case with the *Homo* genus itself, we do not know where exactly our particular species emerged. As we learned earlier, the oldest putative *Homo sapiens* fossils we currently possess are from North Africa, but the oldest fully modern human fossils are the 190,000-year-old relics obtained from the East African country of Ethiopia. To further complicate matters, genetic data suggest that modern *Homo sapiens* could have first arisen in South Africa [32]. Yet, regardless of where modern *Homo sapiens* first appeared, we do know our species was shaped by matings between African populations that had been exposed to a wide range of climates, food sources, pathogens, and predators [33]. Consequently, it is likely that *Homo sapiens* developed a considerable degree of biological diversity because of the various wide-ranging environments that our African ancestors evolved within.

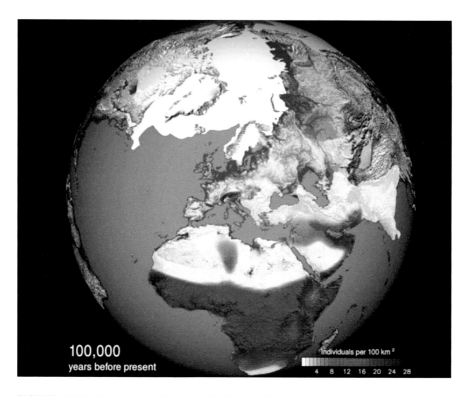

FIGURE 12.5 The geographical distribution of *Homo sapiens* 100,000 years ago [31], Human Origins Program, NMNH, Smithsonian Institution.

If the aforementioned hypothesis is correct, then the numerous and constantly changing climatic zones of Africa played an outsized role in the evolution of our kind. Interestingly, one of the reasons that Africa has so many different climes is that it is orientated along a north-south axis (see Figure 12.6). As Jared Diamond points out in his 1998 book, *Guns, Germs and Steel* [34], continents aligned in this way contain many regions that vary in latitude, and in such places, moving relatively small distances north or south often results in profound climate changes. In fact, areas on the same latitude that are separated longitudinally by four thousand miles are often more climatically like each other than landmasses on the same longitude that vary by just 1,000 miles in latitude. So, ultimately, Africa may have been particularly effective at generating modern humans because of its geographical orientation on the globe, and this suggests that the nature of the *Homo* clade in general (and the *Homo sapiens* in particular) was determined, at least in part, by the plate tectonic movements that positioned Africa and the other continents hundreds of millions of years ago (see Chapter 5).

THE PROBLEM WITH PLAYING THE AWAY GAME

Like Africa, Eurasia is enormous, and it contains many diverse environments. However, unlike humanity's original homeland, the Eurasian supercontinent has

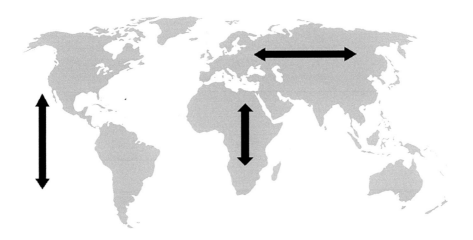

FIGURE 12.6 The orientation of the continents as described by Jared Diamond. (From: Espíritu nocturno, CC BY 4.0 <https://creativecommons.org/licenses/by/4.0>, via Wikimedia Commons.)

a horizontal orientation (see Figure 12.6). Consequently, within Eurasia, there are large tracts of land stretching from east to west that maintain similar climates and comparable environments. In addition, Eurasia also differs from Africa because it has vast regions that lie in the cold and inhospitable northern latitudes, which up until about 20,000 years ago, humans could not occupy because they lacked the technology to build warm shelters and sew clothing. Accordingly, for most of human history, large tracts of Eurasia were off-limits to humanity, and the majority of humans in this part of the world lived south of Kiev and northern Germany [34].

Ultimately, because of its orientation and location, the Eurasian continent likely produced smaller, more isolated, and genetically less diverse populations than did Africa [35]. As a result, humans migrating along the horizontal axis of the Eurasian continent would have encountered few previously established populations, and they would have been less likely to engage in intraspecies and interspecies conflicts. In addition, because they moved and settled along the horizontal axis of the Eurasian supercontinent (Figure 12.7 and https://commons.wikimedia.org/wiki/File:Köppen-Geiger_Climate_Classification_Map.png), these humans had less need for innovation, and species such as the Neanderthals could draw heavily upon their existing cultural, technological, and biological attributes as they migrated across the supercontinent. This would have been beneficial for the early Eurasian humans in the short term, but the relatively slow rate of genetic and cultural evolution that resulted may have placed the Eurasian populations at a disadvantage when they began to regularly encounter the more genetically and culturally diverse *Homo sapiens* interlopers.

Ultimately, in comparison to Africa, Eurasia may not have been as effective at accelerating the genetic and cultural evolution of its indigenous human populations. But, nevertheless, prior to the emergence of the *Homo sapiens'* behaviorally modern suite of traits, the Eurasian humans, and the African *Homo sapiens* may have been

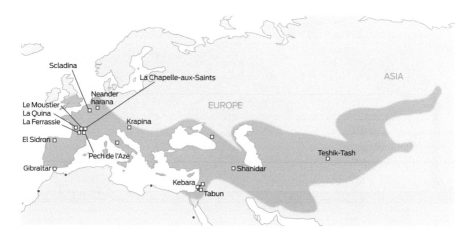

FIGURE 12.7 The geographical distribution of *Homo neanderthalensis*. The main Neanderthal settlement sites are also indicated. (From: Berria, CC BY-SA 4.0 <https://cre-ativecommons.org/licenses/by-sa/4.0>, via Wikimedia Commons.)

close enough in their abilities that the Eurasians may have retained a numerical as well as a "home field" advantage over the early *Homo sapiens* who wandered into their territory. If true, this would explain why the first *Homo sapiens* to leave Africa around 200,000 years ago did not long endure.

While the *Homo sapiens* did not fare well in their initial attempts to leave Africa, they did much better after migrating into the Near East about 50,000–80,000 years ago [24, 36]. By then, the individuals making the trip were likely very similar to modern-day humans, and the behaviorally modern traits that they had evolved may have given this population of *Homo sapiens* the edge they needed to succeed in the new land.

THE EXTINCTION OF THE ORIGINAL EURASIANS

So, it is possible that some modern *Homo sapiens* had a competitive edge over their rivals some 50,000–80,000 years ago, but how this edge manifested is not clear. One possibility is that the modern *Homo sapiens* were better able to adapt to the climate changes that were occurring in Eurasia during this period. We know, for example, that at about this time, some of the woodlands of the Eurasian supercontinent were becoming drier and more savannah-like, and this would have been problematic for the Eurasian Neanderthals because, although they were well adapted to hunting in woodlands, they were not well suited to hunting and gathering in savannah land-scapes [37]. Contrastingly, *Homo sapiens* would have benefited from this climatic alteration because they evolved in savannah landscapes.

Others have suggested that Neanderthals went extinct because their speedy metabolism, coupled with competition from modern *Homo sapiens*, caused them to starve to death. We know that the Neanderthals had a higher rate of metabolism than modern *Homo sapiens* [38], and they therefore needed more food per pound of body weight. However, because they could only use a limited range of species as a food source, and because modern *Homo sapiens* were depleting some of their options, the Neanderthals may not have had much to eat. Moreover, the Neanderthals also lacked the highly efficient social interactions and sophisticated toolset that modern *Homo sapiens* possessed [39], and therefore, it is likely that when they did compete with the modern *Homo sapiens*, they often found themselves outdone [40, 41].

The modern *Homo sapiens* may have also eliminated some members of their sister species by interbreeding with them and essentially absorbing them into their population. It is also possible that they eradicated the original Eurasians by inadvertently infecting them with lethal African pathogens [42].

Still another possible explanation for why we are still here, and our sister species are not, is that our species may have fielded a larger and better armed fighting force when they successfully began to colonize Eurasia. We know that the modern *Homo sapiens* population was more numerous than their competitors, and they were equipped with compound, projectile weapons, which the Eurasians lacked [43]. This combination of factors would have given the modern *Homo sapiens* a significant advantage if warfare between various human species did occur.

In the end, we don't know enough about the interactions between the original Eurasians and the modern *Homo sapiens* to determine exactly what happened when the two groups collided about 50,000–80,000 years ago, but we do know that, by approximately 35,000 years ago, all the original Eurasians were gone. So, after hundreds of thousands of years of sharing the planet with several other species of humans, our sister species vanished, and our species became the sole extant member of the *Homo* clade.

THE GREAT DIASPORA

At the time of our sister species' demise, our *Homo sapiens* ancestors were rapidly spreading across the globe. By about 40,000 years ago, modern *Homo sapiens* could be found throughout Eurasia and Australia, and by 12,000 years ago, they were dispersed throughout the Americas. Our ancestors even settled the remote Polynesian Islands of the South Pacific by 1,000 CE, and by the time the European explorers embarked on their global voyages, only a few small distant outposts, such as the Azores and Bermuda, remained undiscovered.

During modern *Homo sapiens'* great migrations, some populations were largely isolated from others for many thousands of years. However, within a given localized area, many groups routinely intermixed. Evidence of this phenomenon can be observed from the study of Western European genetics. About 10,000 years ago, the Western European population was composed of farmers from Iran, farmers from the Fertile Crescent, hunters and gatherers from Central and Western Europe, and hunters and gathers from Eastern Europe [36]. When these four groups first encountered each other, they were as genetically different as East Asians and Europeans are today. However, all four of these populations interbred and contributed to the modern Western European genome of the time.

Some 5,000 years after these four groups initially intermixed, the Western Europeans' genome changed significantly yet again. At this point, the Yamnaya swept into Western Europe from the Central European and Central Asian steppes. Upon arriving around the year 3000 BCE, these invading herders introduced the earlier occupants of Western Europe to exotic pathogens, domesticated horses, and the wheel. They also imposed their well-developed culture and language as well as their Iranian and Eastern European genetic heritage on those they overwhelmed [36].

Today, there is no one ethnic group that can claim that they were the original Western Europeans. Instead, it is necessary to acknowledge that Western Europeans formed from the interbreeding of a wide range of different peoples. Furthermore, given that similar intermixing occurred throughout the world, it is also true that no one group anywhere on the planet can assert that they were the "original" and "sole inhabitants" of a particular large tract of territory [44].

The fact that human populations are of a genetically mixed nature has interesting implications for those who espouse nationalistic ideologies. Chief among them is the fact that our mongrel heritage falsifies claims predicated on the idea that there are "racially pure" populations. Sadly, this reality has not stopped many from disseminating their inaccurate views about their supposedly "pure" and "unmixed" genomes.

To the dismay of many who continue to support "blood and soil" type dogmas, the integration of the human population has not abated. In fact, due to the recent construction of high-speed transportation systems that link wide-ranging social and cultural networks, it has increased. Recently, groups of non-Africans that had been isolated from each other for over 50,000 years and populations of sub-Saharan Africans that had been separated for as long as 200,000 years are now coming together and interbreeding [36]. However, all this intermingling will not result in anything particularly dramatic. This is true, in part, because even though there are genetic differences between groups, modern humans are nevertheless genetically very similar to each other, and the distinctions that do exist tend to be limited in scope.

Among the first to demonstrate that humans resemble each other genetically was Richard Lewontin. In 1972, after grouping modern humans into seven populations (West Eurasians, East Asians, South Asians, Oceanians, Australians, Native Americans, and Africans), Lewontin examined the variation in blood group proteins and various other biological markers, and he found that approximately 85% of the total variation within these marker sets could be located within any one segment of the population [45]. Based on this work, Lewontin concluded that human populations were remarkably similar, and we now know that any two individuals, regardless of where their recent ancestors came from, are about 99.4% the same at the genetic level [46]. This degree of genetic similarity is truly remarkable, and as a group, we modern *Homo sapiens* are about 10–50 times more genetically similar to each other than are individuals in most other species [46]. Indeed, some have argued that modern *Homo sapiens* are so similar that, within our species, there is not enough genetic variability to form subspecies or races [44, 47]. However, whether there are "subspecies" or "races" of humans depends on how you define these terms, and as you might expect, there are many ways in which to do so [48]. In the end, what can't be debated is that we are all genetically very much alike, and we all have a recent common ancestry. In fact, according to the Yale University statistician, Joseph Chang,

> the most recent common ancestor for the world's current population lived in the relatively recent past—perhaps within the last few thousand years. And a few thousand years before that, although we have received genetic material in markedly different proportions from the people alive at the time, the ancestors of everyone on Earth today were exactly the same [49].

Before moving on from this topic, it is important to note once again that, despite our strong similarities, there remain some genetic distinctions between human populations. For example, the degree of clustering of various genetic traits does vary between groups [50], and it is also likely that there are subtle *average* genetic differences between human populations [36]. As a result, companies like "Ancestry" and "23andMe" can analyze a series of genetic markers and attribute a specific genetic heritage to an individual, and they can also use the differences in the structure of single genes, as well as clusters of genes, to make predictions about an individual's health prospects. However, as previously isolated populations continue to intermix, groups will become even more genetically homogenous, and as that happens, our common history, as well as our shared fate, will become even more obvious.

THE "WINNING HAND" OF THE AFRICAN *Homo sapiens*

As I stated earlier, archaic *Homo sapiens* and all our sister species are gone, but we modern *Homo sapiens* are still here, and we are currently distributed across the globe. So, what made us so successful?

We know that modern *Homo sapiens* have many noteworthy traits, but our ability to use grammatical language is unique. We are also unusual in that we can generate sophisticated abstract concepts, complex logic, and deep reasoning, and we can innovate at high rates. Unfortunately, for those who are interested in these traits, grammatical language, abstract thoughts, logical thinking, and deep reasoning do not fossilize. But innovation does sometimes produce products that remain long after their creators are gone, and one ancient commodity that differentiated us from our sister species was our tools. In comparison to other humans, early modern *Homo sapiens* had better tools, more strategies for producing tools, and more varied styles of tools. In addition, with the help of our tools, early members of our species were able to generate superior weapon systems, more elaborate jewelry and body decorations, and more sophisticated representational art than any other species [36].

To some extent, the apparent disparity in innovative capabilities may have been linked to various human species' population densities. As was mentioned earlier, we know that soon after modern *Homo sapiens* arrived in Eurasia, their group expanded, and they eventually outnumbered their Neanderthal competitors by about 10:1 [35]. Due to this overwhelming demographic advantage, the modern *Homo sapiens* would have had a significant cultural and technological edge, because all else being equal, more people results in more innovation. But, in addition to large numbers, modern

Homo sapiens likely had brains that were wired differently from those of their sister species, and this too may have given our group an innovative edge.

To gain a sense of the differences in the brains of Neanderthals and modern *Homo sapiens*, archaeologists have compared their fossilized skulls. Interestingly, they found that Neanderthals and modern *Homo sapiens* had similar brain sizes, but when the two species' endocranial volumes were adjusted for body mass and the size of their visual system, they determined that the modern *Homo sapiens* had larger relative cranial volumes [51]. Moreover, recent work using techniques developed by computational neuroanatomists indicates that modern *Homo sapiens* also had larger cerebellums, and this may have given our African ancestors superior language processing skills, enhanced cognitive flexibility, and augmented memory capacities and attention spans [52].

Additional evidence suggesting the brains of modern *Homo sapiens* differed from those of Neanderthals was discovered by molecular biologists. These scientists demonstrated that Neanderthal DNA sequences in modern humans tend not to be expressed at high levels in the brain [53], which is what you would expect to see if the brains of the two species were different and there was selection against the expression of Neanderthal genes.

So, modern *Homo sapiens* brains were likely different from those of the other human species, and early modern *Homo sapiens* were apparently more innovative. But which particular neurobiological traits drove our ability to think abstractly and logically, and what specific neurological capacities helped us reason and innovate in such an extraordinary way? The answers to these questions are subject to debate, but according to many, the traits and capacities that are most responsible for our thinking are those that gave rise to our facility for grammatical language.

With grammatical language, we can more readily explore our tactical situation as well as brainstorm practical solutions to the problems we face. In addition, we can explore our feelings, revisit the past, and ponder the future through the employment of an internal monologue.

Using grammatical language, we can generate thoughts that integrate geometric and non-geometric features of our environment [54], and as a result, we can produce ideas that combine time, actors, actions, numbers, spatial dimensions, and specific attributes. We can, for instance, assemble a thought such as, "Every other Wednesday, after the guard leaves the area at about 12 p.m., two large pots of food are placed under the table in the section of the room painted red." Consequently, with our capacity to recall, integrate, and organize geometric and non-geometric features through language, we can silently produce accurate, detailed, and precise strategies and plans. In short, with grammatical language, we can think more efficiently and clearly, and therefore grammatical language is, first and foremost, an extremely powerful cognitive tool [54].

In addition to being an essential tool for cognition, spoken grammatical language is also an indispensable instrument for effective communication. Using language, our ancestors transmitted information to one another with a high degree of precision and fidelity. With this same tool, they also organized and choreographed the activities of large numbers of people, which enhanced our forebearers' ability to construct intricate social networks, complex cultures, and new innovative technologies.

So, because of evolutionary processes, our brains acquired the ability to produce grammatical language, and that, in turn, likely facilitated the evolution of our capacity to think creatively, communicate effectively, and innovate rapidly. If this is true, then our brain's facility for grammatical language may be what most distinguished us from other organisms, and this trait may be what allowed us to become the deep-thinking, civilization-producing creatures that we are.

THE GENETICS OF GRAMMATICAL LANGUAGE

Given the central importance of grammatical language, it is easy to see why there has been immense interest in locating language-enabling genes in humans, and the best studied of all language-facilitating genes is *FOXP2*.

FOXP2 encodes a transcription factor, which is a protein that regulates the expression of many other genes, and it is involved in the wiring of the basal ganglia and prefrontal cortex of the brain. In addition, *FOXP2* influences the brachial arch formation and craniofacial development. Collectively, these data indicate that *FOXP2* plays a role in forming brain structures required for language, and it is also involved in constructing the facial and neck structures necessary for spoken language [55, 56].

When *FOXP2* is mutationally inactivated, affected individuals develop profound speech and language impairments, including language processing difficulties and verbal dyspraxia (which is the inability to carry out specific and sequenced orofacial movements). In addition, individuals with mutated *FOXP2* genes exhibit glaring deficits in their ability to interpret and utilize the rules of grammar.

Remarkably, mutations in *FOXP2* do not usually alter other forms of non-verbal cognition, and even though individuals with *FOXP2* mutations often suffer from severe verbal dyspraxia, they do not typically exhibit difficulties feeding themselves, nor do they present with abnormalities in gross motor development. This information suggests that *FOXP2* primarily affects verbal cognition as well as the physiological processes required for speech production [55].

Intriguingly, the regulatory domain of *FOXP2* in modern *Homo sapiens* differs from that found in the Neanderthals, and therefore the gene itself was likely expressed differently in the two groups [57]. As a result of this genetic difference, the language abilities of modern *Homo sapiens* and Neanderthals were probably distinctive, and it is possible that modern *Homo sapiens* had superior speech and language skills as well as greater verbal cognition at least in part because of this *FOXP2* mutation. If this was the case, then FOXP2 would have granted modern *Homo sapiens* an advantage in generating language-based thoughts, and this same mutation would have allowed them to build more effective social networks. Undoubtedly, these advantages would have contributed significantly to the cultural sophistication of *Homo sapiens* as well as the rate of innovation within modern human societies.

In addition to the alterations in *FOXP2*, there were many other genetic differences that distinguished modern *Homo sapiens* from the Neanderthals [58], and since 80–95% of all human genes are expressed in the brain at some point [56], it is likely that a large number of these genetic differences affected grammatical language, reasoned thought, and rapid innovation [59, 60]. Identifying and understanding how exactly these genetic differences distinguished modern *Homo sapiens* will further enlighten our understanding of the traits that make our species particularly unusual.

THE NATURAL SELECTION OF BEHAVIORALLY MODERN HUMANS

In addition to enumerating the genes required for language and modern human behavior, many biologists would also like to determine when the genes required for these traits first appeared. Currently, we can't answer this question, but we can say that they likely emerged long before modern *Homo sapiens* permanently settled outside of Africa. The geneticist David Reich recently came to this conclusion because he and others demonstrated that there are few, if any, DNA regions (outside of the maternally inherited mitochondrial DNA and the paternally inherited Y chromosome) that all humans inherited from a recent common ancestor. Instead, they found that the segments of the genome that all currently living humans share originated in common ancestors that lived at least 320,000 years ago [36].

Reich's observation is significant because, if modern *Homo sapiens* did arise because of gene mutations that occurred right before behaviorally modern humans emerged, we would expect all modern *Homo sapiens* to have common segments of DNA (other than the mitochondria DNA and the Y chromosome), which they inherited from a recent common ancestor. But, since modern humans do not appear to have such DNA regions, and because some groups of modern humans (such as those in parts of Africa) have been separated from other groups for most of the last 200,000 years, one of the following three conditions must hold: either all of the genes that imparted the modern behavioral suite of traits arose independently and nearly simultaneously in all the non-interbreeding *Homo sapiens* populations during the last 50,000–80,000 years, or some groups of modern humans lacked the ability to generate grammatical language and complex culture, or the genes that allowed the modern behavioral suite of traits to emerge arose in an ancient population of forebearers that all modern humans have in common. Given the extreme improbability of the first condition and given that we know that all groups of modern humans employ grammatical language, it seems likely that the genes needed to produce the behaviorally modern suite of traits arose in an ancient population of common ancestors. But, if this is so, why did behaviorally modern humans only appear about 50,000–80,000 years ago?

In an attempt to answer this question, Reich suggests that, within early *Homo sapiens* populations, there were probably a large number of different genes that influenced the repertoire of characteristics exhibited by behaviorally modern humans, and he further postulates that each of these genes had a small effect on some behaviorally modern human trait [36]. So, for example, there may be versions of a gene within the population that make it more likely that an individual can generate a specific type of vocalization, and there may also be other alleles (i.e., molecular variants of a given gene) that decrease the likelihood that an individual can utter a particular type of vocalization. These genes would constitute one of many genetic regions that influence modern human behavior.

According to Reich, in the groups that lived before behaviorally modern humans first appeared, natural selection likely privileged individuals who had sets of alleles that fostered the development of some behaviorally modern traits. So, natural selection may have, for example, favored people with the alleles that made it more likely for them to produce a wide range of vocalizations, and consequently, the frequency of this specific set of alleles would have increased within human populations. Presumably, other alleles that affected other aspects of the modern behavioral repertoire would have been selected in a similar fashion.

At some point, some individuals ostensibly inherited all the alleles needed for the manifestation of the full suite of behaviorally modern traits. When that happened, the first truly behaviorally modern *Homo sapiens* emerged. Once these individuals existed, the alleles that enabled their survival advantage, i.e., the set of alleles that imparted behaviorally modern characteristics, would have been favored, and these alleles would have become more common in a population.

To support the idea that numerous alleles can undergo differential selection within a population and ultimately give rise to a particular trait, Reich points to the more than 180 DNA regions that influence height in humans. He notes that northern European populations tend to have molecular versions of the genes that increase height, while southern Europeans generally have slightly different variants of these same genes that result in shorter stature. In this case, a wide range of height alleles were very likely present in the founding populations that gave rise to the northern and southern Europeans. However, once the group fractured, there was apparently selection for alleles that increased height in northern Europe, while in southern Europe, alleles that produced a shorter stature were favored. So, Reich points out that natural selection influenced the average height of the northern and southern Europeans, but in this case, it did so not by favoring new alleles (i.e., newly mutated genes) that arose in the separate populations, but rather, by altering the frequency of existing height-affecting alleles within each distinct group. Given this, it is reasonable to assume that in a similar way, natural selection could have favored a combination of alleles that gave rise to the traits exhibited by behaviorally modern humans. So, in the ancient human populations of 320,000 years ago, all the alleles needed for modern human behavioral traits may have existed, but it wasn't until more recently that individuals with the complete set of the specific alleles needed for modern human behavior emerged.

While the aforementioned hypothesis is interesting and plausible, it still does not answer the question of why it took so long for nature to select individuals that had all the alleles needed for modern behavior. To address this issue, some suggest changes in the ecological environment occurred around 50,000–80,000 years ago, and these conditions favored the selection of the set of alleles that gave rise to behaviorally modern traits. However, if natural selection did favor a particular set of alleles, it ostensibly did so nearly simultaneously across the large swaths of the globe where humans lived at this time, and this fact makes it difficult to affirm changing ecological conditions as the primary driver in this allele selection process.

There are other potential answers to the question of why natural selection increased the frequency of the alleles that gave rise to modern human behaviors about 50,000–80,000 years ago. For example, Reich notes some sections of the human genome have still not been fully analyzed. Consequently, it is possible that there still could be unanalyzed regions of the genome that harbor behavior-inducing alleles that arose when all people still shared a recent common ancestor. However, Reich cautions that, as the genomes of various groups of humans continue to be studied, this possibility is becoming increasingly less likely.

Still another potential scenario is that modern human behavior only recently arose because *Homo sapiens* culture only recently created the conditions necessary to promote the selection of the set of alleles that gave rise to these behaviors. For example, perhaps recently emerging human cultures placed great value on grammatical language and sophisticated thought. In such environments, those that possessed these traits would be more likely to survive and reproduce. If this situation did occur, it could have led to the creation of a positive feedback loop, and as a result, the alleles that enhanced a person's ability to generate grammatical language and sophisticated thought would have quickly increased in frequency. If this is indeed what happened, then our common cultural practices influenced our genetic evolution in a very profound manner, and they induced a selection process that led to the emergence of the modern suite of human behavioral traits. Interestingly, if this scenario did occur, it apparently did so with similar kinetics across all groups of *Homo sapiens.*

THE WORLD THAT CAME AFTER THE EMERGENCE OF BEHAVIORALLY MODERN HUMANS

Regardless of exactly how and when the full suite of behaviorally modern traits arose, shortly after it did, *Homo sapiens* developed novel ideas and new ways of subsisting. Within the realm of new ideas, one of the most influential and revolutionary was that of a God. According to some scholars, the concept of a God first entered the modern human psyche about 14,000 years ago [61], and by 4,000 years ago, modern humans had developed the foundations of some of the world's largest religious traditions. Today, we have well-developed concepts of a creator God who expresses an interest in our fate and well-being, and if the theistically minded individuals among us are correct, then our species only very recently accomplished what could arguably be the most significant of all historical and evolutionary achievements, namely, the recognition that God exists.

About 10,000–12,000 years ago, soon after the idea of a God appeared, some modern humans learned that they could survive as agriculturalists. With the genesis of this new subsistence strategy, people started to transition from a nomadic hunter-gatherer lifestyle to that of a village-dwelling farmer or pastoralist, and with this change, the human population began to expand about 100 times faster than it had during much of the late Paleolithic [62]. Our species' ability to harness the power of sunlight more effectively through agriculture led to these increases, and this, in turn, fundamentally changed our fate.

Eventually, farming technology improved to the point where one farmer could produce more food than he needed, and this allowed for the development of specialists of all kinds. These specialists, and the farmers that made their existence possible, formed the first cities, which emerged about 5,500 years ago in the river valleys of the Near East, Egypt, India, and China. Within these metropolises, modern humans developed culture-altering technologies such as the wheel, writing, and metallurgy, and consequently, our kind gained new strategies for organizing each other and controlling the natural world.

If we move forward another several millennia, to about the year 1550 CE, we will observe still another monumental event, namely, the beginning of modern science. With the birth of this discipline, many humans altered how they attempted to understand the world, and soon after modern science began, additional novel ideas and technologies further revolutionized human cultures across the globe.

One particularly important byproduct of modern science was the industrial revolution, which developed around 1800. By the dawn of this era, the human population had expanded from the five million souls that existed at the start of the agricultural period to about one billion in 1804 [62, 63]. However, once humans learned how to use fossil fuels our population expanded at unprecedented rates. By tapping into the power stored in these fuels, we essentially discovered how to harvest the energy of the sunlight tied up in the remains of the ancient plants and animals, and that, in turn, made extraordinary population growth rates possible. By 1927, the number of humans had doubled to two billion, and by October 31, 2011, there were seven billion members of our kind on the planet. So, it took our species over 300,000 years to get to a population of one billion, 123 years to go from one billion to two billion, and only 12 years to go from six billion to seven billion.

As impressive and potentially worrisome as the rate of our population expansion is, the culture and technological innovations of the last two hundred are even more striking. To get a sense of this, it is worth contemplating the life of John Adams, the second president of the United States. By the dawn of the industrial age, Adams and his colleagues had come a long way from the days when our ancestors lived in trees. But, despite this reality, in 1800 the second president still had to huddle around an open fire to keep warm during the winter, and in this regard, he was behaving as our ancestors did a million years earlier. Furthermore, if we were to observe Adams and his colleagues at the beginning of the industrial revolution, we would see that the amenities most of us take for granted, such as indoor plumbing, electricity, effective medicines, and rapid transport were not available to even the most powerful individuals. In fact, in 1800, Adams and his ilk did not even have access to a bicycle, because at that point, this simple two-wheeled machine had not yet been invented.

Today, just over 200 years after Adams left the White House, the communal learning undertaken by the human population has fundamentally altered our lives. As a result, at the beginning of the 21st century, people like me move about on carbon fiber, multi-geared, GPS-guided, high-tech bicycles, and we quickly travel long distances in highly sophisticated planes, trains, and automobiles. Furthermore, once we get to our destinations, we retreat to domiciles replete with indoor plumbing, central heat, central air conditioning, portable phones, televisions, high speed, worldwide-web-connected computers, and laser-powered security systems. In the world today, if we fall ill, we arrange to see physicians who prescribe the latest battery of medicines and pain-free surgeries, and we satisfy our curiosity by making use of tools like atom smashers, DNA sequencers, and rocket ships. After two million years of incredibly slow technological advancement, human innovation suddenly increased at a breathtaking pace.

THE FUTURE OF HUMANKIND

In the year 2022, as we gaze into the future, our species is attempting to build self-driving cars, robots with artificial intelligence, and quantum computers [64, 65]. In addition, we are considering the possibility of engineering our own genomes [66], and as always, we are trying to develop new and more efficient ways to kill each other [67]. Given all this, and given the fact that our planet is constantly changing, what can we say about the prospects for humankind? Will we continue indefinitely, or will humanities' luck run out as we join the ranks of the extinct? Of course, the answer to this query is not known, but the question itself is nevertheless worth exploring. So, in the last chapter of this book, we will ponder the future of life on Earth, and we will see if we can utilize our current knowledge to gain a tentative glimpse of humankind's potential destiny.

REFERENCES

1. Anton, S.C., R. Potts, and L.C. Aiello, *Human evolution. Evolution of early Homo: an integrated biological perspective.* Science, 2014. **345**(6192): p. 1236828.
2. Zink, K.D. and D.E. *Lieberman, impact of meat and lower Palaeolithic food processing techniques on chewing in humans.* Nature, 2016. **531**: p. 500–503.
3. Berna, F., et al., *Microstratigraphic evidence of in situ fire in the Acheulean strata of Wonderwerk Cave, Northern Cape Province, South Africa.* Proceedings of the National Academy of Sciences of the United States of America, 2012. **109**(20): p. E1215–E1220.
4. Adler, J., *Why fire makes us human.* Smithsonian.com, 2013.
5. Richard Wrangham, et al., *The raw and the stolen.* Current Anthropology, 1999. **40**: p. 567–593.
6. Laden, G. and R. Wrangham, *The rise of the hominids as an adaptive shift in fallback foods: plant underground storage organs (USOs) and Australopith origins.* Journal of Human Evolution, 2005. **49**(4): p. 482–498.
7. Garcia, T., et al., *Earliest human remains in Eurasia: New 40Ar/39Ar dating of the Dmanisi hominid-bearing levels, Georgia.* Quaternary Geochronology, 2010. **5**(4): p. 443–451.
8. Zhu, R.X., et al., *New evidence on the earliest human presence at high northern latitudes in northeast Asia.* Nature, 2004. **431**(7008): p. 559–562.
9. Eudald Carbonell, J., et al., *The first hominin of Europe.* Nature, 2008. **452**: p. 465–469.
10. Richter, D., et al., *The age of the hominin fossils from Jebel Irhoud, Morocco, and the origins of the Middle Stone Age.* Nature, 2017. **546**(7657): p. 293–296.
11. Scerri, E., et al., *Did our species evolve in subdivided populations across Africa, and why does it matter?* Trends in Evolution and Ecology, 2018. **33**(8): p. 582–594.
12. Hublin, J.J., et al., *New fossils from Jebel Irhoud, Morocco and the pan-African origin of Homo sapiens.* Nature, 2017. **546**(7657): p. 289–292.
13. Callaway, E., *Oldest Homo sapiens fossil claim rewrites our species' history. Nature* (2017). https://doi.org/10.1038/nature.2017.22114
14. Luskin, C., *The Genus Homo: All in the family.* Evolution News, 2012: p. https://evolution news.org/2012/08/the_genus_homo/.
15. Harvati, K., et al., *Apidima Cave fossils provide earliest evidence of Homo sapiens in Eurasia.* Nature, 2019. **571**(7766): p. 500–504.
16. Reich, D., *Who we are and how we got here: ancient DNA and the new science of the human past.* First edition. ed. 2018, New York, NY: Pantheon Books. xxv, 335 p.

17. Wong, K., *Why did Homo sapiens alone survive to the modern era?* Scientific American, 2018. **319**(3): p. 64–69.

18. Ackermann, R.R., A. Mackay, and M.L. Arnold, *The hybrid origin of "modern" humans.* Evolutionary Biology, 2015. **43**(1): p. 1–11.

19. Stringer, C. and J. Galway-Witham, *The origin of our species.* Nature, 2017. **546**: p. 212–214.

20. Hammer, M.F., et al., *Genetic evidence for archaic admixture in Africa.* Proceedings of the National Academy of Sciences of the United States of America, 2011. **108**(37): p. 15123–15128.

21. Callaway, E., *Hunter-gatherer genomes a trove of genetic diversity.* Nature, 2012. https://doi.org/10.1038/nature.2012.11076.

22. Callaway, E., *Israeli fossils are the oldest modern humans ever found outside of Africa.* Nature, 2018. **554**(15–16).

23. Posth, C., et al., *Deeply divergent archaic mitochondrial genome provides lower time boundary for African gene flow into Neanderthals.* Nature Communications, 2017. **8**: p. 16046.

24. Zimmer, C., *A single migration from Africa populated the world, studies find—see attachments for primary research articles.* New York Times, 2017.

25. Chen, L., et al., *Identifying and interpreting apparent Neanderthal ancestry in African individuals.* Cell, 2020. **180**(4): p. 677–687, e16.

26. Lopez, S., L. van Dorp, and G. Hellenthal, *Human dispersal out of Africa: a lasting debate.* Evolutionary Bioinformatics Online, 2015. **11**(Suppl 2): p. 57–68.

27. Gibbons, A., *Five matings for humans, Neandertals.* Science, 2016. **351**(6279): p. 1250–1251.

28. Vernot, B. and J.M. Akey, *Resurrecting surviving Neandertal lineages from modern human genomes.* Science, 2014. **343**(6174): p. 1017–1021.

29. Teixeira, J.C. and A. Cooper, *Using hominin introgression to trace modern human dispersals.* Proceedings of the National Academy of Sciences of the United States of America, 2019. **116**(31): p. 15327–15332.

30. Warren, M., *Mum's a Neanderthal, Dad's a Denisovan: first discovery of an ancient-human hybrid.* Nature, 2018. **560**: p. 417–418.

31. Timmermann, A. and T. Friedrich, *Late Pleistocene climate drivers of early human migration.* Nature, 2016. **538**(7623): p. 92–95.

32. Schlebusch, C.M., et al., *Southern African ancient genomes estimate modern human divergence to 350,000 to 260,000 years ago.* Science, 2017. **358**(6363): p. 652–655.

33. Gomez, F., J. Hirbo, and S.A. Tishkoff, *Genetic variation and adaptation in Africa: implications for human evolution and disease.* Cold Spring Harbor Perspectives in Biology, 2014. **6**(7): p. a008524.

34. Diamond, J.M., *Guns, germs, and steel: the fates of human societies.* First edition. ed. 1998, New York, NY: W.W. Norton & Co. 480 p.

35. Mellars, P. and J.C. French, *Tenfold population increase in Western Europe at the Neandertal-to-modern human transition.* Science, 2011. **333**(6042): p. 623–627.

36. Reich, D., *Who we are and how we got here: ancient DNA and the new science of the human past.* 2018, New York, NY: Pantheon Books.

37. Hogenboom, M., *Why are we the only human species still alive?* BBC Earth, 2018.

38. Dorey, F., *Homo neanderthalensis—the Neanderthals.* Australian Museum, 2019. https://australian.museum/learn/science/human-evolution/homo-neanderthalensis/.

39. Brown, Kyle S., et al., *An early and enduring advanced technology originating 71,000 years ago in South Africa.* Nature, 2012. **491**: p. 590–593.

40. Mellars, P., *Why did modern human populations disperse from Africa ca. 60,000 years ago? A new model.* Proceedings of the National Academy of Sciences of the United States of America, 2006. **103**(25): p. 9381–9386.

41. Diamond, J.M., *The third chimpanzee: the evolution and future of the human animal.* First edition. ed. 1992, New York, NY: HarperCollins. viii, 407 p.

42. Houldcroft, C.J. and S.J. Underdown, *Neanderthal genomics suggests a pleistocene time frame for the first epidemiologic transition.* American Journal of Physical Anthropology, 2016. **160**(3): p. 379–388.

43. Churchill, S.E., et al., *Shanidar 3 Neandertal rib puncture wound and paleolithic weaponry.* Journal of Human Evolution, 2009. **57**(2): p. 163–178.

44. Rutherford, A., *How to argue with a racist: what our genes do (and don't) say about human difference.* 2020, New York, NY: The Experiment LLC. xviii, 221 p.

45. Lewontin, R.C., *The apportionment of human diversity.* Evolutionary Biology, 1972. **6**: p. 381–397.

46. Collins, F.S., *The language of God: a scientist presents evidence for belief,* in *ISSR Library; variation: ISSR Library.* 2006, New York, NY: Free Press.

47. Smedley, A., *American anthropological association's statement on race,* 1998. www.americananthro.org/ConnectWithAAA/Content.aspx?ItemNumber=2583.

48. Weiss, K. and S. Fullerton, *Racing around, getting nowhere.* Evolutionary Anthropology, 2005. **14**: p. 165–169.

49. Rohde, Douglas L.T., Steve Olson, and J.T. Chang, *Modelling the recent common ancestry of all living humans.* Nature, 2004. **431**(7008): p. 562–566.

50. Reich, D., *How genetics is changing our understanding of 'race'.* New York Times, 2018.

51. Pearce, E., C. Stringer, and R.I. Dunbar, *New insights into differences in brain organization between Neanderthals and anatomically modern humans.* Proceedings of the Royal Society, 2013. **280**(1758): p. 20130168.

52. Kochiyama, T., et al., *Reconstructing the Neanderthal brain using computational anatomy.* Scientific Reports, 2018. **8**(1): p. 6296.

53. McCoy, R.C., J. Wakefield, and J.M. Akey, *Impacts of Neanderthal-introgressed sequences on the landscape of human gene expression.* Cell, 2017. **168**(5): p. 916–927 e12.

54. Berwick, R.C. and N. Chomsky, *Why only us: language and evolution.* 2016, Cambridge, MA: MIT Press.

55. Cecilia, S.L., S.E.F. Lai, Jane A. Hurst, Faraneh Vargha-Khadem, and Anthony P. Monaco, *A forkhead-domain gene is mutated in a severe speech and language disorder.* Nature, 2001. **413**: p. 519–523.

56. Byoung-il Bae, D.J., and Christopher A. Walsh, *Genetic changes shaping the human brain.* Developmental Cell, 2015. **32**(4): p. 423–434.

57. Maricic, T., et al., *A recent evolutionary change affects a regulatory element in the human FOXP2 gene.* Molecular Biology and Evolution, 2013. **30**(4): p. 844–852.

58. Green, R.E., et al., *A draft sequence of the Neandertal genome.* Science, 2010. **328**(5979): p. 710–722.

59. Preuss, T.M., *Human brain evolution: from gene discovery to phenotype discovery.* Proceedings of the National Academy of Sciences of the United States of America, 2012. **109**(Suppl 1): p. 10709–10716.

60. Fitch, W.T., *The biology and evolution of speech: a comparative analysis.* Annual Review of Linguistics, 2018. **4**: p. 255–279.

61. Armstrong, K., *A history of God: the 4000-year quest of Judaism, Christianity, and Islam.* First American ed. 1993, New York, NY: A.A. Knopf: Distributed by Random House. xxiii, 460 p.

62. Christian, D., *Origin story: a big history of everything.* First edition. ed. 2018, New York, NY: Little, Brown and Company. x, 357 p.

63. Bello, D., *Human population reaches 7 billion—how did this happen and can it go on?* Scientific American, 2011. https://www.scientificamerican.com/article/human-population-reaches-seven-billion/

64. Kitman, J.L., *Google wants driverless cars, but do we?* New York Times, 2016.

65. Arute, F., et al., *Quantum supremacy using a programmable superconducting processor.* Nature, 2019. **574**(7779): p. 505–510.

66. Klug, W.S., et al., *Concepts of genetics.* Twelfth edition. ed. 2019, New York, NY: Pearson.

67. Gordon, M.R., *U.S. plans new nuclear weapons.* Wall Street Journal, 2018.

13 The Future of Life on Earth

It's tough to make predictions, especially about the future.

Yogi Berra

The probability of global catastrophe is very high . . . inaction and brinkmanship have continued, endangering every person on Earth.

The authors of the 2017 Bulletin of Atomic Scientists

What's the use of a fine house if you haven't got a tolerable planet to put it on?

Henry David Thoreau

If you're going through hell, keep going.

Winston Churchill

We currently have a substantial global impact on the biosphere, and the changes we introduce are likely causing a mass extinction [1]. Consequently, in the near term, the future of life on Earth will be significantly influenced by what happens to us.

When considering the fate of humanity, there are two sets of threats to contemplate, namely, those that we impose on ourselves and those imposed upon us by nature. To get an idea of the nature of these threats, I will briefly describe examples of each.

WILL WE END OUR LUCKY STREAK?

Our large brains have helped us avoid some extinction threats; however, these same brains expose us to self-imposed dangers that no other creature faces. The list of these potential hazards is lengthy, and there have been many books and articles written on this topic (see [1–4] for some examples). However, it is not my intention to enumerate all the ways in which we could facilitate our own destruction. Rather, I would like to briefly mention just a few of the most probable scenarios for our consideration.

THE POTENTIAL FOR HUMANITY TO DESTROY ITSELF WITH ITS OWN WEAPONRY

Included among the immediate problems we face is the potentially catastrophic use of biological weapons. Sadly, humankind is not above the use of germ warfare, and documented attempts to spread the disease during a conflict can be traced to Hittite combatants of 3,500 years ago [2]. More recently, in 1346, Mongol warriors who

DOI: 10.1201/9781003270294-16

died of plaque were thrown over the walls of besieged Russian cities in an attempt to sicken the population within [3], and in modern times, Europeans used smallpox-infected blankets to unleash lethal viruses among the indigenous Native American populations [4]. Unfortunately, this is not an exhaustive list of history's germ warfare offensives.

Today, many countries can use viruses, bacteria, and fungi as weapons of war, and because of advances in biotechnology, lethal forms of these creatures can be genetically engineered to be even more deadly. Ultimately, the use of this type of weapon would likely yield unforeseen consequences, and those employing it could inadvertently bring about their own destruction along with that of their intended victims.

As I write this chapter, the world is enduring a pandemic brought about by the COVID-19 virus. This virus appears to have originated in China, and in just a few months, it spread around the world. Within one-half of a year, the virus infected over 6.3 million individuals and killed more than 374,000. In the United States alone, there were over 100,000 deaths in just the first six months of the virus' appearance, and in less than two years, over 820,000 people died from this infection. Keep in mind that this happened even though the United States is a highly technologically advanced country, and it is located many thousands of miles from the virus' point of origin.

In addition to compromising public health, the virus decimated the economies of many countries, and it helped potentiate political and civil unrest. While COVID-19 is clearly dangerous, one can only imagine the worldwide devastation that would occur if a much more lethal and rapid spreading pathogen were engineered and purposefully released.

Recent advances in genetic engineering have also made it possible for scientists to genetically alter humans themselves, and today, for the first time in history, we can engineer our genome in a laboratory [5]. With this ability, we can potentially direct our own evolution, but whether we have the wisdom to undertake this endeavor is certainly debatable. If we do engage in human genetic engineering, and it seems likely that we will [6], it is conceivable that we could one day generate a group of people that are so different from the humans of today that they would constitute a new species. History suggests that two species of hominids cannot coexist on this planet (see Chapter 12), so if two different species of humans do emerge, a conflict between them is likely. During this struggle, there is the possibility that the two groups would eliminate each other. However, if the genetically engineered humans gain the upper hand, they would likely assimilate or eliminate the non-engineered individuals, and if this were to occur, humans, as we currently know them, would cease to exist. How the new species of hominoid would affect life on the planet is anyone's guess.

Another possibility is that we will develop a form of artificial intelligence that will overtake the planet and eliminate us. There are some who argue this is a very real risk, while others doubt that a conscious and truly independent form of artificial intelligence could ever arise [7–10]. But, if intelligent machines do ultimately replace us, organic evolution would occur in parallel with inorganic evolution, and "life" would evolve in a very new and unpredictable manner.

Yet another possibility is that we will eventually integrate genetic engineering, nanotechnology, robotics, and computer technology, and in so doing, transform ourselves into organic/inorganic hybrids. Currently, we augment our natural functioning with equipment such as ceramic dental implants, computer-enhanced hearing aids, laser-guided surgery, and artificial organs. In the future, genetically engineered individuals may further augment their physiological and intellectual function, and if we were to come back hundreds of years from now and find such a world, we would barely recognize our descendants.

While the anthropogenic threats and advances described earlier are worth pondering, the most immediate dangers are those imposed by nuclear weapons and environmental destruction. Nuclear weapons have already been used in battle, and the destruction of the environment is occurring right now. Of these two sets of threats, the large-scale use of nuclear weapons would result in the most rapid destruction of our species, so it is with this topic that I will begin.

HOMICIDAL TENDENCIES AND NUCLEAR ARMAGEDDON—TROUBLE FROM WITHIN

The nuclear era was ushered in on July 16, 1945, at 5:29:45 a.m., in the Jornada del Muerto desert in New Mexico. On that morning, the first nuclear device was detonated, and soon after, on the mornings of August 6 and August 9, 1945, nuclear bombs were unleashed on Hiroshima and Nagasaki in Japan. The resulting horror is extremely difficult describe. The bombs destroyed both cities and resulted in as many as 246,000 deaths. Many recoiled at the magnitude of the pain and suffering that was induced in just minutes, and since this time, religious, military, and political leaders have considered how we might rid our world of this existential threat [11]. However, as of this writing, the world has at least 9,200 nuclear warheads ready to be deployed [12–14], and some of these devices have more than 3,000 times the explosive yield of the bombs used in 1945 [15].

In the history of the world, no other species possessed the ability to willfully obliterate itself, and thankfully, for the last 75 years, we have managed to avoid that potential reality. But, 75 years is not a long period, and in that interval, there have been many dozens of incidents during which mishaps and miscalculations could have led to a nuclear holocaust. The Union of Concerned Scientists has an excellent six-page summary of some of these events in a text entitled *"Close Calls with Nuclear Weapons"* [16], and what is clear from that manuscript and other related documents [17–21] is that, in some instances, a nuclear holocaust was avoided because of the prudent judgment of a single individual. One such event took place on October 27, 1962, in the middle of the Cuban Missile Crisis. On that day, which John F. Kennedy referred to as "Black Saturday," the American Navy surrounded a Cuba-bound Soviet patrol submarine and tried to force it to surface. Valentin Savitsky, the captain of the Soviet submarine, was not able to communicate with Moscow, and in an exhausted state, and with low levels of life support, he believed that it was likely that a war between the Soviet Union and the United States had already started. In response, Savitsky wanted his submarine crew to launch its 10-kiloton nuclear missile at the American naval forces. But, to fire the missile, the top two officers on board had

FIGURE 13.1   Vasily Arkhipov: The man who likely saved the world in 1962. (From: Olga Arkhipova, CC BY-SA 4.0 <https://creativecommons.org/licenses/by-sa/4.0>, via Wikimedia Commons.)

to concur with his decision. This policy was not a typical one, and in 1962, most nuclear-armed Russian submarine captains only needed one political officer to concur. However, the flotilla commander, Vasily Arkhipov (see Figure 13.1), was part of Savitsky's crew on that day, and consequently, Arkipov's approval was also required. Ultimately, Ivan Maslennikov, the political officer, agreed with the Savitsky, but Arkhipov opposed the plan, and after an argument, the captain acquiesced. Had Arkhipov not been aboard that Soviet submarine, and had he not maintained his ability to calmly engage in rational thought while under tremendous pressure and the threat of imminent death, there likely would have been a large-scale nuclear war unlike any the world has seen. After learning of this incident, Arthur Schlesinger Jr., a historian, and an advisor to President Kennedy, declared that this situation "was not only the most dangerous moment of the Cold War. It was the most dangerous moment in human history."

Notably, this was not the only time a single individual stood between peace and nuclear Armageddon. Another well-documented instance of this phenomenon occurred 21 years after the Cuban Missile Crisis, on September 26, 1983. On that day, Stanislav Petrov (see Figure 13.2), a 44-year-old lieutenant colonel in the Soviet military, was the duty officer at a Soviet Command Center near Moscow when the Soviet early warning satellites indicated that the American military had launched five nuclear missiles at the Soviet Union. According to the satellite data,

FIGURE 13.2 Stanislav Petrov in 2017—another individual that singlehandedly saved the world from a devastating nuclear exchange. (From: Queery-54, CC BY-SA 4.0 <https://creativecommons.org/licenses/by-sa/4.0>, via Wikimedia Commons.)

the reliability of this information was "very high," and it predicted the U.S. nuclear missiles would hit the Soviet Union in about 25 minutes. Despite this situation, Petrov decided to report the warning as a malfunction, and he later stated that he did not trust the reliability of the early warning system, and he "had a funny feeling." Furthermore, the warning indicated that "only five missiles" were headed toward the Soviet Union and Patrov concluded that, "when people start a war, they don't start it with only five missiles" [21].

Subsequent investigations revealed that the false alarm was triggered when the satellites interpreted the Sun's reflection off the clouds as incoming missiles. Given the analysis that Petrov carried out when making his decision about whether to initiate a counterstrike, it is hard to imagine what the world would look like today if the early warning system had indicated that hundreds of missiles were heading toward the Soviet Union instead of five. It is also difficult to contemplate how different the world would likely be today if a different officer had been on duty at the Soviet Command Center on that fateful day in 1983. In the end, had the situation unfolded in a slightly different fashion, many of us reading this now would probably be dead or never born, and if some of us had somehow survived a nuclear exchange between the Soviet Union and the United States, our lives would have been very different.

Alarmingly, there were many other instances in which the actions of a small number of people made the difference between war and peace. Fortunately, however, when false alarms of impending nuclear attacks have occurred, rational individuals have made choices that profoundly benefited humankind. Furthermore, up until now, we have not encountered a situation during which someone in control of a nuclear arsenal acted on the gravely mistaken belief that he or she could benefit from a nuclear confrontation and ultimately emerge victorious. Additionally, in the last 75 years, we have not experienced a situation in which a suicidal, desperate, and mentally disturbed individual with control of nuclear weapons decided to destroy humanity.

MOVING FORWARD IN THE NUCLEAR ERA

If a nuclear war does occur, the resulting conflict would be unlike any of the past. To begin to understand the potential destruction of such an event, it is worth pondering the fact that, currently, the United States can kill one-quarter of Russia's population (i.e., 36 million people) with just 147 warheads, and it can destroy one-quarter of China's population (325 million people) with only 789 nuclear weapons [13]. Likewise, the Russians or the Chinese could instantly kill one-quarter of the U.S. population (82 million people) with 124 warheads, and they could, in just a few minutes, incinerate all the largest cities in the United States including New York, Los Angeles, Chicago, Houston, Phoenix, Philadelphia, San Antonio, San Diego, Dallas, Seattle, Denver, Boston, and Washington, to name just a few.

Keep in mind that all this hypothetical destruction within the Soviet Union, China, and the United States could be accomplished with a total of just 1,261 warheads, and today, the United States and Russia each have more than 4,000 of these weapons. In addition, the United Kingdom, France, and China have about 200–300 nuclear weapons apiece, and India, Pakistan, North Korea, and Israel each control up to several scores of nuclear warheads [14].

Given this reality, and given that reasonable and sane people understand that the only useful purpose of nuclear weapons is to deter an adversary who would use them, how can humanity decrease the risk that it will destroy itself in a nuclear war? Many suggestions have been put forward, and one of the more obvious propositions is to enhance systems designed to prevent terrorists from obtaining these instruments of mass destruction. We can also improve humanity's prospects by halting nuclear proliferation, decreasing nuclear weapon stockpiles, enhancing safeguards against false alarms, and limiting the power of individuals to unilaterally initiate a nuclear exchange. If we do not take these and other necessary actions to prevent the use of nuclear weapons, we will increase the probability that humanity will suffer in ways that would be difficult, if not impossible, to recover from.

Regrettably, despite having so much at stake, humans have, up until now, largely relied on luck to get us through this nuclear era. But luck is not an effective survival strategy, and sooner or later, our good fortune will run out. If we are to continue to exist, we need to think carefully about how we can significantly decrease the enormous inherent risks associated with nuclear weapons, and we need to enact as many safeguards as possible.

ENVIRONMENTAL DESTRUCTION

In addition to a potential nuclear conflagration, we face the very real prospect that we could severely damage, or even destroy, the environment that we depend upon to survive. Recently, there has been a tremendous amount of attention rightly focused on global warming and its effect on the environment. However, it is important to note that maintaining and protecting the environment requires that we consider more than just global warming.

In his outstanding book, *Collapse: How Societies Choose to Fail or Succeed* [22], Jared Diamond notes that to survive on our planet we need to:

  a. Control our population growth,
  b. Maintain our natural resources, and
  c. Limit the pollution of the biosphere and the degradation of our ecosystems.

Not surprisingly, the difficulties we have maintaining our natural resources and limiting our destructive potential increase as a function of the size of the global population, and in 1798, Thomas Malthus warned that we would eventually reach a point at which Earth could no longer support the human population. However, until now, we have managed to increase our numbers and simultaneously substantially increase the standard of living for a large fraction of the population [23]. But the question that we must address is how long can we continue this feat? Ultimately, we need to know when the number of people on the planet will be greater than that which Earth can support, and we need to be able to limit our growth so that it does not exceed this number. Unfortunately, this tipping point is hard to estimate, and even if we knew what the number was, it is likely that we will not easily be able to keep the world's population below that number. What we do know is that we cannot indefinitely increase our population size and expect to survive on planet Earth [24].

So, how might we control human population growth? The simple truth is, we don't know. Throughout history, for various reasons, societies have attempted to attenuate their expanding numbers, but there have always been countervailing forces that successfully opposed these initiatives [22, 25]. As a result, today there are 7.8 billion people.

With regard to the world's population, the good news is the rate of population increase has slowed. Currently, it is 1.1%, down from a recent high of 2.1% from 1965 to 1970. However, in absolute numbers, the world population is still increasing dramatically. In fact, in the year 2018, the global population *increased* by 82 million, which is a number equal to about ¼ of the population in the United States. Many experts think that the human population will level out at about 11 billion, but we can't be sure they are right, and even if they are, we don't know if the planet can sustain that number of individuals indefinitely.

The ability of the planet to maintain a growing human population depends, in part, on the stability of the small sliver of the biosphere in which most life dwells. But, sadly for us, this domain is not infinitely robust. Natural habitats, as well as flora and fauna, can be irreversibly altered or destroyed, and there are finite supplies of many of the materials that we need such as fossil fuels, freshwater, topsoil,

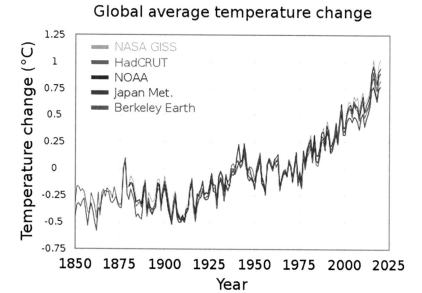

FIGURE 13.3 Earth's biosphere is heating up. Average global temperatures were determined by NASA, NOAA, Berkeley Earth, and the meteorological offices of Japan and the United Kingdom. The data sets are in strong agreement and the pairwise correlations of temperature increases were between 98.09 and 99.04%. (From: RCraig09, CC BY-SA 4.0 <https://creativecommons.org/licenses/by-sa/4.0>, via Wikimedia Commons.)

minerals, and agriculturally suitable land. Today, as we continue to alter the biosphere, we are destroying ancient habitats, and we are facilitating the destruction of species at a speed that is estimated to be at least 100 times the normal extinction rate [26, 27]. In addition, human-induced perturbations in Earth's nitrogen and phosphorous cycles are altering the resilience of our planet's natural and managed ecosystems [28, 29].

As we strive to maintain our resources and live within our means, we must also consider how the byproducts of our activity affect the habitats in which we reside. Toxic materials, particulate matter in the air and water, and the introduction of foreign species have crippled or destroyed ecosystems. Over the last 50 years, we have also discovered that of all the byproducts we generate, gaseous wastes are particularly concerning because they induce global warming (see Figure 13.3). Currently, 97% of the world's climate scientists and most of the world's major scientific organizations have warned that we must dramatically limit our use of fossil fuels if we are to prevent significant increases in global temperatures [30]. If we do not heed these warnings, and if we continue to burn fossil fuels at our current rate, the temperature on the planet will increase by as much as 4.8°F by 2050 and by 4.7–8.6°F by the end of this century [31], and that would be utterly devastating to our biosphere.

WILL NATURE END OUR LUCKY STREAK?

On average, a typical species lives about ten million years, but mammals fare less well with an average "species lifespan" of only about one million years [32]. Our species has been in existence for about 300,000 years, and how long it will continue is obviously not known. What we do know is that in the past, billions and possibly even trillions of species have existed, and Nature itself was responsible for the destruction of more than 99.9% of them.

To efficiently chronicle some of the hazards Nature could generate, I have grouped them into three categories. The first cluster contains events that could happen at any time, but they have a very low probability of occurring. The second set of episodes consists of threats that are more likely to materialize. These events occurred in the past, and they will occur again, but exactly when, we do not know. Finally, the last group of cataclysmic incidents is cosmic in nature, and life ending, but they will almost certainly not take place until sometime in the very distant future. Many of the events in this final cluster of threats are related to the aging of our solar system, and while interesting in an existentially scary way, these dangers are not of immediate concern.

GROUP ONE THREATS: A GRAB BAG OF NASTY EVENTS

There are a large number of low probability naturally occurring events humanity could encounter. Among them are gamma-ray bursts generated by relatively nearby supernovas or kilonovas; supermassive black hole activity; itinerant black holes, stars, and planets; and exuberant solar activity.

In one worst-case scenario, gamma-ray and X-ray bursts from supernovas, kilonovas, and magnetar's could sterilize the planet, or in a less dramatic but equally deadly development, they could destroy our atmosphere. We have, in fact, been bombarded with high-energy radiation from distant cosmic objects in the past, and one recent incident was recorded on August 27, 1998. For about five minutes on that summer day, gamma rays and X-rays from a magnetar in the constellation Aquila slammed into Earth, and although Aquila is 20,000 light years from our planet, the energy released from this region still spiked the radiation sensors on Earth's satellites.

Luckily, the energy from Aquila that did strike Earth was absorbed by our atmosphere, and its destructive punch only made it to about 30 miles from our planet's surface. This occurred because, over its 20,000-year journey, much of the radiation headed our way from Aquila dissipated. But, had this magnetar been "only" 2,000 light years away, the radiation that inundated our planet would have been 100 times greater [33], and the consequences would have been much more significant.

Life-destroying radiation can also emanate from black holes, and recently a truly colossal amount of radiation from a black hole in the central galaxy of the Ophiuchus cluster was detected. This particular burst was the largest explosion ever documented in the universe, and it destroyed a section of the cosmos that was 15 times the size of our Milky Way galaxy (keep in mind that the Milky Way has a diameter of over 100,000 light years) [34]. Sadly, any life that existed around the countless billions of stars in this region was instantaneously wiped out. Thankfully for us, this dreadful event occurred some 390 million light years from here, so life on Earth endured without incident. However, if a similar event occurred in our cosmic realm, we and our entire solar system would join the ranks of the extinct.

Fortunately, Earth has not been in the path of a life-sterilizing energy source during its four-billion-year history, and we obviously avoided any solar system destroying black hole activity. However, our planet has been hit by high-energy radiation many times, and it will be hit again. By one estimate, there are more than 20 potentially life-threatening gamma-ray-producing events that could occur in the next two billion years, and there are currently 12 potential pre-supernova stars within 3,260 light years (or 1 kiloparsec) of Earth [35].

In addition to high-energy radiation, large itinerant objects, such as roaming black holes, stars, and planets, have the potential to destroy our planet if they approach our solar system. But fortunately for us, the galaxy is enormous, and wandering celestial objects appear to be relatively rare [36].

High-energy radiation blasts and wandering cosmic objects fit into the biblical "end of times" genera of cataclysmic events, and there is nothing we could currently do to prevent the damage they would generate. When it comes to this group of catastrophes, our only hope is that we continue to be lucky enough to avoid them.

DANGER NEARBY

Closer to home, there are threats that emanate from our Sun. Among these are giant solar flares that could destroy our atmosphere. Fortunately, however, at least at the moment, our Sun appears to be relatively benign when it comes to the generation of atmosphere-ending solar flares [37, 38]. But, our Sun has generated significant solar flares that are capable of knocking out our power grids, and some may even be capable of erasing our electronic records [39]. Such flares, if they were to occur today, would be extremely destructive and expensive [40].

A HODGEPODGE OF OTHER DANGERS

There are other phenomena that we can include within this group of highly unlikely, life-ending events, including vacillations in the vacuum state of the universe, alterations in the fundamental constants of the universe, and the invasion of Earth by hostile and genocidal aliens [41–44]. However, as far as we know, these events have never occurred in the past, and it is likely that they won't happen in the future either.

Finally, within this group of existential cosmic hazards are the "unknown unknowns." Regrettably, there are likely many events of which we are not yet aware that could pose a threat to our existence. Consider, for example, five hundred years ago, we didn't know that giant rocks could fall from the sky and cause mass extinctions, and we didn't know that the ground could open and spew enough molten lava to destroy a large fraction of the life on the planet. In fact, as recently as 1980, the idea that mass extinctions were caused by these events was considered highly speculative [45]. Consider also that today, carbon dioxide emissions threaten life on Earth, but just 300 years ago, humanity was unaware that this gas even existed. In the future, we will likely encounter currently unknown dangerous phenomena, but there is not much point in worrying about something that you can't yet even imagine. That, after all, would seem to be the very definition of an anxiety disorder.

GROUP TWO THREATS: NATURAL EVENTS THAT WILL THREATEN OUR EXISTENCE SOMETIME IN THE FUTURE

Unlike the events described in group one, the natural events described in this second group are likely to occur at some time in the not-so-distant future, and in at least some instances, we might be able to do something to mitigate their effects. Within this category are catastrophic geological events (such as supervolcanic eruptions and flood basalts) and collisions between our planet and various forms of space debris (such as asteroids and comets). These events have caused the planet's greatest mass extinctions, and they are the most significant non-anthropogenic threats that we face in the future.

VOLCANIC ERUPTIONS AND BASALT FLOODS—TROUBLE FROM BELOW

Between catastrophic geological events and cosmic collisions, the most concerning are the geological catastrophes, and in particular, supervolcanic eruptions. By definition, supervolcanic eruptions have a Volcanic Explosivity Index (VEI) of 8 or higher, and this means they eject 1000 km^3 (which is 240 cubic miles) or more of debris. During the last two million years, there have been 27 of these types of eruptions, but luckily for us, *Homo sapiens* have only experienced two of them. The first was the Toba supervolcanic eruption of 74,000 years ago. This explosion was the second-largest volcanic paroxysm known to have occurred, and with a VEI of 8.8, it dumped 670 cubic miles of ash across Central Asia and the Middle East. By the time it was over, the Toba eruption had released enough debris and gas to profoundly alter the global climate for centuries.

The effect the Toba explosion had on our ancestors has been a matter of vigorous debate [46–48]. Some investigators argue that it nearly eradicated our species, and they point to genetic evidence that indicates that, after the Toba eruption, the number of humans on the planet may have dropped to between 15,000 and 40,000 individuals [47]. However, others believe that groups of humans on faraway landmasses were not greatly affected by the Toba incident, and they point to data that suggest humans did okay in the years immediately following the eruption [48]. What is clear is that the Toba explosion devastated Indonesia and some of the regions around it, and if such an eruption were to occur in today's crowded and highly interconnected world, the consequences would be profound.

The second supervolcanic explosion that humans experienced was the Taupo eruption, which occurred about 27,000 years ago in New Zealand. At the time of this VEI 8.1 event, our ancestors did not live in New Zealand, but it is likely that the Taupo eruption generated tsunamis that devastated inhabited sections of Southeast Asia and Australia, and the explosion also likely produced a volcanic winter that greatly affected those alive at the time.

Today, several supervolcanic sites are still active, and among the deadliest of these is Yellowstone, which has generated three cataclysmic eruptions in the last 2.1 million years. The largest of the three, which took place 2.1 million years ago in what is now the Western United States, was the fourth most violent volcanic blast known to have occurred in the history of the planet. With a VEI of 8.7, this eruption had a force hundreds of thousands of times greater than the one generated by the nuclear bomb that destroyed Hiroshima, and when the eruption occurred, 30 miles of the Rocky Mountains were almost instantaneously annihilated, and hundreds of cubic miles of gases and molten lava were released into the atmosphere and onto the surrounding landmass. The lava spread over an area of more than 6,000 square miles (which is an area greater than the state of Connecticut), and during the explosion, eight cubic miles of the mountain were blown clear into the stratosphere. And that was just the beginning of the damage this monster rendered. The molten lava and falling rock ignited fires across a wide swath of land, and the gases released by the volcano damaged the atmosphere and triggered a multi-century-long volcanic winter. In the end, the resulting climatic changes devastated ecosystems across the planet and severely affected the flora and fauna on what is now the continental United States.

The last major eruption of the Yellowstone volcano occurred 640,000 years ago, and it had a VEI of 8.3. Unfortunately, it is likely that Yellowstone will erupt again, but we can't say when. However, if this supervolcano were to erupt today with a force like that it displayed 2.1 million years ago, over a billion people would likely be dead within months of the explosion, and the entire continent of North America would be devastated. Furthermore, the living standards of people across the globe would plummet, and the event would lead to unimaginable political and civil unrest and chaos.

Regrettably, Yellowstone is not the world's only active supervolcano, and there are other similar behemoths located across the globe. Consequently, the odds that humanity will experience another supervolcanic eruption are not trivial, and it has been estimated that there is a 0.12% probability that a supervolcanic explosion will occur within the next 100 years [49]. Smaller volcanic eruptions are even more likely, and volcanic eruptions with a VEI of 7 (which emit a debris volume of 100 km^3) take place about one to two times every millennium. If a VEI eruption of 7 were to occur today, entire cities would likely be destroyed, and large areas would be devastated and paralyzed by tsunamis and volcanic ash. In addition, global cooling would likely ensue [50], and this climate-changing event could last for decades or even centuries. Under these conditions, the planet's ability to sustain the current human population would be severely strained, and the stability and security of nations across the globe would be compromised.

Clearly, large-scale volcanic eruptions would be disastrous for humanity, but they are not the only geologically based threat we face. Lurking below the surface of the planet are prodigious pools of magma that could be released as part of a flood basalt event. During such an incident, enormous volumes of lava (usually more than 2,000 km^3) flow from a fissure within the planet. These lava flows can occur periodically, over a period of millions of years, and they can form step-like plateaus of volcanic land. Mercifully, flood basalts are rare, and they are orders of magnitude less likely to occur than are supervolcanic eruptions [49]. Over the last 250 million years, there have only been 11 known instances of continental flood basalts, and the last one occurred about 17 million years ago in what is now the northwestern section of the United States [51]. So, at the time of the last continental flood basalt, humanity did not exist, and we can be thankful that flood basalts are very infrequent, because if that were not the case, we would not be here.

ASTEROIDS AND COMETS—TROUBLE FROM ABOVE

In an average year, our planet smashes into about 40,000 tons of little meteors and tiny space particles, but it collides with a 1-kilometer cosmic mass, most often an asteroid, only about once every 600,000 years [52]. Luckily, collisions between our planet and even larger cosmic masses are rarer still, and Earth typically strikes an asteroid with a diameter greater than 10 kilometers only about once every 100 million years [33]. These numbers indicate that collisions between Earth and 1-kilometer pieces of space debris occur less frequently than do supervolcanic eruptions, and in fact, such a collision is 5–100 times less likely to occur than is a supervolcanic eruption [47, 49]. Nevertheless, when these collisions do take place, the outcome is catastrophic, and like VEI 8 volcanic eruptions, collisions with space debris pose an existential threat to humanity.

The potential destructive power of asteroids greater than 10 kilometers in diameter has been well documented by those studying the last great mass extinction (see Chapter 10), but even a much smaller asteroid can wreak havoc of biblical proportions. In his book, *A Short History of Nearly Everything*, Bill Bryson eloquently describes the incredible catastrophe that ensued when an asteroid with a diameter of 1.2 miles slammed into what is now Manson, Iowa 74 million years ago [53]. In short, during the reign of the dinosaurs, a 10-billion-ton rock, which had been wandering through the cold darkness of space for many millions of years, began to drift toward Earth, and at some point, this colossal boulder was trapped by Earth's gravity. Once ensnarled, the hunk of space debris rocketed to our planet's surface at a speed 200 times faster than that of sound, and as it approached the surface of the planet, the air below it compressed and heated to about 108,000°F.

Within one second of entering the atmosphere, the huge asteroid slammed into Earth, and it created a hole that was three miles deep and more than 20 miles wide. The heat and blast wave that resulted instantaneously destroyed every living creature within 150 miles of the impact site, and within seconds, excavated material soared into the atmosphere, earthquakes shook the planet, volcanoes spit debris and gases into the environment, and tsunamis headed for distant shores. Just minutes after the impact, the damage intensified when some of the red-hot debris that careened upward into the sky rained back down and set the landscape ablaze. Within this same brief interval, some of the asteroid and ejected blast material became ensconced within the atmosphere, and this plus the volcanic gases, and the soot and ash generated by the fires, blocked the incoming sunlight for months or possibly even years.

Ultimately, this asteroid, which *only* had a diameter of 1.2 miles, destroyed ecosystems throughout what is now the entirety of the midwestern United States, and it caused damage that likely took the area centuries to repair.

Today, 74 million years after the collision, the asteroid-generated hole in Manson, Iowa is gone. Over the long stretch of time that separates us from this cataclysmic event, glaciers filled and grated the asteroid-generated crater, which was so large that it once dwarfed the Grand Crayon.

Surprisingly, despite the ferocity of the asteroid's impact, the dinosaurs and all the other creatures alive at the time eventually recovered, and apparently, the blast did not cause a single species to go extinct. Nevertheless, this rock did unthinkable damage, and if an asteroid like the one that hit Manson, Iowa, was to hit the midwestern United States today, it would be hard to predict the effects it would have on humanity.

MITIGATING THE MOST LIKELY NEAR-TERM NON-ANTHROPOGENIC THREATS

Unfortunately, we can't predict with certainty when geological catastrophes and asteroid strikes will occur, and it is not certain that we could do anything to stop them, even if we could. There is some hope, however, that we could succeed in saving ourselves. For example, NASA recently suggested that we could pump water at high pressure into a volcano and extract heat, and in so doing, make it less likely that the volcano would erupt. Money and political will would be needed to develop and test this strategy, but these resources may not be that difficult to obtain, given that the extracted heat could be used to produce very competitively priced electricity. Still, while this is a promising idea, we haven't yet tried it, and we don't know with certainty that it would work. In addition, the heat extraction process takes time, so if a volcano is going to erupt, we would need to start the process well ahead of the explosion [54].

Despite the dangers these geological monsters present, there are currently more than 500 million people who live near volcanoes. To protect these individuals, volcanologists have proposed implementing a three-part strategy to prepare for the inevitable eruptions. First, they suggest that governments curtail volcano-induced damage by limiting or prohibiting the use of land that is in a potential hazard zone. Second, they recommend that governments monitor a volcano for signs that it may become dangerous. Finally, they advocate for the implementation of an emergency response plan in the event of an eruption. All this is worthwhile, and it may mitigate the damage generated by a relatively small volcano, but it would be difficult to apply this advice to decrease the potential damage that would be generated by a very large and infrequently erupting volcano. After all, how could a government prohibit the use of all the land that could be threatened by a supervolcanic eruption that might only occur once every 600,000 years? Further, if a volcano-like Yellowstone were to erupt, how would we evacuate all the individuals in the several states that would be immediately affected? Where would all the evacuated people go? Ultimately, an eruption with a VEI of 7 or 8 would be a disaster unlike any that modern humanity has experienced, and even with our best efforts, it would be exceedingly difficult, if not impossible, to effectively prepare for the consequences of such a blast.

In addition to trying to develop a plan to decrease the possibility of a volcanic eruption, NASA is also working on plans to prevent an asteroid impact. The first step in their strategy involves mapping potentially hazardous Near-Earth Objects (NEOs), and in 2005, the United States Congress mandated that NASA map at least 90% of the NEOs with a diameter of 140 meters (460 feet) or more, while simultaneously identifying as many smaller NEOs as possible [55].

As of June 2018, 893 of the estimated 920, or 97% of the near-Earth asteroids with a diameter of 1 kilometer or more, have been mapped. However, in 2017, NASA projected that less than half of the asteroids with a diameter of 140 meters would be mapped by 2033 [56], and they indicated that asteroids under 140 meters in diameter would be even more difficult to identify.

On the good news side of the ledger, NASA determined that none of the currently identified large asteroids (i.e., those with a diameter of one kilometer or larger) are likely to collide with our planet anytime soon. Of course, this statement must be tempered by the fact that we haven't mapped all the large NEOs, and asteroids in the asteroid belt or elsewhere can be jolted out of their current trajectories by future interactions with other asteroids or nearby planets.

If at some point we do identify a NEO on a collision course with our planet, part two of NASA's plan would go into effect. This would involve altering the trajectory of the incoming asteroid so that it would not collide with Earth. One strategy for doing this would be to set one or more "kinetic impactors" on a collision course with the asteroid [57, 58]. If successful, these impactors would alter the asteroid's flight path while leaving the asteroid intact.

To test the utility of this approach, NASA launched the "DART mission" on November 24, 2021. This operation will culminate in late September or early October of 2022, when NASA's 1,210-pound spacecraft will collide with a targeted 525-foot asteroid named Dimorphos [59]. If this mission is successful, humans will have generated the first evidence that we may be able to do what the dinosaurs could not, namely, save ourselves from the destructive impact of giant space rocks. However, while the success of the "DART mission" will be a substantial step forward in our attempts to protect our planet, we still will have work to do before we can demonstrate that we can successfully alter the trajectory of an asteroid with a diameter that is an order of magnitude larger than that of Dimorphos. Furthermore, to rest more comfortably, we will have to map all NEOs quickly, so that if they are on a collision course, we will have the time we need to stop them.

So, given that we can't currently be certain we can stop large-scale geological eruptions or NEOs, we must live with the possibility that the world as we know it could end because of one of these events. The good news is that this type of catastrophic situation is not likely to happen anytime soon, and like the dinosaurs, we could enjoy many tens of millions of years without encountering an extinction-level episode. Nevertheless, one day our luck will run out, and unless we are able to stop these threats, they will likely destroy our civilization.

GROUP 3 THREATS: NATURAL EVENTS THAT WILL DESTROY LIFE ON EARTH SOMETIME IN THE DISTANT FUTURE

The final natural events that I will discuss are those that will destroy life on Earth, as we know it, sometime in the distant future. Included in this list are perturbations in the orbitals of the solar system's planets. According to a recent study, the planets in our solar system will continue to stably rotate around the Sun for the next 40 million years, but after that, there is a 1–2% probability that Mercury's orbital will destabilize sometime in the next five billion years. If this were to occur, the rotation of the planets in the inner solar system would be disrupted, and eventually, Earth could slam into Mercury or Mars. If this did occur, Earth itself might very well be destroyed, and if our planet managed to remain intact, its surface would heat up to temperatures approaching those in a giant red star [60].

Another sad reality for creatures in the distant future is that the Moon is slowly inching away from Earth, and approximately two billion years from now, the Moon will be too distant to stabilize Earth's obliquity. This will result in severe, rapid, and unstable changes in Earth's climate, and the violence of these fluctuations will likely lead to the elimination of terrestrial animal life [33].

Ultimately, in the very distant future, our planet, and the life on it, will be destroyed by the increasing luminosity of the Sun. As the Sun ages, it will eventually run out of hydrogen to fuse within its core, and when that happens, it will begin to fuse helium in a shell around its core. As this occurs, the Sun will expand in size, and the temperature on Earth and the weathering process (see Chapter 5) will become so elevated that most of the carbon will be depleted from the atmosphere within 600 million years. The decreasing levels of carbon will eventually lead to the cessation of carbon-fixing photosynthesis, and as a result, oxygen levels will begin to plummet. When this happens, any plant and animal life that managed to survive the increasingly inhospitable climate will die.

The problems associated with the heating up of the Sun will become even more dramatic with time, and approximately one billion years in the future, the Sun's energy output will be 10% higher than it is now. As a result, Earth's oceans will evaporate, and as these great bodies of water dry up, the water vapor in the atmosphere will increase tremendously, further facilitating the warming of the planet. In addition, as the oceans evaporate, plate tectonic activity will generate increasing amounts of friction, and at some point, the level of friction will become high enough to stop the plate's movements. This development will end geologically induced carbon cycling, and it will impede Earth's internal heat dissipation mechanism.

By two billion to three billion years in the future, it is likely that the planet's magnetic dynamo will also fail, and with that, Earth's magnetosphere would quickly deteriorate. In the absence of this protective shield, the ever-increasing heat would cause any remaining water on the planet to be lost, and without water, even the most tenacious single-cell polyextremophiles will go extinct. According to one model, all the events described earlier will play out, and life itself will perish from Earth, in about 1.6–2.8 billion years [61, 62].

Eventually, Earth itself will be incinerated by the Sun as our star transitions into a red giant. When, exactly, this will happen is not clear, but most estimates suggest that our planet will cease to be in seven billion or eight billion years [63].

PEERING INTO OUR DISTANT FUTURE
THROUGH DARK-COLORED GLASSES

If our descendants are still here 0.5 billion years from now, what will happen to them once the Sun starts to cook Earth? The answer to this question is, of course, not knowable. What we can say, however, is that our descendants at that point in time will be very different from us. This fact should not come as a surprise, given that our ancestors from 0.5 billion years ago were sea sponges. In 0.5 billion years, any creatures that we give rise to will look and act differently than we do, and their potential intellectual and technological abilities are very hard to imagine. Frank von Hippel at Princeton University suggests that, by this point, our descendants might have the ability to rejuvenate the Sun, shade the planet from the Sun's intense energy, or move Earth to a safer orbital [64]. It is also possible that, by this time, our descendants could build giant, space traveling, city ships and leave Earth. Presumably, such vessels would provide a comfortable, viable, long-term, and ultrafast method for space travel, and after what could be many generations, our very distant descendants might find and settle another Earth-like planet in some far away sector of the galaxy. But, at this point, all this is just wild speculation. Five hundred million years is a very long time, and between now and then, there are many more immediate obstacles to surmount.

HUMANITY'S LONG-TERM PROSPECTS

Humans are a relatively hardy bunch, and so far, we have survived fluctuations in climate, wars, plaques, and famines. Our survival wasn't guaranteed and preserving while facing these challenges wasn't easy. As I mentioned in a previous chapter, many of these trials pushed humanity to the brink of extinction [65], and even relatively small environmental disruptions may have altered human history by contributing to overwhelming social unrest. (See reference [66] for a description of how the eruption of Alaska's Okmok volcano in 43 BCE may have led to the fall of the Roman Republic.) Nevertheless, despite all this, we managed to endure. Now, however, our numbers are much greater, and we are much more integrated and dependent on technology and each other [67]. So, today, a nuclear war or devastating climatic event in one area would negatively and profoundly affect individuals across the globe. Whether we like it or not, our fates are linked.

Despite this reality, there are situations in which actions driven by stress and irrational thought could culminate in worldwide destruction. Global warming and its attending environmental and economic degradation coupled to a highly deadly pandemic or a global natural disaster could trigger ruinous calamity and worldwide societal collapse [68]. In such chaos, unsavory individuals could seize power within a nuclear state, and these desperate, possibly pathological and irrational individuals could use nuclear weapons in an attempt to gain an advantage. The resulting effects from such a combination of events could be so severe that the human population would simply cease to exist. Alternatively, a nuclear war could start the extinction process, and the environmental, economic, and societal collapse that would likely follow could lead to rampant, uncontrollable, and highly deadly pandemics. Regardless of the order in which the events occur, multiple stressors acting simultaneously would greatly increase the odds that humanity would perish.

It is worth emphasizing, however, that unlike the dominant lifeforms that came before us, we have a better understanding of our world, and we are aware of at least some of the events that could lead to our destruction. We also have the potential to act logically and cooperatively to solve significant problems, and if we acknowledge the dangers that exist and work in unison, we may be able to survive under conditions that other species could not.

If we do manage to persevere, we can continue to build upon the incredible progress that we have made since our emergence. In just the last few centuries alone, humanity has decreased the percentage of people living in poverty, increased life expectancy, improved health care, increased the literacy rate, developed science and technology, and enhanced personal freedoms [69–71]. In short, during a very brief moment in the history of the universe, we have greatly augmented the quality of life for a large fraction of the human population. These developments represent great and substantial achievements, of which humanity should be proud, and if we continue this trend, our descendants may make existence much more enjoyable for an ever-increasing fraction of the human population (for an optimistic and refreshing look at humanity's potential future, see Steven Pinker's *Enlightenment Now: the case for reason, science, humanism, and progress* [71]). It is even within the realm of possibility that one day in the distant future, humans will move out into the galaxy and seed it with life. At that point, human survival would likely continue for a very long time. But all this depends upon our continuing to have good fortune, and it depends on us using our intellect wisely. It also is contingent on us seeing each other as the close relatives that we are and realizing that we must cooperate, on at least some issues, if we are to continue to survive.

If we refuse to acknowledge the reality of the existential threats we face, our 300,000-year-old species, will likely not even reach the average species lifespan for a mammal. Sadly, if this happens, we will be just one more group in the long list of the extinct, and that would be a sad outcome for such an unusual and promising species.

If we do eliminate ourselves or if nature does so, other lifeforms will endure, and although the planet will be a very different place, new species will appear (see *The World Without Us,* for more on this [72]). Ultimately, if a new type of intelligent being were to one day appear on the planet, we might end up being a cautionary tale of the dominant species that once was.

THE IMPORTANCE OF HOPE AND PERSEVERANCE

Those of you who know me know that I am a *Star Trek* fan. I am especially fond of these fictitious sets of television series because they emphasize the positive qualities of humanity, and they focus on the possibility that our species could evolve culturally, technologically, and even biologically to become greater than it is now.

So, in the spirit of *Star Trek*, I want to end this book on a positive note, because although we need to be aware of the threats we face, we also need to know that we can build on the progress of the enlightenment, and we can create a better world in ways that no creature before us could. To do so, a significant fraction of us must want this better world, and we must be willing to move forward with hope, perseverance, and courage. To quote Winston Churchill, "Success in not final, failure is not fatal, it is the courage to continue that counts."

REFERENCES

1. Ceballos, G., et al., *Accelerated modern human-induced species losses: entering the sixth mass extinction.* Science Advances, 2015. **1**(5): p. e1400253.
2. Moore, M., *Hittites 'used germ warfare 3,500 years ago'.* The Telegraph, 2007. https://www.telegraph.co.uk/news/worldnews/1571927/Hittites-used-germ-warfare-3500-years-ago.html.
3. Wheelis, M., *Biological warfare at the 1346 siege of Caffa.* Emerging Infectious Diseases, 2002. **8**(9): p. 971–975.
4. Patterson, K.B. and T. Runge, *Smallpox and the Native American.* American Journal of the Medical Sciences, 2002. **323**(4): p. 216–222.
5. Ma, H., et al., *Correction of a pathogenic gene mutation in human embryos.* Nature, 2017. **548**(7668): p. 413–419.
6. Connor, S., *First human embryos edited in U.S.* MIT Technology Review, 2017. https://www.technologyreview.com/2017/07/26/68093/first-human-embryos-edited-in-us/.
7. Shermer, M., *Apocalypse AI.* Scientific American, 2017. **316**(3): p. 77.
8. Russell, S., *Should we fear supersmart robots?* Scientific American, 2016. **314**(6): p. 58–59.
9. Koch, C., *Proust among the machines.* Scientific American, 2019. **321**(6): p. 46–49.
10. Hawking, S., et al., *Brief answers to the big questions.* First U.S. edition. ed. 2018, New York, NY: Bantam Books, an imprint of Random House. xxiii, 230 p.
11. Bennett, D., *No nukes.* Boston Globe, 2008.
12. McQueen, A., *How to be a prophet of doom.* New York Times, 2018.
13. Editors, *Trump's nuclear arsenal.* New York Times, 2017.
14. Davenport, K. and K. Reif, *Nuclear weapons: who has what at a glance.* Arms Control Association, 2019.
15. Bennett, J., *Here's how much deadlier today's nukes are compared to WWII A-Bombs.* Popular Mechanics, 2016. https://www.popularmechanics.com/military/a23306/nuclear-bombs-powerful-today/.
16. *Close calls with nuclear weapons.* 2015, Cambridge, MA: Union of Concerned Scientists. www.ucsusa.org/weaponsincidents.
17. Cook, J., *Eight times the world narrowly avoided a potential nuclear disaster.* Huffington Post, 2017.
18. Stevens, M. and C. Mele, *Causes of false missile alerts: The Sun, the Moon and a 46-cent chip.* New York Times, 2019.
19. Postol, T., *How a nuclear near-miss in '95 would be a disaster today—opinion.* Boston Globe, 2015.
20. Shuster, S., *Interview: Stanislav Petrov, Russian who averted nuclear war.* Time Magazine, 2017.
21. Chan, S., *Stanislav Petrov, soviet officer who helped avert nuclear war, is dead at 77.* New York Times, 2017.

22. Diamond, J.M., *Collapse: how societies choose to fail or succeed*. 2005, New York, NY: Viking Press.

23. Rosling, H., O. Rosling, and A.R. Ronnlund, *Factfulness: ten reasons we're wrong about the world—and why things are better than you think*. First edition. ed. 2018, New York, NY: Flatiron Books. x, 342 p.

24. Dovers, S. and C. Butler, *How many people can Earth actually support?* Australian Academy of Science, 2017. https://www.livescience.com/16493-people-planet-earth-support.html.

25. LeBlanc, S.A. and K.E. Register, *Constant battles: the myth of the peaceful, noble savage*. 2003, New York, NY: St. Martins Press.

26. Ceballosa, Gerardo, Paul R. Ehrlich, and P.H. Raven, *Vertebrates on the brink as indicators of biological annihilation and the sixth mass extinction*. Proceedings of the National Academy of Sciences, 2020. **117**: p. 13596–13602.

27. Nuwer, R., *Mass extinctions are accelerating, scientists report*. New York Times, 2020.

28. Johan Rockström, et al., *A safe operating space for humanity*. Nature, 2009. **461**(2009): p. 472–475.

29. Penuelas, J., et al., *Human-induced nitrogen-phosphorus imbalances alter natural and managed ecosystems across the globe*. Nature Communications, 2013. **4**: p. 2934.

30. Shaftel, H., et al. *Scientific consensus: Earth's climate is warming*. NASA, 2019; Available from: https://climate.nasa.gov/scientific-consensus/.

31. Editor, *Climate change: evidence and causes*. 2018. Washington, DC: National Academies Press.

32. Newman, M.E., *A model of mass extinction*. Journal of Theoretical Biology, 1997. **189**(3): p. 235–252.

33. Ward, P.D. and D. Brownlee, *Rare Earth: why complex life is uncommon in the universe*. 2000, Göttingen: Copernicus Publications.

34. Overbye, D., *This black hole blew a hole in the cosmos*. New York Times, 2020.

35. Beech, M., *The past, present and future supernova threat to Earth's biosphere*. Astrophysics and Space Science, 2011. **336**(2): p. 287–302.

36. Frogel, J.A. and A. Gould, *No death star—for now*. Astrophysical Journal, 1998. **499**: p. L219—L222.

37. Timo Reinhold, et al., *The Sun is less active than other solar-like stars*. Science, 2020. **368**: p. 518–521.

38. Grossman, L., *The Sun is less magnetically active than other similar stars*. Science News, 2020.

39. O'Callaghan, J., *Solar 'superflares' rocked Earth less than 10,000 years ago—and could strike again*. Scientific American, 2021. **325**(325): p. 58–69.

40. Stromberg, J., *What damage could be caused by a massive solar storm?* Smithsonian. com, 2013.

41. Tegmark, M. and N. Bostrom, *Is a doomsday catastrophe likely?* Nature, 2005. **438**: p. 754.

42. Rosenband, T., et al., *Frequency ratio of Al+ and Hg+ single-ion optical clocks; metrology at the 17th decimal place*. Science, 2008. **319**(5871): p. 1808–1812.

43. Turner, M.S. and F. Wilczek, *Is our vacuum metastable?* Nature, 1982. **298**: p. 633–634.

44. Scharf, C.A., *How to predict a hostile alien invasion*. Scientific American, 2017. https://blogs.scientificamerican.com/life-unbounded/how-to-predict-a-hostile-alien-invasion/.

45. Brannen, P., *What ancient mass extinctions tell us about the future*. Scientific American, 2020. **323**(3): p. 74–81.

46. Ambrose, S.H., *Did the super-eruption of Toba cause a human population bottleneck? Reply to Gathorne-Hardy and Harcourt-Smith*. Journal of Human Evolution, 2003. **45** p. 231–237.

47. Breining, G., *Super volcano: the ticking time bomb beneath Yellowstone National Park*. 2007, McGregor, MN: Voyageur Press.

48. Smith, E.I., et al., *Humans thrived in South Africa through the Toba eruption about 74,000 years ago*. Nature, 2018. **555**: p. 511–515.

49. Papale, P. and W. Marzocchi, *Volcanic threats to global society*. Science, 2019. **363**(6433): p. 1275–1276.

50. Newhall, Chris, Stephen Self, and A. Robock, *Anticipating future Volcanic Explosivity Index (VEI) 7 eruptions and their chilling impacts*. Geosphere, 2018. **14**(2): p. 572–603.

51. Rampino, M. and R. Stothers, *Flood basalt volcanism during the last 250 million years*. Science. **241**: p. 663–668.

52. Editor, *Asteroid of one-kilometer or larger strikes Earth every 600,000 years*. MIT News, 2003.

53. Bryson, B., *A short history of nearly everything*. 2003, Portland, OR: Broadway Books.

54. Cox, D., *NASA's ambitious plan to save Earth from a supervolcano*. BBC Future, August 17, 2017.

55. Editors, *Twenty years of tracking near-earth objects*. NASA, 2018. www.jpl.nasa.gov/news/news.php?feature=7194.

56. Elderkin, B., *America isn't ready to handle a catastrophic asteroid impact, new report warns*. Gizmodo, 2018. https://gizmodo.com/america-isnt-ready-to-handle-a-catastrophic-asteroid-im-1827014709.

57. Kornei, K., *Deflecting an Earth-bound asteroid may take multiple bumps*. New York Times, 2021.

58. Li, M., et al., *Enhanced kinetic impactor for deflecting large potentially hazardous asteroids via maneuvering space rocks*. Scientific Reports, 2020. **10**(1): p. 8506.

59. Roulette, J., *NASA's DART mission launches to crash into killer asteroid and defend Earth*. New York Times, 2021.

60. Shiga, D., *Solar system could go haywire before the Sun dies*. New Scientist, 2008. https://www.newscientist.com/article/dn13757-solar-system-could-go-haywire-before-the-sun-dies/

61. O'Malley-James, J.T., et al., *Swansong biospheres: refuges for life and novel microbial biospheres on terrestrial planets near the end of their habitable lifetimes*. International Journal of Astrobiology, 2012. **12**(2): p. 99–112.

62. Franck, S., C. Bounama, and W.v. Bloh, *Causes and timing of future biosphere extinctions*. Biogeosciences, 2006. **3**: p. 85–92.

63. Emspak, J., *What will happen to Earth when the Sun dies?* Live Science, 2016. https://www.livescience.com/32879-what-happens-to-earth-when-sun-dies.html.

64. Grinspoon, D., *Deep time survival*. Scientific American, 2016(3): p. 76–79.

65. Inglis-Arkell, E., *Close calls: three times when humanity barely escaped extinction*. Gizmodo, 2015. https://io9.gizmodo.com/close-calls-three-times-when-the-human-race-barely-esc-1730998797.

66. McConnell, J.R., et al., *Extreme climate after massive eruption of Alaska's Okmok volcano in 43 BCE and effects on the late Roman Republic and Ptolemaic Kingdom*. Proceedings of the National Academy of Sciences, 2020. **117**(27): p. 15443–15449.

67. Wright, R., *NonZero: the logic of human destiny*. 2000, New York, NY: Pantheon Books.

68. Phillips, C.A., et al., *Compound climate risks in the COVID-19 pandemic*. Nature Climate Change, 2020. **10**: p. 586–588.

69. Pinker, S., *The better angels of our nature: why violence has declined*. 2011, New York, NY: Viking. p. xxviii, 802.

70. Kristof, N., *This has been the best year ever*. New York Times, 2019.

71. Pinker, S., *Enlightenment now: the case for reason, science, humanism, and progress*. 2018, New York, NY: Viking, an imprint of Penguin Random House LLC. xix, 556 p.

72. Weisman, A., *The world without us*. 2007, New York, NY: Thomas Dunne Books/St Martin's Press. p. 432.

Epilogue
Reflecting on Our Long Journey

We cannot solve our problems with the same thinking we used when we created them.

Albert Einstein

The life of the individual only has value [insofar] as it aids in making the life of every living thing nobler and more beautiful.

Albert Einstein

Do what you can, with what you have, where you are.

Teddy Roosevelt

Our journey has come to an end, but before we part ways, it is worth reflecting once more on why we are here. During our exploration, we saw that for us to exist, an unfathomable amount of energy and spacetime first had to emerge from nothing, and it apparently had to do so via a process that we do not understand. Furthermore, as all this energy and spacetime was emerging, it did so in an exquisitely precise manner. As a result, our universe acquired a very long litany of specific characteristics, and these traits, which manifest in part as the laws and constants of physics, shaped all our universe was, is, and ever will be. However, as significant as the process of the universe's trait acquisition was, it is important to remember that we do not have the faintest inkling as to how or why these attributes emerged. What we do know is that, in very many instances, had the fundamental features of our universe been even ever so slightly different, we would not be here. What are the odds of all this occurring in such a way that life could eventually emerge within the universe? No one knows, but they appear to be exceedingly remote.

Once the universe was underway, stars capable of carrying out nucleosynthesis eventually came forth, and in the process of carrying out their alchemy, these mind-bogglingly large structures functioned under the guidance and constraints of the physical laws and constants established at the beginning of time. As they did so, they produced elements whose existence depended on the manifestation of very precise resonance states, and in the presence of dark matter and dark energy, these stars collectively formed colossal galactic structures.

The galaxies that arose also had specific characteristics, and some had compositions and regions that allowed for the formation of Sun-like stars and Earth-like planets. So, here we see that, within this exceedingly improbable universe, we live in an uncommon galactic region near an unusual star. Furthermore, we reside in a solar system whose composition and physical dynamics made it possible for our highly atypical planet to appear.

As time unfolded, the rocky Earth, which was initially molten and barren, gave rise to life. For this to happen, yet another very long list of specific conditions had to be in place, and while these conditions ultimately did come to be, it is not at all obvious that this was inevitable.

With the genesis of life, Earth changed. Within our planet's biosphere, which was shaped by both constructive and destructive forces, the physical laws and the constants of the universe made it possible for life to evolve. The various forms of life that arose as a result ultimately interacted in complex and unpredictable ways, and eventually many of these seemingly extraordinarily unlikely events converged, and they produced us.

At each step in this 13.8-billion-year long process, there were an incalculably large number of ways in which life could have been terminated. There were also an immeasurably large number of ways in which humanity's march into existence could have ended. However, despite this reality, we are here.

So, does our existence, and perhaps that of other intelligent, sentient beings elsewhere in the universe, signify that the universe has meaning and purpose, or is the universe, and all that is within it, just a conglomeration of meaningless entities moving about in a vast, cold, seemingly endless expanse? Unfortunately, this is not a question that science can definitively answer, but it is worth noting that science does reveal our universe to be a very special place with a truly remarkable set of characteristics, and that fact alone can aid those in search of answers related to meaning and purpose [1, 2].

Regardless of one's thoughts about the meaning and purpose of our universe, I hope this book helped you appreciate just how lucky we are to be part of this seemingly most improbable cosmos. I also hope the text helped you to become more aware of the deep connections we share with other lifeforms in general, and with our fellow human beings in particular.

If this book accomplished what I hoped it would, I'm sure you will agree that it is never too late to help a few of our fellow travelers live "nobler and more beautiful" lives, and in closing, I wish you success in doing "what you can, with what you have" to achieve that end. I also wish you happiness and the best of luck during your "fleeting moment in the sun."

REFERENCES

1. Templeton, J., *Evidence of purpose: scientists discover the creator.* 1994, New York, NY: Continuum. 212 p.
2. Meyer, S.C., *Return of the God hypothesis: three scientific discoveries that reveal the mind behind the universe.* First edition. ed. 2021, New York, NY: HarperOne, an imprint of HarperCollins Publishers. 568 p.

Index

Note: Page numbers in *italics* indicate a figure and page numbers in **bold** indicate a table on the corresponding page.